Solving The Pulses Crisis

Solving The Pulses Crisis

edited by
Anil Kumar Singh
B. Gangwar

2013

New India Publishing Agency
Pitam Pura, New Delhi-110 088

Published by

Sumit Pal Jain for

New India Publishing Agency

101, Vikas Surya Plaza, CU Block, L.S.C. Mkt.,
Pitam Pura, New Delhi-110 088, (India)
Phone : 011-27341616, Fax : 011-27341717
Mobile : 09717133558
E-mail : info@nipabooks.com
Web : www.nipabooks.com

ISBN : 978-93-81450-48-2

Printed and Laser Type Setting at : Jai Bharat Printing Press, Delhi.-110032

M.S. Swaminathan
Chairman

M.S. SWAMINATHAN RESEARCH FOUNDATION

3rd Cross Street, Taramani Institutional Area
Chennai (Madras)-600 113, India
Ph.: +91-44-2254 2790, 2254 1698 Fax: 91-44-2254 1319
Email: chairman@mssrf.res.in, swami@mssrf.res.in

FOREWORD

I congratulate the Indian Council of Agricultural Research, Dr. A.K. Singh and the authors and editors of this timely publication, which provides a roadmap for enduring the demand- supply gap in pulses production. A Panchsheel for farmers producing Pulses is urgently needed. The five components are:

- Soil health enhancement
- Water harvesting and efficient use
- Technology, including appropriate seeds, and inputs
- Credit and insurance
- Assured and remunerative marketing

On my suggestion, the Finance Minister of India provided funds in Union Budget of 2011-12, for initiating 60,000 Pulses Villages based on integrated attention to conservation, cultivation, consumption and commerce. This programme is being continued during 2012-13. The impact of such an end-to-end approach is evident for the fact that Pulses production reached a level of 18 million tonnes during 2010-11, thus breaking the era of stagnation in pulse production. Thus, there is hope for ending the protein hunger in our country.

There is no time to relax. It has been estimated that pulses production should reach 27 million tonnes by 2015 and 32 million tonnes by 2030. This will call for an annual growth rate 4% in pulses production. Fortunately, we have a large untapped yield reservoir considering the gap between actual and potential yields even with the currently available varieties and yields. "A bridge the yield gap movement" needs to be launched in the 60,000 Pulses villages. This publication shows how this can be done.

Year 2012 marks the 150th anniversary of the establishment by President Abraham Lincoln of the US Department of Agriculture and the US Land Grant University System. This marked the beginning of the scientific transformation of agriculture in the US. 2012 also marks the 50th anniversary of the publication of Rachel Carson's book, *"The Silent Spring"*, which stimulated the development and adoption of environmentally sound integrated pest management procedures. Science is the prime mover of change. Mahatma Gandhi said over 80 years ago that agricultural stagnation can be ended if intellect and labour are united. Abraham Lincoln also said in 1859, "no other human occupation opens so wide a field for the profitable and agreeable combination of labour with cultivated thought, as agriculture". This book shows how we can marry intellect and labour in pulses production, so that our small farm families can derive, to quote Abraham Lincoln "a comfortable subsistence from a small area of soil".

Our gratitude goes to the editors and authors of this book for their dedication to the cause of ending protein hunger in our country.

M.S. Swaminathan

PREFACE

As a result of concerted multi-agency efforts focussing on R&D and technology with effective policy and institutional mechanism, India could harvest a record production of foodgrains of about 250 million tonnes during 2011-12. However, low production/ productivity and almost stagnant production of pulses remained a matter of concern for the researchers and planners of the country. Backed up with higher support prices and technological backstopping by developmental agencies, pulses production could also be achieved to a record levels of 18.24 million tonnes in 2010-11 and 17.28 million tonnes in 2011-12. In spite of the record production, the country still imported 3.5 million tons of pulses (2010-11) costing about Rs.7,000 crores in foreign exchange, being the biggest importer with close to 20 percent share of the world. Whenever availability of pulses fell short of domestic demand, there has been a spurt in prices causing widespread concerns among the planners and policy makers because pulses are considered a major source of proteins and play an important role in improving food and nutritional security particularly for the poor and small and marginal farmers in India. To alleviate protein-energy malnutrition in the country, a minimum of 50 gm of pulses per capita per day is required in addition to other sources of proteins such as cereals, milk, meat, egg, etc. To make up this shortfall, about 27 million tonnes of pulses are required by 2015 and 32 million tons by 2030. To meet this target, the overall pulses production must increase at the rate of around 4 percent per annum. Though the latest data shows that pulses productivity has reached a record level of approximately 700 kg/ ha, the present productivity growth is still not sufficient enough to meet the diversified pulses demand of the rising population. This seems a herculean task in the light of the past trends in pulses production. India's dependence on global market to meet the domestic demand for pulses would increase in future. Completely relying on imports would not be a good strategy for a country with a large vegetarian population. Therefore, a well thought strategy is urgently needed to meet future demand of pulses and avoid crises situations. The key elements of such a strategy should aim at providing technological backstopping for managing biotic and abiotic stresses and achieving a breakthrough in pulse productivity on sustainable basis alongwith remunerative price support for the farmers. It must be remembered that pulses are going to play crucial role not only for nutritional security but also for improving human- livestock- soil health and also for saving water resource.

Experiences from farming system research reveal that opting for specific cropping pattern for different regions and adopting improved agronomic practices provide us with substantial scope for incorporating pulses in the prevailing farming systems, and realising the target of pulses production is not difficult if location-specific strategies are implemented effectively. The content of the present book addresses all the relevant issues by dovetailing the region-specific strategy and options for solving pulses crises in the country. It is expected that the book would serve to sensitise the planners and policy makers as well as the local agricultural development officials to explore and implement region-specific strategy for enhancing pulses production by incorporating these in the prevailing farming systems by harnessing the available technological potential. Moreover, various technological options and strategies presented in this book would also be useful for all the stakeholders to realise the target of pulses production in near future.

Anil Kumar Singh

B. Gangwar

SYNONYMS

1. **Pigonpea/Redgram (*Cajanus cajan* (L. *Millsp.*):** *Arhar dal* (Bengali/Hindi/Kashmiri), *Tuvara parippu* (Malayalam), *Thugare bele* (Kannad), *Tur dal* (Marathi), *Harada* (Oriya), *Tuvaram paruppu* (Tamil), *Tuver* (Gujarati).

2. **Greengram/Mungbean (*Vigna radiate* L. R. Wilczek):** *Mug* (Gujarati/Bengali), *Hesare kalu* (Kannad), *Muang* (Kashmiri), *Moongi* (Punjabi), *Cheru pararu* (Malayalam), *Mung* (Hindi/Marathi), *Muga* (Oriya), *Pasipayir* (Tamil), *Pesalu* (Telugu).

3. **Blackgram/Urdbean (*Vigna mungo* L. Hepper):** *Mashkalair dal* (Bengali), *Uddina bele* (Kannad), *Kaha* (Kashmiri), *Mah-di-dal* (Punjabi), *Uzhunnu parippu* (Malayalam), *Urd dal* (Hindi), *Biri* (Oriya), *Ulutham paruppu* (Tamil), *Minapa pappu* (Telugu), *Uddachi dal* (Marathi), *Aalad* (Gujarati).

4. **Chick pea (*Cicer arietinum* L.):** *Chana* (Gujarati/Hindi), *Kadale* (Kannad), *Kadala* (Malayalam), *Chanu* (Kashmiri), *Buta* (Oriya), *Harbara* (Marathi), *Chole* (Punjabi), *Kothu kadalai* (Tamil), *Sanagalu* (Telugu), *Chola* (Bengali).

5. **Lentil (*Lens culinaris*):** *Masoor* (Bengali), *Masur dal* (Hindi/Gujarati/Marathi), *Masur bele* (Kannad), *Masur parippu* (Malayalam), *Musur* (Kashmiri), *Masura* (Oriya), *Mysore paruppu* (Tamil), *Misur pappu* (Telugu).

6. **Field pea (*Pisum sativum* L.):** *Matar* (Bengali/Hindi), *Vatana* (Gujarati/Marathi, *Batani* (Kannad/Telugu), *Pattani* (Malayalam/Tamil), *Matara* (Oriya), *Kara* (Kashmiri), *Kabli chole* (Punjabi).

7. **Lathyrus/Grass pea (*Lathyrus sativus* L.):** *Khesari* (Hindi/Bengali/Oriya), *Lakh* (Marathi), *Khesari parippu* (Tamil), *Lanka pappu* (Telugu).

8. **Cluster bean (*Cyamopsis tetragonoloba* L. Taub):** *Gaur* (Hindi/Bengali), *Gori Kayi* (Kannad), *Guwar* (Marathi), *Kothavarnkai* (Tamil).

9. **Mothbean/Dew gram (*Vigna aconitifolia* Jacq. Marechal):** *Moth* (Hindi), *Matki* (Marathi), *Kancha muga* (Oriya), *Thanni kaalu* (Kannad), *Bobbarlu* (Telugu), *Thatta payaru* (Tamil).

10. **Cowpea (*Vigna unguiculata* L. Walp.):** *Barbati* (Bengali), *Lobia* (Hindi), *Alasande* (Kannad), *Payteluguar* (Malayalam), *Chavli* (Marathi), *Chani* (Oriya), *Karamani* (Tamil), *Bobbarlu* (Telugu).

11. **Horsegram (*Macrotyloma uniflorum* L. Verdc.):** *Kulthi-kalai* (Bengali), *Kuleeth* (Gujarati/Marathi), *Hurule* (Kannad), *Muthira* (Malayalam), *Kulthi* (Hindi), *Kolatha* (Oriya), *Kollu* (Tamil), *Ulavalu* (Telugu).

12. **Rajmash/Common Bean (*Phaseolus vulgaris* L.):** *Rajma* (Hindi), *Rajmah* (Assamese), *Raajma* (Bengali, Gujarati, Kannad), *Rajama* (Oriya), *Rajma phalli* (Urdu), *Kidney Beans* (Tamil), *Van Payar* (Malayalam).

ABBREVIATIONS

BNF	Biological Nitrogen Fixation
Mt.	Million tonne
HI	Harvest Index
INM	Integrated Nutrient Management
ICRISAT	International Crop Research Institute for Semi-Arid Tropies
RSI	Relative Spread Index
RYI	Relative Yield Index
AICRP_CS	All India Coordinated Research Project on Cropping Systems
RWCS	Rice-Wheat Cropping System
TGP	Trans-Gangetic Plain
NPK	Commission for Agricultural Costs and Prices
NEPZ	North-East Plain Zone
RCT	Resource Conservation Technologies
SAUs	State Agricultural Universities
IIPR	Indian Institute of Pulses Research
R & D	Research & Development
ICAR	Indian Council of Agricultural Research
NFSM	National Food Security Mission
ATMA	Agricultural Technology Management Agency
RKVY	Rastriya Krishi Vikas Yojana
MSP	Minimum Support Prices
HaNPV	Nuclear Polyhedrosis Virus
YMV	Yellow Mosaic Virus
PM	Powdery Mildew
FW	Fusarium Wilt
PB	Phytophthora Blight
SM	Sterility Mosaic
ALB	Alternaria Leaf Blight
MYMV	Moongbean Yellow Mosaic Virus
BGM	Botrytis Gray Mould
AB	Ascochyta Blight
VAM	Vesicular Arbuscular Mycorrhizae
NWPZ	North Western Plain Zones
NEPZ	North Eestern Plain Zones

CZ	Central Zone
SZ	Southern Zone
U.P.	Uttar Pradesh
WB	West Bengal
M.S.	Maharashtra State
MP	Madhya Pradesh
AP	Andhra Pradesh
NHZ	Northern Hill Zone
PSB	Phosphorus Solubulizing Bacteria
PSM	Phosphorus Solubulizing Mycorhizae
GDP	Gross Domestic Product
PDFSR	Project Directorate for Cropping Systems Reasearch
PDFSR	Project Directorate for Farming Systems Research
IGP	Indo-Gangetic Plain
K	Kharif
R	Rabi
S	Summer
RYE	Rice Equivalent Yield
RPE	Relative Production Efficiency
REE	Relative Energy Efficiency
FIRB	Furrow Irrigated Raised Bed
AER	Agro-Eco-Region
NPOP	National Programme on Organic Production
FTDR	Foreign Trade Development and Regulation ACT
USDA	United States Department of Agriculture
APEDA	Agricultural and Processed Foods Export Development Authority
AMA	Agricultural Marketing Advisor
NAB	National Accreditation Body
FYM	Farm Yard Manure
YEMA	Yeast Extract Mannitol Agar
AICRP-IFS	All India Co-ordinated Research Project on Integrated Farming Systems.
REY	Rise Equivalent Yield
GR	Green Gram Residue
LTE	Long-Term Experiment
PRB	Permanent Raised Bed
FB	Flat Beds
IPN	Integrated Plant Nutrient
DAP	Di-Ammonium Phosphate
SSNM	Site-Specific Nutrient Management
AICRPWM	All India Coordinated Research Project on Water Management
IW/CPE	Irrigation Water/Cumulative Pan Evaporation
WUE	Water Use Efficiency

DAS	Days After Sowing
BBF	Broad Bed Furrow
BLW	Broadleaved Weeds
GATT	General Agreement for Trade and Tariffs
WTO	World Trade Organization
HPR	Host Plant Resistance
SMD	Sterility Mosaic Disease
IPM	Integrated Pest Management
EIQ	Environment Impact Quotient
SSF	Solid-State Fermentation
CIPHET	Central Institute of Post Harvest Enginnering & Technology
CFTRI	Central Fisheries Technology Research Institute
CIAE	Central Institute of Agricultural Engineering
PKV	Phule Krishi Vishvavidhyalaya
OGL	Open General License
DAC	Department of Agriculture & Cooperation
NFSM	National Food Security Mission
ATMA	Agricultural Technology Management Agency
RKVY	Rashtriya Krishi Vikas Yojana
MSP	Minimum Support Prices
MPS	Mechanism of Price Support (MPS)

CONTENTS

Part - I
Issues, Strategies and Technologies

Chapter - 1

Solving the Pulses Crisis : A Road Map

Anil Kumar Singh, B. Gangwar and Harbir Singh
Indian Council of Agricultural Research, New Delhi-110012
Project Directorate for Farming Systems Research, Modipuram, Meerut-250110

India has witnessed a phenomenal increase in foodgrains production since Independence and from a food deficit state producing about 51 million tons in 1950-51 it become food surplus country with a record harvest of more than 250 million tons in 2011-12. This particular year also saw a remarkable record production of pulses at about 18.1 million tons. However, to meet the increasing demand of the continuously rising population about 2-3 million tons of pulses have to be imported every year costing the country about Rs.3000 crores in foreign exchange. Despite the changes in the dietary habits in which the large middle class has moved away from traditional cereals towards high value dairy, meat, fruits and vegetable products, pulses in India are still considered an important source of proteins, particularly for the poor. Pulses are not only an excellent source of protein (20-40%), carbohydrates (50-60%) and minerals (2-4%) but also provide calories almost equal to that of cereals energy to the tune of 300 to 540 Kcal/ 100g (Ofuya and Akhidue, 2005). Thus, pulses play an important role in achieving food and nutritional security for small and marginal farmers. However, the per capita availability of pulses has been declining over the time – it has declined from 60.7 gm/ day in 1951 to about 32 gm/day in 2010 – mainly due to stagnant production (except in recent years), rising demand for consumption and other uses. Pulses not only are used for consumption, but also play an even greater role in maintaining and enhancing soil fertility through nitrogen (N_2) fixation. Though the estimates of N fixed by pulses depends on crop species, soil type, agro-climatic zone and crop management practices, it was reported that early, medium and long duration pigeonpea varieties fix 7, 55 and 69 kg/ N_2 ha^{-1}, respectively, in Vertisols of Patancheru (Kumar Rao and Dart, 1987). Pulses are grown not only as sole crops but also as intercrops and the biological nitrogen fixation (BNF) of pulses substantially benefits the associated or succeeding crops in terms of nitrogen contribution (16-68 kg/ N_2 ha^{-1} for succeeding cereals). Thus on a conservative estimate, the BNF (@46 kg/N_2 ha^{-1}) by pulses, excluding the inter-crop area, alone amounts to 1.06 million tons annually.

In spite of diverse roles of pulses in improving soil fertility, nutritional security and environmental sustainability, their production hovered around 13-15 million ton during last two decades or so. Whenever pulses availability fell short of demand, there has been a spurt in prices causing widespread concern among the policy makers. The issue is what can be done to further increase the pulses productivity and what are the constraints which need to be addressed. Therefore, for meeting the future demand of pulses, a well thought strategy is urgently needed to avoid crises situations. Specific crops and regions would obviously require location-specific strategies to address various issues in pulses production. These have been discussed in the subsequent chapters of the book. This chapter gives a brief overview and highlights the main elements of the strategy needed for solving the pulses crisis in the country.

Pulses scenario

Pulses in India are grown under a wide range of agro-climate conditions. Historically, the pulses production hovered around 13 to 15 million tons during the last decade although India's share in global pulse production increased to 27-28 per cent owing to a record harvest in 2010-11 (18.09 million tons). To alleviate protein-energy malnutrition in the country, a minimum of 50 gm of pulses per capita per day are required in addition to other sources of proteins such as cereals, milk, meat, egg, etc. To make up this shortfall, about 27 million tonnes of pulses are required by 2015 (TMOP, 2000). The long term trend in area, production and yield of pulses shows that though productivity of pulses has been fluctuating around 600 kg/ha, an increasing trend has been visible, and the latest data shows pulses productivity touching a record 697 kg/ha in 2010-11. However, the present productivity growth is not sufficient enough to meet the diversified pulses demand of the rising population. An analysis carried out by the Indian Institute of Pulses Research (IIPR), Kanpur indicated that the demand for pulses in 2030 would be 32 million tons (Table 1). To meet this target, the overall pulses production must increase at the rate of around 4 percent per annum (IIPR, 2011).

The major producers of pulses in the country are Madhya Pradesh (24%), Uttar Pradesh (16%), Maharashtra (14%), Rajasthan (6%), Andhra Pradesh (10%) followed by Karnataka (7%) which together contribute to about 77% of total pulses production while the remaining 23% is produced by Gujarat, Chhattisgarh, Bihar, Orissa and Jharkhand. The maximum share in pulses production is that of chickpea (40-50%) followed by pigeon pea (15-16%), black gram (10-12%) and lentil (9-10%) in that order. With both area under cultivation and production being stagnant, the country has permitted unrestricted imports of pulses with low or nil duties for about 20 years.

Table 1: Crop-wise demand projections of pulses from 2012 to 2030 (million tons)

Pulse crops	Year				
	2012	2017	2022	2025	2030
Chickpea	6.47 (2.00)	7.87 (4.00)	9.12 (3.00)	9.68 (2.00)	10.22 (4.00)
Pigeonpea	2.75 (2.25)	3.59 (5.50)	4.58 (5.00)	5.04 (3.25)	5.52 (4.75)
Mungbean	1.52 (5.25)	1.89 (4.50)	2.30 (4.00)	2.53 (3.25)	2.75 (3.50)
Urdbean	1.97 (5.75)	2.43 (4.25)	2.96 (4.00)	3.23 (3.00)	3.58 (4.00)
Lentil	1.31 (4.00)	1.73 (5.75)	2.21 (5.00)	2.41 (3.00)	2.62 (3.80)
Peas	1.03 (4.00)	1.30 (4.75)	1.58 (3.75)	1.70 (3.00)	1.98 (4.35)
Others	1.97	2.91	3.70	4.83	5.33
Total pulses	17.02 (3.50)	21.72 (5.00)	26.43 (4.00)	29.43 (3.65)	32.00 (4.20)

Figures in parentheses are the growth rate during the period in percentages.

Though the country harvested a record 18.09 million tons of pulses in 2010-11, it is still the biggest importer of pulses, with close to 20 percent share in world import of pulses. India imports its requirements from various countries, such as Myanmar, Canada, Australia and various other countries. Import of pulses has increased from 3.13 lakh tons to 34.48 lakh tons during the period 1991 to 2010. Though the quantum of imports dipped during the mid 2000, it reached a record high level of 3.5 million tons in 2010-11 in spite of a bumper harvest. In value terms, the import of pulses has increased continuously during most of the years, except during 2002-03 to 2004-05. Thereafter, the value of imports has risen consistently and reached around Rs.9,700 crores in 2009-10. It is worth noting that though India imported a record 3.5 million ton of pulses in 2010-11, the value of imports were lower (Rs.7,000 crores) in comparison to that in 2009-10. This may be attributed to the highest production in the country which helped in meeting the pulse demand for human consumption, while the low-cost imports helped in meeting the feed demand. Thus, despite being the world's largest producer of pulses, the country has to resort to large quantities of imports to meet her domestic demand. Despite a ban on exports and free imports flowing in to the country, the wholesale price index of pulses has risen sharply in recent years in comparison to the price index of cereals, which is a major cause of concern.

Due to stagnant production and increasing population, it seems India's dependence on global market to meet the pulse demand would increase in future. Thus, it can be inferred from the discussion that the pulses scenario is a matter of serious concern and calls for adoption of a broader strategy which focuses not only on higher price support and free flow of imports of pulses, but also provides technological backstopping for managing biotic and abiotic stresses is achieving a breakthrough in pulse productivity on sustainable basis and providing effective institutional arrangements for marketing of pulses.

Constraints in pulses production

Generally, pulses are grown in marginal environments which only adds to the risk in pulses production. For example, more than 87 percent area under pulses is rainfed. Moreover, several other biotic and abiotic stresses make cultivation of pulses even more risky. This section briefly lists the major constraints in pulses production in India.

Susceptibility to Diseases and Insect-Pests : Diseases and pests are a major problem in pulse crops. More than 250 insect species are reported to affect pulses in India. Among these, nearly one dozen cause heavy crop losses. On an average 2-2.4 million tons of pulses are lost annually due to ravages of insect pests. Among them, pod borer *(Helicovera armigera)* causes the most harm, followed by pod fly, wilt and root rot. Scientific efforts have been able to address this problem to some extent. For example, *Bacillus thruringiensis* has been tested on pod borer, and field demonstrations resulted in 10-12% reduction in avoidable losses as against 11% by the use of a chemical pesticide. However, the successful release of Bt chickpea/pigeonpea varieties from either public or private research will take some more years.

Yield fluctuations in rainfed cultivation : Moisture stress is the oft-cited reason for crop failures. The mean rainfall of major pulse growing states such as Madhya Pradesh (MP), Uttar Pradesh (UP), Gujarat and Maharashtra is about 1,000 mm and the coefficient of variation of the rainfall is 20-25%. Terminal drought and heat stress results in forced maturity with low yields. Drought stress alone may reduce seed yields by 50% in the tropics. A quantum jump in productivity can be achieved by applying life saving irrigation especially in *rabi* pulses grown on residual moisture. Irrigated area under pulses has virtually remained stagnant at 13% of the total area. In south India, water holding capacity of the soil often limits grain yield to the extent of 50% of what is attained under irrigation on Alfisols. On the contrary, on vertisol soils, higher water holding capacity causes growth reduction up to 5-20%. Higher evapo-transpiration in south India during the *rabi* season causes severe constraints to chickpea yield under drought. Temperature is another factor which influences grain yield. Cold is an abiotic stress, limiting the grain yield of pulse crops. All hot season pulses are sensitive to low temperatures, but generally these are not exposed to low temperatures. Since most pulse crops are drought tolerant, most of the research efforts have been confined to develop genotypes and associated production technologies to suit dry land conditions. Consequently, germplasm suited to high rainfall and irrigated conditions are lacking.

Physiological limitations : There is a general feeling that pulses (C3 plants) suffer from inherently low yield potential and are physiologically inefficient group of plants compared to cereals (C4 plants) such as sorghum and maize. However, Aggarwal *et al.* (1997) reviewed the comparative advantages of C3 and C4 group of plants and argued that C3 and C4 plants seem to compete on fairly even terms in hot dry environments. The fact that C3 plants usually do better in cool climates suggests that C3 plants are better for *rabi* season. However, the disturbing feature is that the harvest index (HI) in pulses is low compared to cereals. Harvest Index is defined as seed yield per unit of recoverable biomass. In pulses, it is only 15-20% compared to 45-50% in case of cereals such as wheat and rice. Low HI results from excessive vegetative growth, but can be overcome by early partitioning of dry matter into seeds (Saxena and Johansen 1990).

Low input use : Even though good results have been obtained at research stations, the adoption of integrated nutrient management (INM) technologies by farmers is dismal at the farm level. Out of 137 pulse growing districts, 87 districts show 20-60 percent sulphur deficiency. About 50 percent pulse growing districts are having Zn deficiency. Therefore, application of micro-nutrients in deficient soil is important for increasing pulse production in the country. About 40 percent of the pulses growing regions have low to medium population of native *Rhizobium.* Seed inoculation with biofertiliser *(Rhizobium)* being a low cost input can increase pulses productivity by 10-12 percent.

Irrigation is an important input in any cropping system. Pulses are generally cultivated as rainfed crops as they do not respond to irrigation to the same magnitude as other competing crops like cereals. However, experimental results at Indian Insititute of Pulses Reserch (IIPR) show that one irrigation in chickpea at a critical stage could substantially improve productivity. At International Crop Research Institute for Semi-arid Tropies (ICRISAT), grain yield increased in response to irrigation by nearly 50% in sole pigeonpea and more than doubled in intercropped pigeonpea with two irrigations given at the vegetative and flowering stages after the harvest of the cereal crop in years of scarce rainfall (Sharma and Jodha 1982).

Post-harvest technology : Post-harvest losses account for 9.5% of total pulses production. Among post-harvest operations, storage is responsible for the maximum loss (7.5%). Processing, threshing and transport cause 1%, 0.5% and 0.5% losses, respectively. Among storage losses, pulses are also most susceptible to damage due to insects (5%) compared to wheat (2.5%), rice (2%) and maize (3.5%) (Deshpande and Singh, 2001). Therefore, lack of appropriate storage structures is also one of the constraints discouraging pulse growers.

Gaps in transfer of technology : Farmers have been applying sub-optimal doses of fertilisers, pesticides and number of irrigations for pulses after meeting the requirements of wheat, rice and vegetable crops. Therefore, wide gaps also exist between yields realized in experimental plots, front-line demonstration plots and farmers' fields. Large scale on-farm demonstrations conducted in the last five years have clearly shown superiority of new technologies over the local practices. Adoption of these technologies can increase pulse production by at least 20 to 50% in the country.

Lack of seed availability : Inadequate supply of quality seeds of improved varieties remains one of the greatest impediments in bridging the wide gap between potential and harvested yields. State-level research evidences show 40-60 percent realizable gaps in yield potential of major pulses. This is a very vital issue which needs serious attention because even with the present technologies (including varieties) available with us, substantial productivity gains can be achieved with the use of proper agronomic practices.

Roadmap for solving pulses crises

Due to stagnant production and increasing population, it seems India's dependence on global market to meet the pulse demand would increase in future. The demand projections show doubling of demand for pulses (like chickpea and pigeon pea) in Africa and Asia (Rao *et al.*, 2010). However, the scope for meeting increasing demand for pulses is limited because of increased competition for farm land use from competing crops. Completely relying on imports also is not a good strategy for a country as large as India with 1.21 billions mouths to feed. Therefore, there is need to evolve an effective strategy for enhancing pulses production. The following may be the major elements of the strategy for solving pulses crises in the country.

1. Area expansion

The scope for bringing more area under cultivation of pulses is limited. However, some areas remain fallow after rising of a single crop, and such fallow areas hold the key for raising pulses production. About 12 million hectares of rice-fallows is available in India. At least half of the area can be conveniently brought under cultivation of pulses. Even assuming the present productivity levels, utilization of this area would result in additional production of 4 million tons per annum. Similarly, a vast area is acidic in nature and remains fallow during post rainy season. With soil amelioration, this area can be brought under pulse cultivation. In some pockets in North Bengal crops, like lentil and black gram are grown during post-*kharif* season after harvest of jute and upland rice. But, there is still ample scope of increasing the acreage under pulses comprising the districts of *Uttar* Dinajpur, *Dakshin* Dinajpur, Jalpaiguri and Cooch Behar, provided the supply of soil ameliorant like dolomite and lime are assured.

For identifying the suitable niche areas for enhancing pulses production in major pulse producing states, **relative spread index (RSI) and relative yield index (RYI)** were estimated. While RSI shows relative intensification of production in a particular state, RYI indicates the yield level in that state relative to the yield at the national level. While a combination of RSI (>100) and RYI (>125) would indicate most efficient pulses producing state, their lower values (<75) would imply that the area/state does not have much potential for pulse production (Table 2). An analysis of relative spread and yield index of pulses indicates some niche areas where intensification of pulses production holds good potential. A closer look on the estimates of relative spread index and yield index highlights the scope for area expansion and yield enhancement in various states (Table 2).

A perusal of table 2 clearly brings out the relatively higher efficiency of pulses production in Andhra Pradesh, Jharkhand and Madhya Pradesh. Though the states of Chhattisgarh, Karnataka, Maharashtra and Rajasthan also show higher intensity of area under pulses but these have lower RYI implying that more focus on increasing the productivity and realizing the yield gaps would substantially enhance the pulses production in these states. If additional area can be brought under pulses cultivation in

Bihar, Gujarat, Haryana, Uttar Pradesh and West Bengal, it would result in comparatively higher contribution to total production as these states show higher level of relative yield index. These scenarios are summarized in table 3.

Table 2: Computed values of relative spread index and relative yield index of pulses in major producing states of India (10 years' average)

States	Relative Spread Index (RSI)	Relative Yield Index (RYI)
1. Andhra Pradesh	127.3	108.7
2. Bihar	63.5	129.0
3. Chhattisgarh	129.7	87.0
4. Gujarat	92.5	110.5
5. Haryana	22.8	124.3
6. Jharkhand	113.4	106.5
7. Karnataka	147.1	67.5
8. Madhya Pradesh	185.6	118.3
9. Maharashtra	132.5	97.7
10. Orissa	72.3	69.0
11. Rajasthan	141.3	66.9
12. Tamil Nadu	83.3	62.5
13. Uttar Pradesh	86.7	132.6
14. West Bengal	18.6	122.5

Table 3: Extent of relative spread index (RSI) and relative yield index (RYI) of pulses in major producing states of India

Relative Spread Index (RSI)	Relative Yield Index (RYI)	
	High	Low
High	Andhra Pradesh, Jharkhand, Madhya Pradesh	Chhattisgarh, Karnataka, Maharashtra, Rajasthan
Low	Bihar, Gujarat, Haryana, Uttar Pradesh, West Bengal	Orissa, Tamil Nadu

2. Technological options

While there is limited scope for area expansion, technological options available with the R&D system provides a ray of hope. The agricultural research system of the country has developed close to 600 improved varieties of pulses. However, only one-fourth of these are being grown on farmers' field. Even if half of all the available technologies could reach farmers' field, these would result in significant contribution in production. In addition to varieties, other technological options for increasing pulses production are explored in this section.

Improved varieties : Increase in the production and productivity of pulses can be brought about by the wider availability and adoption of improved varieties. Nearly 570 improved varieties of different pulse crops having high yield, resistance to key diseases, appropriate duration, larger seed size etc. have been released for cultivation since the inception of coordinated Pulses Improvement Programme in 1967. But at present, only 137 varieties are in the production chain. Many of them are popular among farmers because of their adaptability and resistance against one or other prevalent diseases.

A critical review of average yields and attainable yields of pulses shows that it is possible to increase the production by 40 percent by bridging the yield gaps. For example, yield gap-I in different pulses in Himachal Pradesh ranged from 20-47 percent while yield gap-II varied from 17- 90 percent (Table 4). Similarly, the yield gaps in major pulses in Gujarat ranged from about 100% (in black gram) to about 250% in chickpea. Even if about 50 percent of the existing yield gap is harnessed, it would add about 7 million tons of pulses to the national bowl.

Table 4: Average yields and yield gaps in major pulse crops in Himachal Pradesh

Particular	Crop						
	Rajmash	Pigeon pea	Mung	Kulthi	Gram	Lentil	Cowpea
Experimental yield* (kg/ha)	3800	-	-	1065	2155	-	16.8
"On-farm" yield (kg/ha)	2150	780	600	1750	1500	1450	14.0
State average yield (kg/ha)	1080	480	500	170	647	300	-
Yield gap-I (%)	43.4	-	-	34.3	30.4	-	20.0
Yield gap-II (%)	49.8	38.5	16.7	90.3	56.9	79.3	-

Remunerative cropping systems : With the advent of high yielding and short duration varieties, new cropping systems with high productivity and monetary returns have been developed both for rainfed and irrigated areas which have become popular with the farmers. Among the important cropping systems, rice-wheat-mungbean for intensification of rice-wheat system and short duration pigeonpea-wheat and rice-late sown chickpea for the irrigated areas of Punjab, Haryana, Delhi, central and western U.P., rice-lentil for the eastern U.P., Bihar, West Bengal and Madhya Pradesh, and rice-*rabi* urdbean/ mungbean in rice fallows of coastal peninsula have been found efficient and remunerative. In double-crop systems, even under rice-wheat cropping system, it is possible to introduce crops like green-gram on at least 50% area (5 million ha). This would result in addition of 3.5 million ton to the present production level of pulses.

Diversification through pulses : The diversification strategies vary with available resources, local needs, social economic interactions and market. A large number of experiment carried out under All India Coordinated Research Project on Cropping Systems (AICRP_CS) and other Programmes in different agro-climates have indicated the beneficial effect of legumes (grown as intercrops, catch crops, substitute crops, green fodder or green manure) on the production profitability and soil fertility. The potential avenues for cultivation of pulses are briefly discussed below:

Pulses as summer/catch crop : The most feasible way of including legumes in cereal based systems without decreasing land area of the cereal crops is to grow legume as catch crop. In RWCS, medium duration rice varieties followed by wheat varieties suited for normal sowing or late sowing in north-eastern plain did not leave any scope for growing legume between rice harvest and wheat sowing. However, the period between wheat harvesting and rice planting can be utilized for growing short duration (60-70 days) summer legumes. Experiments at Masodha, Varanasi and Ludhiana indicated that inclusion of mung bean in RWCS enhances the total productivity and net returns as compared to RWCS alone (AICRP_CS report). In other studies, the yield of rice in rice-wheat system generally remained higher when preceded by mung bean (Singh and Sharma, 2002).

Pulses as intercrop : Intercropping is one of the important ways to increase the productivity and provide income stability under limited soil moisture conditions. Extensive studies carried out under AICRP-CS have helped in identification and standardization of several highly profitable inter-cropping systems. Some of the promising inter-cropping system are maize+black gram at Palampur, Ranchi and Banswara, maize+cowpea at Karjat, sorghum+pigeon pea at Indore, pigeon pea+mungbean at Bichpuri and Hanumangarh, and rice+soybean at Kalyani and Jabalpur. Most of these inter-cropping systems have been evaluated on farmers' fields and found to be highly remunerative (15-200%) over the sole crops.

Pulses as substitute crop : The substitution of predominant crops like rice or wheat largely depends on nature of constraints in different agro-ecologies. For instance, in Trans-Gangetic Plain (TGP), where water table depletion is a serious concern, there is scope for substituting rice with pigeon pea. Similarly, in eastern parts of India, where wheat productivity is generally low because of climatic constraints particularly higher thermal regimes, wheat needs to be substituted with chickpea, lentil, pea or groundnut. Extensive studies on comparative performance of rice-wheat vs. rice-pulses or pulses-wheat at different locations of AICRP_CS reveal that growing legumes in the systems with recommended rate of NPK helped to improve system productivity and profitability compared to non-legume based systems.

Pulses in fallow lands and new areas : There is a scope for introduction of pulses in new areas such as wastelands, reclaimed soils and rice fallow lands as part of integrated watershed management and as a replacement of less remunerative crops or intensification. To sustain and improve rice productivity, pulses have to be introduced in rice based cropping systems in different rice growing areas in Peninsular India including the deltaic areas of important rivers *viz,* Krishna, Godavari, Kaveri, etc. An innovative approach of crop intensification in the Bundlkhand and other regions, with respect to rice-pulse system, is cropping on rice bunds. Presently most of the bunds are left fallow.

Population management : Frontline demonstrations of ridge/raised bed planting in North-East Plain Zone (NEPZ) have clearly shown 17-20% yield advantage over flat bed sowing. Most of the *rabi* pulses like chickpea, lentil and lathyrus are grown on residual soil moisture. For optimum population management, the moisture in seed zone remained a limiting factor due to early withdrawal of monsoon. Under such conditions, seed priming (soaking seeds for 6-8 hours in water) has proved quite beneficial in improving plant stand and early vigour which results in improving production. This is a very simple technology with encouraging results in frontline demonstrations and needs to be advocated among the farmers.

Resource conservation technologies : Adoption of resource conservation technologies (RCT) has opened a vista for resource conservation at farm level. The RCTs of zero, strip and rotary till drilling, and bed planting saved 64 to 85% resources (time, labour, cost, fuel and energy). In zero-tillage cultivation, farmers reported about 25-30% savings because zero tillage is possible with the residual moisture of previous crop and untilled soil has less infiltration than tilled soil so farmers can apply irrigation much faster. Since zero tillage takes immediate advantage of residual moisture from the previous crop, as well as cutting down on subsequent irrigation, water use is reduced by about 10 cm-hectare, or approximately 1 million liters per hectare. One additional benefit is less water logging and yellowing of the plants after the first irrigation that is a common occurrence on normal ploughed land. The widespread adoption of RCTs would also bring significant environmental benefits. For example, use of zero till drill on one hectare of land would save 60-70 liters of diesel and approximately 1 million liters/ha of irrigation water. Using a conversion factor of 2.6 kilograms of carbon dioxide per liter of diesel burned, this represents about a quarter ton less emissions per hectare of carbon dioxide, the principal contributor to global warming. These benefits increase dramatically if extended on larger area. Adoption of zero-till on, say, 5 million hectares would represent a savings of 5 billion cubic meters of water each year. In addition, annual diesel fuel savings would come to 0.5 billion liters (including the pumping)- equivalent to a reduction of nearly 1.3 million tons in CO_2 emissions each year.

Integrated nutrient management : In rainfed areas, foliar spray of 2% urea/ DAP at flowering/pod filling stage has benefitted the crops in terms of higher yield. A large number of frontline demonstrations carried out on chickpea, lentil, mungbean and urdbean in different regions have shown a spectacular increase of 13-20% in productivity of these pulses. This could prove to be one of the simplest and low cost technologies for improving production of pulses in the rainfed areas. Of late, deficiencies of the secondary and micronutrients such as sulphur, zinc, boron and iron are emerging and magnitude of response to such nutrients is increasing. A large number of field experiments and frontline demonstrations on biofertilizers have clearly shown that seed inoculation with biofertilizers like *Rhizobium* and PSB improve productivity of pulses by 11-21%. The technology involves inoculation of seeds with *Rhizobium* culture @ 25 g/kg of seed and PSB 10-12 hours before sowing. These are the least expensive inputs which need to be popularized on a large scale among farmers.

Integrated weed management : Cultural practices like hoeing and hand weeding coupled with application of herbicides are ideal for weed management. Pre-emergence application of pendimethalin @ 1.0 to 1.25 kg *a.i.* per ha has been found quite effective in controlling seasonal weeds. Frontline demonstrations of this technology have shown 24-33% yield advantage at farmers' fields and therefore, needs popularization immediately.

Integrated management of diseases and insects: Insulation of crops against biotic stresses is most critical in enhancing crop productivity. *Fusarium* wilt is a serious problem in pigeonpea, chickpea and lentil causing substantial yield losses. Host plant resistance is the best option. However, non--availability of seeds of wilt resistant varieties and existence of pathogenic variability call for integrated approach of wilt management involving summer cultivation, seed dressing with fungicides like carbendazim, thiram and *Trichoderma* which are compatible with *Rhizobium,* promoting intercrops like pigeonpea with sorghum, chickpea with linseed and appropriate crop rotations. An excellent example of containing pigeonpea wilt has been well demonstrated in Maharashtra state. There is need to ensure fungicide treatment of seeds being sold to farmers, adequate production and distribution of *Trichoderma* and creating awareness about the wilt resistant varieties.

Post-harvest technology: Besides losses during harvest, transportation and threshing, pulses also encounter insect pest damage during storage, which is estimated to be around 7.5%. A mass awareness programme to educate farmers on scientific storage along with distribution of seed storage bins can check the post-harvest losses to the tune of 0.45 million tons and increase the shelf life of pulses. Improved production and protection technologies, including quality seeds, timely irrigation, farm operations, and effective control measures of biotic and abiotic stresses have the potential to double the pulses production. The present status of seed production in pulses indicates that indent for breeder seed is low because of poor seed replacement ratio (2-3%) as against the desired level of 10-15% or more. It is assessed that availability of quality seed in time can enhance production by 20% with the available varieties.

Thus it is possible to realize additional production of 14.5 million tons by harnessing the potential of available technologies and better agronomic management practices. Assuming that the present level of pulses production (18.09 million tons) could be maintained, and the roadmap can be implemented in next 5-7 years, we can look forward for realizing the required pulses production in 2030 (32 million tons) a decade earlier (Table 5).

Table 5: Potential of additional production of pulses from areas in India

Niche areas of intervention	Available resources/technologies	Estimated Production gains (million tons)
Rice fallow area	12 million ha (50% area with present technology or productivity levels)	4.0
Bridging yield gap	Gaps in farm yields and attainable yields (40%- 60% to be harnessed)	7.0
Inter-cropping & Interisification through (summer pulses)	Introduction of green gram at 50% of the rice-wheat area under cultivation	3.5
	Achievable additional production	**14.5**

Policy and institutional support

There are a number of key socio-economic and political factors that would be crucial for successful implementation of the roadmap. For technological back up, many ICAR research institutes as well as various State Agricultural Universities (SAUs) are engaged in R&D on pulses in the country. Particularly, the Indian Institute of Pulses Research (IIPR), Kanpur with a major focus on basic and applied research in pulses such as chickpea, pigeon pea, *mung* bean, *urad* bean, lentil, etc. has developed a number of high yielding varieties in respect of these crops. The R&D efforts of the (ICAR) and SAUs have paved the way for Green Revolution in pulses as a group of thirty-one Indian scientists have recently (2 November 2011) decoded the genome of pigeon pea (*Arhar*), the second most important pulse crop of India.

In addition, the Government of India has launched several programmes including National Food Security Mission (NFSM), Agricultural Technology Management Agency (ATMA), Rastriya Krishi Vikas Yojana (RKVY) and 60,000 pulses and oilseeds villages for enhancing pulse production in the country. The NFSM aims at increasing the production of rice, wheat and pulses by 10 million tons, 8 million tons and 2 million tons, respectively, by the end of the Eleventh Plan (2011-12). The planned efforts have borne fruits in the form of achieving target production of pulses even before the end of XIth Plan. However, many more concerted efforts would be required to achieve the projected targets in XII plan. In fact, apart from R&D and extension efforts, the government has adopted two-pronged policy strategy of price incentives and imports. While incentives are given in the form of minimum support prices (MSP) for increasing production, the import of pulses are freely allowed for meeting overall domestic requirements. However, the mechanism of price support (MSP) has not been very effective in the past as the prices declared by the Government always remained lower than the ruling prices. A quick look on the MSP for pulses during the period 2000-01 to 2007-08, shows that the MSP were raised only marginally (maximum 11.8%) and in some years for few crops it remained even constant. Therefore, it is not surprising that pulses production also stagnated during this period and it hovered around 14 million tons. For the first time in 2008-09, MSP for green gram, black gram and pigeon pea were enhanced substantially (29-48%), and that played major role in achieving a record pulse production (18.09 million ton) in 2010-11. Thus, it can be inferred that a proper mix of price incentives, infrastructure, institutional support and technological back up will be the key elements of the strategy for solving pulse crisis through increased production levels of pulses.

Future thrusts

To meet the increasing demand of 32 million tons of pulses by 2030, there are many issues related to technology, policy and institutional set up which need to be addressed. The focus of the strategy for solving pulses crisis in the country is considered to be in the following thrust areas:

- Intensive efforts are needed to increase the diversity of pulses varieties on farmers' fields.
- Harnessing potential yield levels using innovative package of practices
- Application of biotechnological and other modern tools holds the potential to develop plant types which would exhibit good response to higher doses of inputs resulting in higher productivity.
- While relatively efficient pulse producers (Andhra Pradesh, Jharkhand and Madhya Pradesh) should focus on innovative agronomic practices for future improvement in pulse production, states with lower productivity (Chhatisgarh, Karnataka, Maharashtra and Rajasthan) should give more emphasis on technological support for bridging the yield gaps in these states.
- Bringing additional area under pulses cultivation in Bihar, Gujarat, Haryana, Uttar Pradesh and West Bengal would result in comparatively higher contribution to total production as these states show higher level of RYI.
- All round efforts should be made for utilizing the residual moisture in paddy fallows, as a substantial part of paddy fallows in eastern India (about 12 million ha) can be targeted for cultivation of pulses during the *rabi* season.
- Replacement of upland paddy with pulse in another viable option which ahs th potential to give better net returns to farmers.
- To enhance land use efficiency, multiple cropping using short-duration pulses or extra-short duration pulses should be focused upon.
- Introduction of sprinkler irrigation has vast scope for higher productivity of pulses.
- Efforts should be made for bringing more area under micro-irrigation system for enhanced water use efficiency. Incentives may be given to promote pulses as intercrops in newly planted orchards under National Horticultural Mission.
- Watershed management approach showed a potential for resource conservation and increasing the yield of pulses.
- Integrated nutrient, pest and weed management is the need of the day for achieving higher productivity in pulses.
- Innovative water management approaches consisting of optimum irrigation schedules, appropriate method of irrigation and judicious use of other production inputs have to be demonstrated through 'on-farm demonstration' and at farmers' fields.
- The complimentary bio-intensive intercropping systems involving diverse crops need to be studied both under irrigated and rainfed situations focusing on pulse-based cropping systems.

- Short-duration and disease resistance varieties of pulses with better synchrony for fruiting and maturity needs to be developed for summer cultivation particularly for rice-wheat system.
- Resource conservation technologies (RCTs) like furrow irrigated raised bed and broad bed furrow systems have the potential for introducing pulses under various cropping systems.
- Inadequate use of quality seed of pulses is one of the major constraints in pulse production. Hence, the issues related to quality seed and low seed replacement rate must be addressed on priority.
- Past experiences show that production of pulses did not fetch remunerative prices for the farmers which is considered as one of the major factors responsible for stagnation in production. During the last two years, MSP for pulses was enhanced keeping in view the market situation. There have been encouraging results for such policy incentives. Therefore, access to market, crop insurance and minimum support prices to farmers would play a crucial role in making pulses production more profitable and sustainable.

References

Deshpande, S D and Singh, G 2001: "Long Term Storage Structures in Pulses", National Symposium on Pulses for Sustainable Agriculture and Nutritional Security, Indian Institute of Pulses Research, New Delhi, 17-19 April.

Directorate of Agriculture 2008-09. Final forecast report of area, production and productivity, Gujarat state.

IIPR (Indian Institute of Pulses Research), 2011. Vision 2030, IIPR, Kanpur.

Kumar Rao, JVDK and Dart, PJ 1987. Nodulation, nitrogen fixation and nitrogen uptake in pigeonpea (Cajanas cajan) of different maturity groups, Plant and Soil, 99: 255–266.

Ofuya, ZM and Akhidue V 2005, The role of pulses in human nutrition: A review, Journal of Applied Science and Environmental Management, 9 (3): 99-104.

Rao, Parthasarathy, Birthal P S, Bhagavatula S and Bantilan M C S 2010, Chickpea and Pigeonpea Economies in Asia: Facts, Trends and Outlook. Patancheru, Andhra Pradesh, India: International Crops Research Institute for the Semi-Arid Tropics. pp. 76

Saxena, N P and Johansen C 1990: "Realised Yield Potential in Chickpea and Physiological Conside-rations for Further Genetic Improvement" in S K Sinha, P V Sane, S C Bhargava and PK Aggarwal (eds), Proceedings of the International Congress of Plant Physiology, Vol.1, Indian Society for Plant Physiology and Biochemistry, Indian Agricultural Research Institute, New Delhi, pp:279-288.

Sharma, D and Jodha, N.S. 1982: "Pulses Production in Semi-arid Regions of India: Constraints and Opportunities", Economic & Political Weekly, Issue No 52, Vol No XVII. Share of different states in total production

Singh, V.K. and Sharma, B B 2002. Economic evaluation of rice based cropping systems in foothills of Himalayas. Indian Journal of Agronomy 47 (1) : 12-19.

TMOP (Technology Mission on Oilseeds and Pulses), 2000. Expert Committee Report onPulses, DARE, New Delhi.

❑❑❑

Chapter - 2

Technological Options for Increasing Pulses Production

N. Nadarajan and Sanjeev Gupta
Indian Institute of Pulses Research, Kanpur, Uttar Pradesh.

India is the largest producer of pulses in the world with 24% share in the global production. During 2010-11, the total production was 18.09 million tonnes from an area of 25.6 million hectare (Table 1). The important pulse crops grown in India are chickpea (48%), pigeonpea (15%), mungbean (7%), urdbean (7%), lentil (5%) and fieldpea (5%). The major pulse producing states are Madhya Pradesh, Maharashtra, Rajasthan, Uttar Pradesh, Karnataka and Andhra Pradesh, which together account for about 80% of the total production. The pulses production does not keep pace with the domestic requirement and consequently, the nation has to import 2-3 million tonnes and the *per capita* availability of pulses has declined progressively from 41.6 g in 1990 to 37 g in 2008 (Indian Council of Medical Research recommends 65 gm/day/capita). Thus, increasing production from the current level poses a serious challenge for the nation.

Issues in pulses production

Even with the best of efforts, pulses production and productivity has more or less remained stagnant. Due to the low productivity-low input nature, pulses are grown as residual/alternative crops on marginal lands after taking care of food/income needs from high productivity-high input crops like rice and wheat by most farmers. Also, they grow as rainfed crops with little or no modern yield enhancing inputs. The low priority accorded to pulse crops may be related to their relatively low status in the cropping system. As a crop of secondary importance, in many of these systems, pulse crops do not attract much of the farmer's crop management attention. In deficient soils, the application of micronutrients has also been found to be remunerative, yet the farmers hardly use these inputs. Seeds of newly developed varieties are not available to the farmers in the required quantities. As a result the seed replacement rate is hardly about 2 to 5%. In addition to this, these crops are adversely affected by a number of biotic and abiotic stresses, which are responsible for reduced yield. Lack of on-farm storage facilities and the vulnerability of pulses to store grain pests also results in considerable losses. Stray cattle and blue bulls are responsible for a lot of damage to the pulse crops especially during off-season, thus affecting adversely the cultivation of crops like mungbean, urdbean, pigeonpea, fieldpea and lentil in Bihar, Uttar Pradesh, Madhya Pradesh, Rajasthan and Haryana also resulting in instability and low yields.

Table 1: Scenario of Pulses in India

Year	Area (mha)	Production (million tonnes)	Yield (kg/ha)	Imports (million tonnes)	Export (million tonnes)	Per capita availability (g/day)
1990-91	24.6	14.3	578	0.23	0.02	41.6
1991-92	22.5	12.1	533	0.31	0.03	34.3
1992-93	22.4	12.8	573	0.48	0.03	36.2
1993-94	22.2	13.3	598	0.62	0.04	37.2
1994-95	23.0	14.0	610	0.55	0.05	37.8
1995-96	22.3	12.3	552	0.49	0.06	32.7
1996-97	22.5	14.2	635	0.65	0.05	37.1
1997-98	22.8	13.0	567	1.08	0.16	32.8
1998-99	23.5	14.9	634	0.56	0.10	36.5
1999-2000	21.1	13.4	635	0.25	0.19	31.6
2000-01	20.4	11.1	544	0.35	0.24	30.0
2001-02	21.7	13.4	609	2.21	0.16	35.4
2002-03	20.5.	11.1	543	1.99	0.14	29.1
2003-04	23.4	14.9	637	1.72	0.15	35.8
2004-05	22.8	13.1	577	1.33	0.27	31.5
2005-06	22.4	13.1	585	1.69	0.44	32.5
2006-07	23.2	14.2	612	2.27	0.25	35.5
2007-08	23.6	14.8	688	2.79	0.16	41.8
2008-09	23.0	14.2	617	2.30	0.16	37.0
2009-10	23.3	14.66	630	NA	NA	NA
2010-11	25.6	18.1	707	NA	NA	NA

Source: Department of Agriculture and Cooperation (2010-11). NA – Not available.

Constraints in pulses production

The following section deals with different technological constraints in pulses production.

1. Susceptibility to diseases and insect-pests

Yield loss in pulse crops are very high due to the high incidence of diseases and insect pests. The losses may be as high as complete destruction of the crops in certain cases. More than 250 insect species are reported to affect pulses in India. Among these, nearly one dozen cause heavy crop losses. On an average 2-2.4 million tonnes of pulses are lost annually due to ravages of insect pest complex. Among them, pod borer *(Helicovera armigera)* causes the most harm, followed by pod fly, wilt and root rot. Recently many successful trials have been conducted to control pod borer through using nuclear polyhedrosis virus (HaNPV), which has been found to be more efficacious in bringing about higher and quick mortality. In on-farm trials, *Bacillus thruringiensis* has also been tested on pod borer, and field demonstrations resulted in 10-12% reduction in avoidable loss as against 11% in the use of a chemical pesticides. However, the successful release of Bt chickpea/pigeonpea varieties from either public or private research will take some more years. Among important diseases, wilt in chickpea and pigeonpea, yellow mosaic virus (YMV) and powdery mildew (PM) in mungbean and urdbean are common and more damaging (Nadarajan and Chaturvedi 2010). Important biotic and abiotic stresses affecting pulses production in India are given in Table 2.

Table 2 : Biotic and abiotic stresses affecting pulses production in India

Crop	Seasons	Stress	
		Biotic	Abiotic
Chickpea	Timely sown	FW, root rot, chickpea stunt, BGM, pod-borer	Low temperature
	Early sown	FW, root rot, ABor chickpea stunt, pod borer	Salt stress
	Late sown	FW, pod borer	Terminal drought, cold
Pigeonpea	Early	FW, PB, pod-borer complex	Water logging
	Medium late	F W, SM, pod-borer complex	Cold, terminal drought, waterlogging
	Pre-*rabi*	FW,ALB, Podfly	Cold, terminal drought
Mungbean	*Kharif*	MYMV, CLS, WB, thrips	Terminal drought
	Summer/ Spring	MYMV, thrips	Pre-harvest sprouting, Temperature stress, drought
	Rabi	PM, CLS	Terminal drought
Urdbean	*Kharif*	MYMV, anthracnose, WB, LCV, Thrips	Terminal drought
	Summer/-Spring	MYMV, Thrips	Pre-harvest sprouting, temperature stress, drought
	Rabi	PM	Terminal drought
Lentil		FW, root rot, rust	Moisture, temperature

FW= Fusarium wilt, PB= Phytophthora blight, SM= Sterility mosaic, ALB= Alternaria leaf blight, MYMV= Moongbean yellow mosaic virus, BGM= Botrytis gray mould, AB= Ascochyta blight

2. Yield fluctuations in rainfed cultivation

There has been a high degree of risk in pulses production as more than 87% of the area under pulses is presently rainfed. The mean rainfall of major pulse growing states such as Madhya Pradesh (M.P.), Uttar Pradesh (U.P.), Gujarat and Maharastra is about 1,000 mm and the coefficient of variation of the rainfall is 20-25%. Moisture stress is the of-cited reason for crop failures. Terminal drought and heat stress results in forced maturity with low yields. Drought stress alone may reduce seed yields by 50% in the tropics. A quantum jump in productivity can be achieved by applying life saving irrigation especially in rabi pulses grown on residual moisture. Irrigated area under pulses has virtually remained stagnant at 13% of the total area. Availability of adequate soil moisture for crop growth depends on rainfall, water holding capacity and depth of soil in rainfed areas. In south India, water holding capacity of the soil often limits grain yield to the extent of 50% of that possible under irrigation on Alfisols. On the contrary, on vertisol soils, higher water holding capacity causes growth reduction up to 5-20%. Higher evapo-transpiration in south India during the *rabi* season causes severe constraints to chickpea yield under drought. Another major problem is salinity and alkalinity of soils. Salinity and alkalinity is high both in semi-arid tropics and in the Indo-Gangetic plains in irrigated areas, which is a cause for concern, as most pulses are susceptible to salinity and alkalinity.

Grain yield is mainly influenced by temperature. Cold is an abiotic stress, limiting the grain yield of pulse crops. All hot season pulses are sensitive to low temperatures, but generally these are not exposed to low temperatures. On the other hand, cool season pulses (chickpea) are often subjected to chilling temperatures especially in areas of north India. However, not much progress has been made in the development of chilling and frost tolerant varieties. Poor drainage/water stagnation during the rainy season causes heavy losses to pigeonpea on account of low plant stand and increased incidence of phytophthora blight disease, particularly in the states of Uttar Pradesh, Bihar, West Bengal, Chhattisgarh, Madhya Pradesh and Jharkhand. Ridge planting has been found very effective in ensuring optimal plant stand and consequently higher yield. Since most pulse crops are drought tolerant, most of the research efforts have been confined to develop genotypes and associated production technologies to suit dry land conditions. Consequently, germplasm suited to high rainfall and irrigated conditions is lacking.

3. Physiological limitations

There is a general feeling that pulses (C3 plants) suffer from inherently low yield potential and are physiologically inefficient group of plants compared to cereals (C4 plants) such as sorghum and maize. However, Aggarwal *et al.* (1997), reviewed the comparative advantages of C3 and C4 group of plants and argued that C3 and C4 plants seem to compete on fairly even terms in hot dry environments. The fact that C3 plants usually do better in cool climates suggests that C3 plants are better for rabi season. However, the disturbing future is that the harvest index (HI) in pulses is low compared to cereals. HI is defined as seed yield per unit of recoverable biomass. In pulses, it is only 15-20% compared to 45-50% in case of cereals such as wheat and rice. Low HI results from excessive vegetative growth, but can be overcome by early partitioning of dry matter into seeds (Saxena and Johansen 1990).

Pulses in general have a high rate of flower drop. In pigeonpea, over 80% of the flowers produced in a plant are shed; by decreasing flower drop, yield can be increased considerably. This can be done either by breeding lines which retain a large proportion of flowers producing pods or through physiological manipulations, such as spray of hormones which reduce flower drop.

4. Low input use

Even though good results have been obtained on research stations, the adoption of integrated nutrient management (INM) technologies by farmers is dismal at the farm level. Farmers have been applying sub-optimal doses of fertilisers for pulses after meeting the requirements of wheat, rice and vegetable crops. Experimental result shows that deficiency of micronutrients, especially sulphur and zinc, is widespread among pulse growing regions. Out of 137 pulse growing districts, 87 districts show 20-60% sulphur deficiency. About 50% pulse growing districts are having Zn deficiency. Therefore, application of micro-nutrients in deficient soil is important for increasing pulse production in the country. About 40% of the pulses growing regions have low to medium population of native *Rhizobium.* Seed inoculation with biofertiliser *(Rhizobium)* (low cost input) can increase pulses productivity by 10-12%. *Vesicular arbuscular mycorrhizae* (VAM) offer promise for improving supply of phosphate and micronutrients like zinc for a variety of pulse crops. Phosphate solubilisers are yet another group of hydrotropic organisms, which may have applicability in rainfed pulses production systems, particularly in soils with poor phosphorus availability. Lack of quality culture in adequate quantity is one of the major constraints.

Irrigation is an important input in any cropping system. Pulses are generally cultivated as rainfed crops due to less irrigation response than competing crops like cereals. However, experimental results at IIPR show that one irrigation in chickpea at a critical stage could substantially improve productivity. At the ICRISAT centre, grain yield increased in response to irrigation by nearly 50% in sole pigeonpea and more than doubled in intercropped pigeonpea with two irrigations given at the vegetative and flowering stages after the harvest of the cereal crop in years of scarce rainfall (Sharma and Jodha 1982). However, in northern parts of India where temperatures and evaporation are low with good amount of winter rainfall, responses to irrigation have not been very large. In view of good response for supplemental irrigation to pulse crops, government should encourage policies to provide supplemental irrigation.

5. Post-harvest technology

Post-harvest losses account for 9.5% of total pulses production. Among post-harvest operations, storage is responsible for the maximum loss (7.5%). Processing, threshing and transport cause 1%, 0.5% and 0.5% losses, respectively. Among storage losses, pulses are also most susceptible to damage due to insects (5%) compared to wheat

(2.5%), rice (2%) and maize (3.5%) (Deshpande and Singh 2001). Therefore, lack of appropriate storage structures is also one of the constraints discouraging growers for pulse production.

6. Gaps in transfer of technology

Farmers have been applying sub-optimal doses of fertilisers, pesticides and number of irrigations for pulses after meeting the requirements of wheat, rice and vegetable crops. Therefore, wide gaps also exist between yields realized in experimental plots, front-line demonstration plots and farmers' fields. Large scale on-farm demonstrations conducted in the last five years have clearly shown superiority of new technologies over the local practices. Adoption of these technologies can increase pulse production by at least 20 to 46% in the country (Table 3). Improved varieties of different pulse crops hold promise to increase productivity by 20 to 41% while package technology comprising of improved varieties and integrated management of nutrients and pests has shown 26 to 46% yield advantage over the farmers' practices in large number of frontline demonstrations conducted across the country. Improved production technologies like raised bed planting, ridge furrow planting, seed treatment with *Rhizobium*, application of sulphur @ 20 kg/ha, pre-emergence application of pendimethalin @ 1-1.25 kg a.i./ ha, foliar spray of 2% urea/ DAP, and bio-intensive IPM modules have been advocated after experimentation and large scale frontline demonstrations. Under complex rainfed areas, the farmer participatory research (FPR) needs to be developed and involve farmers more closely in on-farm research. Farmer-participatory testing will help refine the technologies, pinpoint and eliminate adoption constraints.

Table 3: Yield gaps observed in frontline demonstrations of improved technologies and Farmers' practices.

Crop	Technology	Number of FLDs conducted	Yield (kg/ha)			Yield advantage (%)
			In FLDs	In farmers practices	Gaps	
Chickpea	Improved varieties	650	1383	1158	225	19.5
	Package technology	183	1447	1150	297	25.8
Pigeonpea	Improved varieties	69	1591	1346	245	18.0
	Package technology	234	1287	967	320	33.0
Lentil	Improved varieties	114	1221	992	229	27.0
	Package technology	59	1266	918	348	37.8
Mungbean	Improved varieties	72	801	612	189	30.9
	Package technology	27	565	387	178	46.0
Urdbean	Improved varieties	150	728	556	172	30.8
	Package technology	36	801	631	170	26.9
Fieldpea	Improved varieties	75	1468	1197	271	22.6
	Package technology	40	1687	1185	502	42.4

7. Lack of seed availability

Inadequate supply of quality seeds of improved varieties remains one of the greatest impediments in bridging the wide gap between potential and harvested yields. Last year, the ICAR produced 12807 quintals Breeder seed of 137 improved varieties of different pulse crops, more than the DAC indent. However, the efforts have been still inadequate in meeting the seed requirement mainly due to poor conversion of breeder seeds into foundation and certified seeds, resulting in the poor availability of quality seed. Presently, the total supply of quality seed of pulse varieties is around half of the present requirement of 13 lakh quintals, assuming 15% seed replacement rate. There is thus need to improve seed production and distribution system across the country involving different stakeholders. The intention should be to double the national seed production while making seed readily available to farmers by virtue of the decentralized approach.

Obviously, for any meaningful pulse production programme, the above issues will have to be addressed through research and development options. Also, the farmers will have to appreciate the new improved technologies for increasing pulse production in the country.

Technological options for increasing pulses production

1. Improved varieties

Increase in the production and productivity of pulses can be brought about by the wider availability and adoption of improved varieties. Nearly 570 improved varieties of different pulse crops having high yield, resistance to key diseases, appropriate duration, larger seed size etc. have been released for cultivation since the inception of coordinated pulses improve-ment programme in 1967. But at present, only 137 varieties are in the production chain. Many of them are popular among farmers because of their adaptability and resistance against one or other prevalent diseases. The list of varieties with specific traits is given in Table 4. A large number of frontline demonstrations conducted in different agro-climatic zones have shown mean yield gain of 19.5%, 27%, 18%, 22.6%, 31.9% and 30.8% in chickpea, lentil, pigeonpea, fieldpea, mungbean and urdbean, respectively due to improved varieties over local cultivars.

2. Remunerative cropping systems

With the advent of high yielding and short duration varieties, new cropping systems with high productivity and monetary returns have been developed both for rainfed and irrigated areas which have become popular with the farmers. Among the important cropping systems, rice-wheat-mungbean for intensification of rice-wheat system and short duration pigeonpea-wheat and rice - late sown chickpea for the irrigated areas of

Punjab, Haryana, Delhi, central and western Uttar Pradesh, rice-lentil for the eastern Uttar Pradesh, Bihar, West Bengal and Madhya Pradesh, and rice- *rabi* urdbean/ mungbean in rice fallows of coastal peninsula have been found efficient and remunerative. The efficient and rumenerative cropping systems involving pulses, which have potential not only to increase pulses production in the country but also to improve production sustainability, are as follows :

- Sugarcane + urdbean/ mungbean (1:2)
- Sunflower + urdbean/ mungbean (2:6)
- Potato + rajmash (3:2)
- Sorghum/ pearlmillet + pigeonpea (1:1)
- Pigeonpea + urdbean/ mungbean (1:1)
- Cotton + urdbean/ mungbean (1:1)
- Chickpea + mustard/ linseed (6:2)
- Lentil + linseed (2:1)
- Pigeonpea + groundnut (1:1)
- Pigeonpea + soybean (2:4)

3. Population management

Poor plant population gives reduced yield in pulse crops. *Kharif* pulses, especially pigeonpea, suffer heavy losses due to water stagnation, resulting in poor plant population and seed yield. Ridge planting of pigeonpea has been advocated for its successful cultivation especially in North- East Plain Zone (Eastern U.P., Bihar, Jharkhand, West Bengal) not only to ensure desired plant stand of the crop but also to minimize the incidence of Phytophthora stem blight. The technology involves making ridge and furrows by tractor/bullock drawn ridge planter at 60 cm spacing and sowing seeds in the centre of the ridge. Ridge planters, which are now available, need to be popularized amongst farmers through community services. Raised bed planting is also getting favour of the farmers with the availability of low cost raised bed planters which ensures quick disposal of run-off water and economizes use of supplemental irrigation, if needed. Frontline demonstrations of ridge/raised bed planting in North –East Plain Zone have clearly shown 17-20% yield advantage over flat bed sowing. Most of the *rabi* pulses like chickpea, lentil and lathyrus are grown on residual soil moisture. For population management, the moisture in the seed zone has remained as a limiting factor due to early withdrawal of monsoon. Under such conditions, seed priming (soaking seeds for 6-8 hours in water) has proved quite beneficial in improving plant stand and early vigour which results in improving production. This is a very simple technology with encouraging results in frontline demonstrations and needs to be advocated among the farmers.

Table 4. Improved varieties of major pulse crops

Variety name	Yield (q/ha)	Suitability	Major characteristics
CHICKPEA			
Pusa 256 (BG 256)	18-20	NWPZ, NEPZ, CZ	Medium tall, semi-spreading plants, attractive and bold seeds, tolerant to wilt and *Ascochyta* blight
Pusa 372 (BG 372)	18-22	NWPZ, NEPZ, CZ	Semi-spreading plants, moderately resistant to wilt, *Ascochyta*blight and root rot, small seeds
RSG 888	21	NWPZ	Semi-spreading, small seeded, drought tolerant. twin podded genotype, tolerant to dry root rot
Gujarat Gram 4 (GCP 105)	18-20	NEPZ	Medium tall, semi-erect, tolerant to wilt, pinkish flower, smooth, round and brown seeds
HK 2 (HK 94-134)	22	NEPZ	Tolerant to wilt and root diseases, suitable for irrigated areas.
Pusa Kabuli 1003 (BG 1003:K)	18	NEPZ	White large seeds, tolerant to wilt, medium tall
Uday (KPG 59)	20	NWPZ, NEPZ	Semi-erect, medium tall plants, small seeded, tolerant to wilt, root rot, stunt and pod borer
Ganguar (GNG 1581)	23-24	NWPZ	Medium height (≈ 50 cm) with semi erect plant type, tolerant against diseases like wilt, *Ascochyta* Blight, Stunt and root rot and possesses a good level of protein (21.88 percent) also.
Rajas(Phule G 9425-9)	19	NWPZ	Suitable for normal sown conditions.
Pusa 547 (BGM 547)	18	NWPZ	Suitable for normal sown conditions.
Anvita (RSG-931)	20	NWPZ	Double podded high yielding chickpea variety , resistance to dry root rot, wilt and nematodes in addition to less damage by pod borer
Karnal Chana 1 (CSG 8962)	20-22	NWPZ	Tolerant to mild salinity, resistant to wilt, small brown seeds
DCP 92-3	19	NWPZ	Genotype is semi erect, medium tall, lodging resistant, wilt resistant seeds yellowish brown and medium large
Samrat (GNG 469)	21	NWPZ	Plant erect and tall, seeds brown and large, tolerant to wilt and resistant to *Ascochyta* blight
KWR 108	20-23	NWPZ, NEPZ	Medium tall and semi-erect plants, seeds are brown and large

Contd.

Pusa 362 (BG 362)	24	NWPZ	Medium tall, tolerant to wilt and large seeds
Pusa Chamatkar (BG 1053)	17	NWPZ	Large seeded kabuli, tolerant to wilt
Aadhar (RSG-963)	20	NWPZ	Although it was released for late sown condition but performs well under normal sown as well as rainfed condition. It has good level of resistance to wilt, dry root rot and pod borer.
Pusa Shubhra (BGD 128)	18	CZ	Suitable for normal sown conditions.
SAKI 9516 (JG 16)	20	CZ	Medium seed size, resistant to wilt, tolerant to collar rot, BGM and stunt
Gujarat Gram 1 (GCP 101)	18	CZ	Medium tall, semi-erect plants, resistant to wilt, medium large and dark brown seeds
BGD 72	18	CZ	Semi-erect plants, medium large seeds, tolerant to wilt
Pusa 391 (BG 391)	17-18	CZ	Seeds are large and dark brown, moderately resistant to wilt and root rot
Vijay (Phule G 81-1-1)	20	CZ	Resistant to wilt, tolerant to low moisture stress, small brown seeds
JAKI 92-18	18	CZ	Semi-spreading, bold seeded and resistant to wilt
Digvijay	16-18	Maharashtra	Yellowish brown large seeds, wilt resistant
KAK 2 (PKV Kabuli 2)	18	CZ	Semi-spreading plants, large and white seeded
IPCK 02-29 (Shubra) (kabuli)	19-22	Central India	Moderately resistant to *F. udum*, medium tall and typically erect, medium maturity duration, seeds are large (34-35 g/100 seed wt.)
JG 11	15	SZ	Semi-spreading, large seeds, resistant to wilt and moderately resistant to root rot
Vihar (Phule G 95311)	18	SZ	Large seeded kabuli, wilt resistant
PIGEONPEA			
UPAS 120	11-15	Punjab, Haryana, North Raj., West U.P.	Indeterminate. semi-spreading, synchronous in maturity, tolerant to pod borer
Bahar	25-30	U.P., Bihar, W.B.	Compact resistant to sterility mosaic
Amar (KA 32-1)	16-20	U.P.	Compact, resistant to SMD
Narendra Arhar 1	20-22	U.P.	Resistant to SMD and tolerant to wilt and Phytophthora blight

Contd.

Pusa 9	22-26	U.P., Bihar, W.B.	Indterminate resistant to SMD and Alternaria suitable for Pre-rabi also
MAL 13	2465	U.P., Bihar, W.B.	Tolerant to wilt, PB, SME
NDA 98-1(NDA2)	2443	U.P., Bihar, W.B.	Resistant to wilt
MA 6	2281	Bihar, U.P.	Resistant to wilt and SMD
Paras (H 82-1)	15-20	Haryana	Indeterminate
Al 201	15-16	Punjab	Indeterminate
Pusa 992	14-18	Delhi, Punjab, Haryana	Suitable for normal sown conditions.
TJT501	18-60	MS, Guj. & MP.	Tolerant to pod borer pod fly
Malviya Vikalp (MA 3)	20-22	M.P., Gujrat, Maharashtra	Spreading, constricted pod resistant to pod fly
Jawahar (JKM 7)	18-20	M.P., Gujrat, Maharashtra	Tolerant to wilt and Phytophthora blight
BSMR 175	11-12	Maharashtra	White seeded, wilt and sterility mosaic resistant
JA 4	16-18	Madhya Pradesh	Pod borer tolerant
Asha (ICPL 87119)	16-18	M.P., Gujrat, Maharashtra	Indeterminate, spreading, bold seeded, wilt and SMD resistant
BSMR 736	12-18	Maharashtra	Red seeded, wilt and SMD resistant
Co 6	9-10	Tamil Nadu	Indeterminate, tolerant to pod borer
Laxmi (ICPL 85063)	18-20	Andhra Pradesh	Semi-spreading Suitable for *rabi* planting
GT 100	16-18	Gujrat	Determinate, white bold seeded, tolerant to pod borer and pod fly
AKT 8811	13-14	Maharashtra	Moderately resistant to wilt & pod borer
CORG 9701	11-12	A.P., Tamil Nadu, Karnataka,Orissa	Short duration
Durga (ICPL 84031)	8-10	Andhra Pradesh	Determinate
Vamban 2	10-12	Tamil Nadu	Medium duration, indeterminate
Vamban 3	10-12	Tamil Nadu	Short duration
GTH 1	14-15	M.P., Gujrat, Maharashtra	Medium duration
MUNGBEAN			
Pant Mung 6	12 - 15	NEHZ	Resistant to MYMV, PM, CLS and LCV
Shalimar Mung 1	9 - 10	Jammu & Kashmir	Shining seed, medium large
Sweta	10-11	NHZ	Resistant to MYMV
Pusa 672	9-10	NHZ	Medium large shining green seed, tolerant to MYMV
Pant Mung 5	12-15	U.P & Uttarakhand	Resistant to MYMV, Suitable for all seasons

Contd.

HUM 16	10-12	NEPZ	Resistant to MYMV, suitable for Summer season
Sukumar (WBM 29)	9-10	West Bengal	Suitable for *rabi* season, small seed
OBGG 52	8-9	Orissa	Tolerant to MYMV, suitable for all seasons
Pratap (SG-1)	10-12	Assam	Suitable for early and late situation
Pant Moong 4 (UPM 92-1)	12-15	NEPZ	Resistant to MYMV, dull green seed
NarendraMung 1	10-12	U.P.	Tolerant to MYMV
TMB 37	11	NEPZ	Tolerant to MYMV, suitable for spring season
IPM 99-125 (Meha)	10-12	NEPZ	Resistant to MYMV, suitable for spring season
HUM 12	11-13	NEPZ	Suitable for summer season
HUM 6	10-11	U.P.	Tolerant to MYMV, suitable for spring season
HUM 2	10 - 12	U.P.	Resistant to MYMV, suitable for spring season
Malviya Jankalyani	10-12	NEPZ	Erect, shining green large seed, tolerant to MYMV, suitable for summer season
Samrat	10-12	U.P.	Erect, synchronous shining green seed, highly resistant to MYMV, suitable for spring season
MUM 2	11-12	NWPZ	Resistant to MYMV, shining green seed
IPM 02-3	10	NWPZ	Resistant to MYMV, suitable for *kharif* seaons
Satya	12-14	NWPZ	Erect, resistant to MYMV, suitable for *kharif* seaosn
MH 96-1	10-11	Haryana	Resistant to MYMV
ML 818	10-12	Punjab	Tolerant to MYMV, BLS and CLS
RMG 268 (Durga)	8- 9	Rajasthan	Photothermo-insensitive , shining green seed
ML 613	10-12	Punjab	Tolerant to MYMV and CLS
Ganga 1	10-12	Rajasthan	Moderately resistant to MYMV
PAU 911	12-14	Punjab	Resistant to MYMV, suitable for *kharif* season
Basanti	10-12	Haryana	Resistant to MYMV, suitable for *kharif* and spring seaosn
SML 668	11-13	Punjab	Tolerant to MYMV, suitable for spring / summer season
Ganga-8 (Gangotri)	9-10	NWPZ	Suitable for *kharif* season
Pusa Bold 1	11-12	NWPZ	Erect, Synchronous, large seed, tolerant to MYMV, suitable for spring season
SML 134	11-12	Punjab	Suitable for spring/summer

Contd.

AKM 9911	10-12	Maharashtra	Tolerant to PM
GM 4	15	Gujarat	Tolerant to MYMV
PKV Mung 8802	12-14	Maharashtra	Suitable for *kharif* season, tolerant to MYMV and MB
TARM 18	10-12	Maharashtra	Resistant to PM
TARM 1	10-12	Maharashtra	Resistant to PM, suitable for *rabi* season
JM 721	12-14	Madhya Pradesh	Resistant to PM and MYMV
Gujarat M 3	10-12	Gujarat	Erect, shining green small seed, tolerant to MYMV
TARM 2	10-12	Maharashtra	Small seed, tolerant to PM
BM 4	10-12	CZ	Large green seed, semi-erect bushy type
AKM 9904			
HUM 1	10-12	CZ & SZ	Suitable for Spring and *kharif* season
Pusa 9531	9-10	CZ	suitable for summer season
KKM 3	8-10	Karnataka	Resistant to MYMV, suitable for rice-
TM 96-2		Andhra Pradesh	fallow
COGG 912	8-9	SZ	Suitable for *kharif* season
OUM 11-5	7-8	SZ	Suitable for *kharif* season
Vamban-2 & 3	8-10	Tamilnadu	Tolerant to drought, glaborous pod
WGG 37	6-8	A.P.	Erect, tall, resistant to MYMV
CO 6	10-11	Tamilnadu	Resistant to MYMV, Suitable for all season
LGG 460	10-12	Andhra Pradesh	Resistant to MYMV
Madhira 295	12-14	Andhra Pradesh	Tolerant to PM
Madhira Pesara 347	11-12	A.P.	Tolerant to MYMV, suitable for *kharif* and *rabi* season
Warangal 2	10-12	A.P.	Suitable for *kharif*, *rabi* and summer
Pusa 9072	9-10	SZ	tolerant to PM, suitable for *rabi* season
LGG 410	10-12	A.P.	Resistant to leaf spot, suitable for rice-fallow
LGG 450	11-13	Andhra Pradesh	Tolerant to MYMV and pre-harvest sprouting
LGG 407	12-14	Andhra Pradesh	Resistant to MYMV
URDBEAN			
NDU 99-2	10-12	NHZ	Resistant to MYMV, suitable for *kharif* season
Pant Urd 31	10-12	Uttarakhand and Himachal Pradesh	Main stem bearing, resistant to MYMV, suitable for intercropping in *kharif* and spring season

Contd.

Pant Urd 40	10-12	For Plain and lower hill of Uttarakhand	Main stem bearing, resistant to MYMV, suitable for *kharif* season
Uttara	12-13	NWPZ and NEPZ	Resistant to MYMV, Suitable for *kharif* season
Ujala	8-10	Orissa	Resistant to MYMV and CLS, suitable for *kharif* and *rabi* season
KU 309	10-12	U.P.	Tolerant to MYMV
BirsaUrd 1	10-12	Bihar and Jharkhand	Tolerant to MYMV
WBU 108	10-12	NWPZ, SZ and NEPZ	Resistant to MYMV
NarendraUrd 1	10-12	U.P.	Black, large seed, resistant to MYMV
WBU 109	10-12	NEPZ	Resistant to MYMV, suitable for spring season
KU 91-2	10-11	NEPZ	Resistant to MYMV, suitable for spring season
Mash 338	8-10	Punjab	Tolerant to MYMV
Mash 414	9-10	Punjab	Resistant to MYMV, suitable for spring season
Mash 1008	10-12	Punjab	Resistant to MYMV, suitable for spring season
KU 300	10-12	NWPZ	Resistant to MYMV, suitable for spring season
KU 96-3	8-10	CZ	Resistant to YMV, suitable for *kharif* season
AKU 15	10-12	Maharashtra	Tolerant to PM
GU 1	10-12	Gujarat	Tolerant to PM and CLS
Barkha	9-10	CZ	Large seed, tolerant to CLS
AKU 4	10-11	Maharashtra	Tolerant to MYMV and PM, suitable for *rabi* season
TAU 2	10-11	Maharashtra	Medium large seed, tolerant to PM
TPU 4	7-8	CZ	Erect, medium large seed
JawaharUrd 3	10-12	M.P.	Medium seed size, tolerant to MYMV and CLS
TAU 1	12-14	Vidarbha region of Maharashtra	Purplish black, large seed, tolerant to powdery mildew
IPU 2-43	10-12	SZ	Resistant to MYMV and PM, suitable for *kharif* season
Vamban 5	8-10	Tamil Nadu	Resistant to PM and MYMV
LBG 625 (Rashmi)	10-12	A.P.	Suitable for rice fallows
Vamban 4	10-12	Tamil Nadu	Tolerant to MYMV and PM
Vamban 3	8-10	Tamil Nadu	Dull black seed, resistant to MYMV and tolerant to Powdery mildew

Contd.

WBG 26	9-10	SZ	Resistant to PM
LBG 685	10-12	A.P.	Resistant to wilt
KU 301	10-12	SZ	Resistant to MYMV and tolerant to PM and suitable for *rabi* season
LBG 623	10-12	A.P.	Photo insensitive, large shining black seed, tolerant to MYMV
KBG 512	7-8	Tamil Nadu	Tolerant to stem fly and leaf spot
Prasad	8-9	Orissa	Tolerant to MYMV, suitable for *kharif* and *rabi* season
TU 94-2	12-13	SZ	Resistant to MYMV, tolerant to PM, suitable for *rabi* season
LBG 611	12-14	A.P.	Tolerant to wilt
LBG 20	12-14	A.P.	Tolerant to MYMV and PM, suitable for rice fallows
LBG 402	10-12	SZ (*Rabi*)	Tolerant to wilt and PM, suitable for rice fallows
LBG 17	10-12	SZ	Erect, greenish black large seed, tolerant to PM
LBG 709	10-12	A.P.	Photosensitive resistant to wilt and tolerant to YMV, suitable for *rabi* season in uplands and rice fallows
FIELD PEA			
HFP 9426	25-30	Haryana and NHZ	Resistant to powdery mildew
IFD 1-10	25-30	CZ and NHZ	Resistant to powdery mildew and rust
Pant P 25	20-25	Plains and Hills of Uttarakhand	Resistant to Powdery mildew and tolerant to rust
VL 45	20-25	NHZ	Tolerant to wilt and rust
HUDP 15	25-30	NHZ and NEPZ	Semi leafless, resistant to powdery mildew and tolerant to rust
Pant P 13	20-25	Plains and Hills of Uttarakhand	Resistant to Powdery mildew and tolerant to rust
DDR 23	18-20	NEPZ	Extra early, resistant to powdery mildew
Swati	22-25	U.P.	Semi leaf less, tolerant to powdery mildew
Sapna (KPMR 144-1)	18	U.P.	Dwarf, semi-leafless, large seeds
VL Matar 42	18-22	NEPZ	Resistant to powdery mildew and tolerant to rust
HFP 9907 B (Hariyal)	25-30	NWPZ	Resistant to powdery mildew and tolerant to rust
Pant P 42	25-30	NWPZ	Resistant to powdery mildew and tolerant to rust
KPMR 522	20-25	NWPZ	Semi leafless, resistant to powdery mildew

Contd.

DDR 27	18-20	NWPZ	Very early, resistant to powdery mildew
HFP 4	20-25	NWPZ	Semi leafless, creamy white round and large seeds, resistance to powdery mildew
HFP 8712	20-25	Haryana	Large seed, resistant to powdery mildew
HFP 8909	20-25	NWPZ	Semi leaf less, large seed, resistant to powdery mildew
Shikha	20-25	NWPZ	Large seeds, resistant to powdery mildew
IFD 99-13	25-30	CZ	Resistant to powdery mildew
IPF 99-25	25-30	CZ	Resistant to powdery mildew
KPMR 400	20-25	CZ	Semi leafless resistant to powdery mildew
Ambika	15-20	CZ	Tall, resistant to PM
JP 885	18-22	CZ	Erect plant, large seed
Paras	20-25	Chhatisgarh	Resistant to powdery mildew
LENTIL			
VL Masoor 507	15-17	NHZ	Large seed, tolerant to wilt
HUL 57	18-20	NEPZ, NHZ	Small seed, resistant to rust
Shalimar Masoor 1	14-16	J&K	Large seed
VL Masoor 1	16-18	Uttarakhand	Small and black seed, tolerant to rust and wilt
VL Masoor 4	16-18	Uttarakhand	Small and black seed, tolerant to rust and wilt
VL Masoor 103	16-18	Uttarakhand	Small seed, tolerant to rust
VL Masoor 125	16-18	NHZ	Small black seed, tolerant to wilt and root rot
VL Masoor 126	15-17	NHZ	Small seed, resistant to rust
VL Masoor 507	15-17	NHZ	Large seed, tolerant to wilt
VL Masoor 129	15-17	Uttarakhand	Small seed, tolerant to wilt and root rot
Pant Masoor 6	16-18	Uttarakhand	Large seed, resistant to rust, wilt, *Ascochyta* blight
Pant Masoor 7	16-18	Uttarakhand	Large seed, light gray seed colour, resistant to rust, wilt, *Ascochyta* blight and pod borer
Arun	15-18	Bihar	Medium large seeds, tolerant to rust
Narendra Masoor 1	18-20	U.P.	Large seeds, resistant to rust, tolerant to wilt
Subrata	18-20	NEPZ	Large seeds, tolerant to wilt and rust
Pant L 5	18-22	U.P.	Large seed, resistant to rust, tolerant to wilt and *Ascochyta* blight
KLS 218	15-17	NEPZ	Small seed
WBL 77	16-18	NEPZ	Small seed, resistant to rust

Contd.

IPL 406	16-18	NWPZ	Extra large seed, resistant to rust
LH 89-48	16-18	Haryana	Small seed, resistant to rust
LL 699	12-15	Punjab	Tolerant to rust and blight
DPL 62	18-20	NWPZ	Extra large seed, resistant to rust and tolerant to wilt
Garima	16-20	Haryana	Large seed, tolerant to rust
PusaVaibhav	18-22	NWPZ	Small seed, resistant to rust and tolerant to wilt
DPL 15	16-20	NWPZ	Large seed, tolerant to wilt and rust
L 4076	18-22	NWPZ and CZ	Large seed, tolerant to rust and wilt
Pant Lentil 4	18-22	NWPZ	Small seed, resistant to rust, tolerant to wilt
IPL 81	18-20	CZ	Large seed, tolerant to rust and wilt
Jawahar L 3	18-20	CZ	Large seed, wilt tolerant
JL 1	16-18	M.P.	Medium large seed, early maturity
RAJMASH			
PDR 14 (Uday)	16-20	NEPZ	Tol. to BCMV. 100 seed wt. 44 g white red variegated seeds
HUR 15 (Malviya Rajmash 17)	16-20	NEPZ	Mod. Susceptible to BCMV .Erect determinate white seeded
IPR 96-4 (Amber)	15-16	NEPZ	Res. to BCMV & leaf crinckle. 100 seed 39 g
IPR 98-5 (Utkarsh)	17	NEPZ	Tol. To BCMV. Gulf red seed
IPR 98-3-1 (Arun)	24-25	CZ	Attractive dark red seed , Medium seed size (38 g/100 seeds)

4. Integrated nutrient management

Pulses have shown varying degrees of response to fertilizers in different agro-climatic situations. While response to nitrogen is limited only to starter dose because of atmospheric nitrogen fixing ability, the pulses have shown significant response to phosphorus in terms of yield. In rainfed areas, foliar spray of 2% urea/ DAP at flowering/ pod filling stage has turned out to be rewarding in terms of higher yield. A large number of frontline demonstrations carried out on chickpea, lentil, mungbean and urdbean in different regions have shown spectacular increase of 13-20% in productivity of these pulses. This could prove to be one of the simplest and low cost technologies for improving production of pulses in the rainfed areas. Of late, deficiencies of the secondary and micronutrients such as sulphur, zinc, boron and iron are emerging and magnitude of response to such nutrients is increasing. Sulphur is one of the essential secondary nutrients required for proper plant growth. In recent years, introduction of high yielding varieties under intensive cropping systems coupled with use of sulphur-free fertilizers like urea and DAP and progressive decline in the use of organic manures/compost, has led to widespread deficiency of sulphur in soils. Recognizing this, a comprehensive research programme is initiated to determine response of different pulse crops to sulphur and find out an efficient and low cost source of sulphur. It has been observed that most of the pulse crops respond well to 20 kg of sulphur/ha and in some areas with light textured soils, the response is noticed up to 40 kg/ha. Gypsum, Pyrite or SSP has proved to be low cost source of sulphur. Balanced and integrated nutrient management coupled with timely weed control need to be promoted. A large number of field experiments and frontline demonstrations on biofertilizers have clearly shown that seed inoculation with biofertilizers like *Rhizobium* and PSB improve productivity of pulses by 11-21%. The technology involves inoculation of seeds with *Rhizobium* culture @ 25 g/kg of seed and PSB 10-12 hours before sowing. These are the least expensive inputs which need to be popularized on large scale among farmers.

5. Integrated weed management

Amongst various production inputs, weed management has been found as the most critical input for *kharif* pulses and second most important in *rabi* season as weeds cause substantial yield losses. Cultural practices like hoeing and hand weeding coupled with application of herbicides are ideal for weed management. Pre-emergence application of pendimethalin @ 1.0 to 1.25 kg a.i. per ha has been found quite effective in controlling seasonal weeds. Frontline demonstrations of this technology have shown 24-33% yield advantage at farmers' fields and therefore, needs popularization immediately. Efforts are being made to identify the post-emergence herbicides which will help in containing weed menace in pulses.

6. Integrated management of diseases and insects

Insulation of crops against biotic stresses is most critical in enhancing crop productivity. *Fusarium* wilt is a serious problem in pigeonpea, chickpea and lentil causing substantial yield losses. Host plant resistance is the best option. However, non--availability of seeds of wilt resistant varieties and existence of pathogenic variability call for integrated approach of wilt management involving summer cultivation, seed dressing with fungicides like carbendazim, thiram and *Trichoderma* which are compatible with *Rhizobium,* promoting intercrops like pigeonpea with sorghum, chickpea with linseed and appropriate crop rotations. A glaring example of containing pigeonpea wilt has been well demonstrated in Maharashtra. There is an urgent need to ensure fungicide treatment of seeds being sold to farmers, adequate production and distribution of *Trichoderma* and creating awareness about the wilt resistant varieties. Gram pod borer (*Helicoverpa armigera*) is the key pest of chickpea and pigeonpea causing heavy losses. IPM modules for management of this dreaded pest have been developed and field-tested. Use of sex pheromone trap at 3-5 traps/ha for monitoring the pest is recommended. The economic threshold level (ETL) is 1-2 larva per metre row length. The integrated pest management strategy comprises timely sowing to exploit host avoidance phenomenon, intercropping with mustard, barley and linseed in chickpea and sorghum in pigeonpea, use of trap crops like *Vicia sativa* and African giant marigold, use of Nuclear Polyhedrosis Virus (NPV) @ 250 LE/ha or *Bacillus thuringensis* (Bt) @ 1-1.5 kg/ha, and erection of perches @ 20-30/ha to attract insectivorous birds. Spraying with 5% Neem Seed Kernel Extract (NSKE) or Achook @ 1.25 l/ha or need based use of chemical control like 0.004% fenvelarate at 15-20 days interval is recommended. Integrated management of diseases and pest including use of resistance cultivars, use of healthy seeds, modification of cultural practices, judicious use of fungicides and biocontrol agents may contribute substantially in stabilizing the yields. A comprehensive IPM module for different crops is given in Table 5.

Table 5: Integrated pest management modules for major pulse crops

Crop	IPM interventions		
	Before sowing	**At the time of sowing**	**Standing crop**
Chickpea	Deep summer ploughing, application of neem seed kernel powder @ 50kg/ha in nematode infested soils.	Timely sowing, seed treatment with carbendazim + thiram (1+2 g/kg or Trichoderma (4 g) + carboxin (1 g)/ kg, in nematode infested soils, seed treatment with carbosulfan @ 1%.	Use of pheromone trap @ 4- 5 per ha, installation of bird perches (35-40/ha), start insecticidal spray if the pod borer reaches economic threshold leveli.e. 1-2 larvae per linear meter- First spray of NSKE (5%)- Second spray of NPV 250 LE- Third spray(need based) of NPV @ 250 LE.
Pigeonpea	Deep summer ploughing, application of neem seed kernel powder @ 50kg/ha in soils infested with	Sowing on ridges, intercropping with sorghum (2:1 or 4:2), seed treatment with	Monitoring of the crop for the occurrence of diseases and pests, use of pheromone trap @ 4-5/ha, carbendazim @ 1g/kg moth trap or trichoderma trap for 3-4 consecutive (4 g) and vitavax (2 g). nights *Early maturing crop*- First spray of monocrotophos (0.04%) Second spray of NSKE (5%).- If required, third and fourth spray of either NSKE or NPV @ 250 LE. *Long duration crop*- First spray of dimethoate 0.03% (for pod fly).- Second and third spray similar to early maturing crop.
Mungbean & urdbean	Deep summer ploughing	Seed treatment with carbendazim+ thiram 1+2 g/kg.	In summer sown crops, timely irrigation minimizes thrips infestation, foliar spray of monocrotophos 0.08% or dimethoate 0.03% against pod borers and sucking pests, foliar spray of carbendazim (0.05 %) or

Contd.

			thiophanate methyl (0.025%) at 30 DAS.
Lentil	Deep summer ploughing.	Timely sowing to avoid rust, seed treatment with carbendazim + thiram (1 + 2 g/kg seed), or Trichoderma(4 g) + carboxin (1 g)/kg seed, intercropping (2:1) or mixed cropping with linseed (1:1).	Foliar spray of dimethoate 0.03% in case of aphid infestation, foliar spray of *wettablesulphur* (0.2-0.3%) or mancozeb (0.2%) against rust, removal of diseased plant debris.
Fieldpea	Deep summer ploughing	Timely sowing to avoid powdey mildew and foliarspray of rust, seed treatment with carbendazim + thiram (1 + 2 g/kg seed), or Trichoderma(4 g) + carboxin (1 g)/kg seed.	Regular monitoring of the crop, foliar spray of crop, endosulfan 0.07%, foliar spray of wettable sulphur (0.2-0.3%) against rust and powdery mildew, Carbend-azim (1g/litre of water) or dinocap, karathane 48EC (.05ml/l of water) are also used against powdery mildew, , removal of diseased plant debris.
Rajmash	Deep summer ploughing	Seed treatment with carbendazim + thiram (1 + 2 g/kg seed), or Trichoderma(4 g) + carboxin (1 g)/kg seed.	Foliar spray of metasystox or monocrotophos (0.04%) for aphids that are vectors of bean common mosaic virus, foliar spray of carbendazim (1g/litre of water) against stem blight, removal and destruction of stem blight affected plants, avoiding stagnation of water in the field for limiting the buildup of stem blight and root rot diseases.

7. Post-harvest technology

Besides losses during harvest, transportation and threshing, pulses also encounter insect pest damage during storage, which is estimated to be around 7.5%. A mass awareness programme to educate farmers on scientific storage along with distribution of seed storage bins can check the post-harvest losses to the tune of 0.45 million tons and increase the shelf life of pulses. Installation of efficient and modern *Dal* mills in the production hubs is expected to increase dal recovery by 11%, leading to a saving of at least 0.90 million tonnes of pulses. At village/panchayat level, the mini dal mills developed by IIPR, CIAE, CFTRI and other institutes need to be popularized so that the farmers can get higher return from their produce, store them for a longer period and generate self employment.

Improved production and protection technologies, including quality seeds, timely irrigation, farm operations, and effective control measures of biotic and abiotic stresses have the potential to double the pulses production. The present status of seed production in pulses indicates that indent for breeder seed is low because of poor seed replacement ratio (2-3%) as against the desired level of 10-15%. It is assessed that availability of quality seed in time can uplift about 20% production even with the available varieties. Therefore, a massive seed production programme is a must involving both public and private seed production agencies, NGOs, village panchayats, etc. Mass production of quality biofertilizers and bio-agents needs to be taken up for wide distribution among pulse growers. In the low productivity regions, yield can substantially be raised by alleviating biophysical constraints like lack of water and unfavourable temperature through better management or by matching technologies to withstand the constraint without much reduction in yield. Even a modest gain of 100 kg per ha from 10 million ha low productivity area under pulses would have a far-reaching effect on pulse supply in the country.

References

Deshpande, S.D. and Singh, G 2001: "Long Term Storage Structures in Pulses", National Symposium on Pulses for Sustainable Agriculture and Nutritional Security, Indian Institute of Pulses Research, New Delhi, 17-19 April.

Nadarajan, N and Chaturvedi, S.K. 2010. Genetic options for enhancing productivity of major pulses-retrospect, issues and strategies. J. Food Legumes 23 (1): 1-8

Saxena, N P and Johansen, C 1990: "Realised Yield Potential in Chickpea and Physiological Conside-rations for Further Genetic Improvement" in S K Sinha, P V Sane, S C Bhargava and P K Aggarwal (ed.), Proceedings of the International Congress of Plant Physiology, Vol 1, Indian Society for Plant Physiology and Biochemistry, Indian Agricultural Research Institute, New Delhi, pp 279-88.

Sharma, D and Jodha, N S 1982: "Pulses Production in Semi-arid Regions of India: Constraints and Opportunities", Economic & Political Weekly, Issue No 52, Vol No XVII. Share of different states in total production

❑❑❑

Chapter - 3

Issues and Technologies for Doubling Pulses Production in NEH Region

S.V. Ngachan, G.C. Munda, S.P. Das, Sanjay Gupta, Anup Das and D.P. Patel
ICAR Research Complex for NEH Region, Umiam, Meghalaya, 793 103, India
e-mail:svngachan@rediffmail.com

Pulse production in NE states in still too low as compared to whole India. However, there is very good potential for enhancing pulse producity in the region and contribute to food grain production of the country. In view of these constraints, prospects for increasing pulse production from this region have been visualized.

Pulse production scenario

Despite immense importance of pulses in human and animal nutrition, the share of north-eastern region in the total pulse are of India is too low (0.76%) and this region contributes only 176.44 thousand tons (0.99%) in the all India pulse production (Table 1). Within NE Region, most of the pulse area is occupied by Assam and 47% of total N.E. Region's pulse production is contributed by this state alone. After Assam, Nagaland and Tripura are major pulse growing states with 33.5 and 14.5 thousand hectares of area and 36.46 and 24.20 thousand tons of production, respectively. In Meghalaya, pulses were grown on 4 thousand ha area and 4.2 thousand tonnes of pulses were harvested in 2010-11. The fact that the average pulse productivity of NE region (883.0 kg/ha) is much higher than that of whole India (689 kg/ha) offers opportunity for increasing area and production of pulses in the region.

Table 1 : Area, Production and Productivity of Pulses in NEH Region of India (2000-01)

State	Area (000, ha)	Production (000, tonnes)	Productivity (kg/ha)
Arunachal Pradesh	8.9	9.05	1089.888
Assam	115.3	82.4	560.2775
Manipur	14.5	24.20	496.5517
Meghalaya	4	4.2	875
Mizoram	3.9	4.32	1666.667
Nagaland	33.5	36.46	1035.821
Sikkim	13.2	36.46	1035.821
Tripura	6.4	7.06	703.125
Total North East	199.7	176.44	883.0
All India	26.28*	18.09*	689
% contribution of NEH region	0.76	0.97	128.1

Source : Publication of Various State Governments and NEDFI, Gauhati (2010-11) & Directorate of Economics and Statistics, Department of Agriculture and Cooperation (2011).* area & production in million ha & million tonnes respectively.

The requirement and deficit of pulses of North Eastern region, considering per capita consumption at 18.25 kg have been worked out and presented in Table 2. The region as a whole is deficit if 78.79% pulse grain, ranging from 0.86% surplus in Nagaland to 92.33% deficit in Meghalaya.

Table 2 : Area, Production and Productivity of Pulses in NEH Region of India (2000-01)

States	2011 Population	Production (000, tonnes)	Requirement (2011 Population)	Deficit/Surplus (000, tonnes)	Deficit/Surplus (%)
Arunachal Pradesh	1382611	9.05	25.23	-16.18	-64.13
Assam	31169272	82.4	568.84	-486.44	-85.51
Manipur	2721756	24.20	49.67	-25.47	-51.27
Meghalaya	2964007	4.2	54.09	-49.89	-92.23
Mizoram	1091014	4.32	19.91	-15.59	-78.30
Nagaland	1980602	36.46	36.15	0.31	0.86
Sikkim	607688	8.75	11.09	-2.34	-21.10
Tripura	3671032	7.06	66.10	-59.93	-89.46
Total North East	45587982	176.44	831.98	-655.54	-78.79

Source : Publication of Various State Governments and NEDFI, Gauhati (2010-11)

Cropping intensity in the region is also very low (120 %) because of monocropping of cereal crops like maize, rice and rhizomatus crops like ginger and turmeric. Cultivation of pulses in north east region assumes greater significance in view of increasing demand for pulses and their role in soil fertility build up and reducing soil erosion. Most of the pulses are of short duration will also be useful in increasing the cropping intensity. Higher productivity of pulses could be obtained by adopting appropriate package of practices.

Major pulses of North Eastern Region

The major pulses grown in the region are green gram (*Vigna radiata*), black gram (*V. mungo*), pigeon pea (*Cajanus cajan*), rice bean (*Vigna umbellata*) and cowpea (*V. unguiculata)* in kharif and french bean (*Phaseolus vulgaris*), chickpea (*Cicer arietinum*), lentil (*L. culinaris*) and field pea (*Pisum sativum*) in *rabi*. In hills, various other beans such as faba bean (*V. faba*), adzuki bean (*Vigna angularis*), moth bean (*V. aconitifolia*) and broad bean (*Dolichos lablab*) are also used as pulse. In Meghalaya, pigeonpea is major pulse in *kharif* season and chickpea is a major pulse in *rabi* season. In *jhum* areas, rice bean is a predominant crop in *kharif* season.

Pigeonpea

Pigeonpea is the most widely grown pulse crop in the region. Pod borer and pod weevil are major threats to its cultivation in the region. A wide array of diversity in *Atylosia sp.*, a wild relative of pigeonpea occurs in India with maximum concentration in the Western Ghat in the North Eastern region. Out of 22 species of *Atylosia,* 4 species viz. *A barbata, A villosa, A volubilis* and *A scarabaeoides* are often found in NE region. In the NE region perennial types are commonly grown especially in garo hills of Meghalaya and Karbi Anglong district of Assam. There are variations in seed size and colour in local perennial germplasm.

Arhar is found growing up to 1000 m altitude. The promising arhar genotypes for low altitude (less than 100 m) are: Tripura local, T 21 and Pant A-3, and for medium altitude are: UPAS 120, H 76-44, AL-1 and H 76-11. Tripura local is a late maturing genotype while others are of early maturity group (140-150 days). These genotypes can be grown at lower altitude as well as on dry terraces and hill slopes. These can be sown with the onset of monsoon in April/May. These genotypes are also suitable for intercropping with maize, soybean, groundnut etc. These varieties are having an yield potentials of 1.5-2.5 t/ha in the region.

Mungbean and Urdbean

Mung bean is grown as *kharif* pulse in NE region and pod decay due to rains is the most severe problem. Sowing time of this pulse is critical for successful harvest. It should be sown in first week of August so that at the time of pod maturity the rains are almost over. Local collections of mungbean are semi bushy to bushy types exhibiting much variation in leaf size, inflorescence nature and pod type. Considerable variations prevail in grain colour (green, yellow, brown, black and mottled), the common colour being green.

The sowing of urdbean, should also be completed in first week of August. This crop is better adapted to this region than mungbean. It is one of the most extensively grown pulses in NE region especially in Assam. Local collections are semi spreading types. Much variability in days to maturity occurs usually in medium to late maturity group. The indigenous types are observed to be photo period sensitive. One of its forms 'Pahelo Dal' characterized by its green seed coat is very common in this region.

Mungbean and urdbean have been found suitable for low as well as medium altitudes. Mung bean genotypes, namely, MH 309, PDM 11, PDM 54, ML 5, PDM 14, ML 326, PDM 62 and urdbean genotypes, namely, T 9, UG 203, B 3-8-8, Pant U 19, UG 301, PDU 2, LU 4, JU 77-41 have been found suitable in low and mid altitudes of Meghalaya. The yield potential of these genotypes is 9-10 q/ha at low altitude and 6-9 q/ha at medium altitude (up to 1300 m). The mungbean matures in 70-80 days whereas urdbean matures in 80-90 days.

French bean (Rajmash)

French bean (Rajmash) is an important and highly profitable vegetable/pulse crop of north-eastern hill region of India. It occupies an important position among various *kharif* pulses and vegetable crops grown in temperate hills of India at an altitude ranging from 700 m to 2500 m. Generally, it is cultivated for vegetable purposes round the year except winter months (October – January), in foothills as well as in mid altitude in all the North Eastern States including Sikkim. For vegetable purposes, the round pod type with more flesh and less string is preferred. Flat pod type is mainly grown for whole seed. The increase in productivity of French bean/Rajmash will meet the increasing demand of green vegetables/pulses of the region. A yield potential of 10 t/ha of green pod yield of frenchbean, Naga local (pole type), has been achieved in mid altitude of Meghalaya. Because of its shorter duration (90 days) frenchbean – maize, Maize-frenchbean, rice-french bean cropping systems are possible with high yield potential. Another significant aspect is that the crop has less pest and disease problems and very much suitable for organic farming. Suitable bush type varieties are: VL 63, PDR 14, Arka Komal and contender with yield potential of 10-20 q/ha and pole type varieties are Nagala local, Meghalaya local, Manipuri etc with yield potential of 10-15 q/ha seed yield.

Rice bean

In the North Eastern Region of India, rice bean is grown predominantly under the rainfed conditions in a mixed farming system under the shifting cultivation or in kitchen gardens. Rice bean is used as food, fodder, green manure and cover crop. The dried seeds are usually eaten boiled or in *dal* form. Young immature pods are used as vegetables. Rice bean seeds are also used as animal feed. It contains high amount of protein and carbohydrates, starch, calcium, iron and phosphorus.

Rice bean assumes considerable importance as an underutilized and under exploited legume and has recently gained attention as a supplementary food crop with potential economic importance for the future. Rice bean possesses high nutritional quality and grain yield. It has resistance to yellow mosaic virus, good storage quality, rich genetic diversity and grain yield and is a hardy crop with wider acceptability. Rice bean has adapted to humid conditions of north east through its highly photosensitive nature. It is a short day plant (Sarma *et al.*, 1995) and irrespective of sowing it flowers when day length is reduced in September. This ensures pod formation and maturation in the post rainy season.

In addition to its use as a pulse crop, it is also used as a forage crop. The seed protein of rice bean varies from 14-25%, the amino acid composition specially methionine and tryptophane, is considerably high compared to other *Vigna* species. It contains high quality of vitamins, thiamine, niacin and riboflavin. Singh *et al.* (1980) reported that calcium and iron contents are also appreciably high. Phytin phosphorus content which generally inhibits the phosphorus availability and lowers protein digestibility in most of the Asian pulses is low in rice bean.

In view of the enormous variability, high yield (15-20 q/ha), tolerance to disease and pest and wider adaptability, this under exploited legume, grown predominantly in mixed farming system, has great potential to become an important pulse crop of the region. A number of suitable genotypes of rice bean have been identified for Meghalaya. among these RCRB 1-6 (19-23 q/ha), RBL 1 (15-18 q/ha), MNPL 1 (13-16 q/ha), PRR 1 (10-12 q/ha), RBL 6 (12-16 q/ha) and MNPL 2 (9-10 q/ha) are promising ones. These genotypes mature in 120-150 days.

Pea

Pea is grown in all the states of North East India but mainly used as vegetable and only a part of it is used as dal. Pea is the most successful *rabi* crop in the region. Lowland rice fallows with ample water offer tremendous potential to increase its area in the state. In *rabi* crops there is a general problem of maturity because of low temperature and rains during March / April but pea has the capability to mature under these conditions also. Open grazing during winter is the major obstacle in raising pea. This crop suffers heavy losses due to rust and powdery mildew and there is a need to breed resistant varieties suitable for this region.

Varieties DMR 7, DDR 50, TRCP 8, TRCP 9, Arkel, Rachana and Ambika have been found suitable for both mid and low altitudes. TRCP 8 and TRCP 9 are late maturing genotypes (130-140 days) and produce 18-20 q/ha while other varieties mature in 100-125 days and produce genotypes 15-17 q/ha.

Lentil

Lentil is the most favourite pulse among the people of the NE region but currently it is not grown at a larger scale. Trials conducted have shown that it ia a promising crop for the region. Like pea, this crop also has the potential to mature in low temperature and cloudy conditions.

Varieties, L 4147, PL 639, PL 4, VL 1 and VL 125, have been found suitable for mid altitude. These varieties mature in 120-130 days and produce 8-10 of ha. Varieties, Pant L 406 and Sapana, have been found suitable for low altitude. These varieties mature in 135-140 days and produce 14-15 of q/ha.

Chickpea

Chickpea is also a favourite pulse among the local people of the region. Acidic soils of this region make phosphorus insoluble which is not available to plants. Chickpea exudates malic and other organic acids through their roots, which dissolve insoluble P compounds in the soil that are normally not available to other plants (Bolland *et al.*, 1999). This crop exhibits excellent vegetative growth, flowering and pod setting but unlike pea and lentil, it fails to mature when temperature is less and rains are there at the time of maturity. Its seed has poor shelf life and high humid conditions prevailing throughout the year in the state markedly reduce the seed germination.

Cowpea

Cowpea is rarely grown as a pulse crop in NE region. It is mostly grown for vegetable purposes. The native genetic resources possess much wider variability for their cultivation in the plains, hills and even in stressed environments. Good variation is exhibited in growth habit, photoperiodic and photothermic sensitivity, seed size, pod shape and colour. Varying selection pressure had resulted in enriching the variability of grain, vegetable and fodder types of cowpea in NE region.

Minor pulses

Adzukibean (*Vigna angularis*), more akin to rice bean, is collected from Arunachal Pradesh. Viney forms prevail that possess thinner pods like mungbean with red, green and creamish grain colour. Large variation in pod and grain size of broad bean (*Vicia faba*) from NE region is observed. Local collection of moth bean (*Vigna aconitifolia*) possesses good variation in growth habit (spreading types predominant), leaf location, grain size, pod and seed colour. Local *khesari* (*Lathyrus sativus*) germplasm largely has a spreading growth habit and is well adapted to drought. *Vigna vaxillata* is another less known potential pulse cum tuber crop of this region which resembles cowpea and produces both edible seeds and tubers. At the younger stage, the pods are used as vegetable and at maturity, the seeds are used as *dal* by tribals. The tubers are also used as a substitute for tapioca in animal feed. Both the seeds and tubers are very rich in protein content.

Constraints of pulse production in NEH Region

A number of biotic and abiotic factors limit the pulse production in this region. The important ones are discussed here.

A. Biotic factors

Weeds : As a result of conducive weather conditions, prolific growth of weeds causes serious problems in pulse crops of the region. The soil is also rich in unhumified organic matters. Weed problem thus assumes an alarming proportion and is a major hindrance in crop production. The undulating land surface with large variation in slopes (0-100%), altitude and rainfall pattern leads to a diversification of weed flora. The high rainfall during April to September and sparse rainfall during winter is a major factor supporting the diverse weed flora. The weeds compete with pulse crops for the nutrients and harbour insect pests. Because of excessive rains during cropping season, it has become difficult to control weeds either manually or by spraying herbicides.

Insect pests : Insect pests are serious yield reducers of pulses in NE region and cause 20% loss of crop yield. Since the climatic conditions favour growth of different types of agricultural crops including pulses, it is also congenial for the multiplication of insect- pests. Therefore, insect pest problems in the region are many and serious. Several species of the insect pests attack pigeonpea from germination to maturity. The most important insect pest of pigeon pea and chickpea is pod borer. On an average, 20 – 40% crop is annually lost due to damage caused by pod borers in these two crops. The other pests of pigeon pea are pod- boring weevil (*Apion clavipes*), pod fly (*Agromyza obtuse)*, blister beetle (*Mylabris phalerata, M. pustulata*), aphids (*Aphis craccivora*) and flea beetle (*Chaetocnema basalis*). Leaf folders, aphids, Bihar hairy caterpillar and blister beetle are important pests of mung and urd. Cut- worms, flea beetle, pod borer, aphids and blister beetle infest gram, lentil, rajmash and other beans.

Diseases : Plant diseases and insects cause economic losses in the yield to a great extent. It has been found that diseases cause 25% loss of crop yield. This region receives mean annual rainfall of more than 2000 mm and remains humid almost throughout the year. A survey conducted in NEH region for a number of years revealed that leaf spot, powdery mildew in green gram and black gram, wilt and anthracnose in pigeon pea, leaf spot, anthracnose, rust, pod blight in french bean, leaf spot in cowpea, rust, powdery mildew and cercospora leaf blight in rice bean are the major diseases of pulses.

B. Abiotic factors

Soil factors : About 95% of the soils of the region are acidic in reaction. Besides H^+, about 20% of the exchangeable complex is saturated with exchangeable aluminium which results in aluminium toxicity to plants. Generally, the soils of high altitudes under high rainfall areas are strongly acidic imposing tremendous limitation to high productivity. Phosphorus deficiency is wide spread in the region. It results in stunted growth and decreased grain formation leading to poor yields.

Topography and climatic factors : The altitude in the region varies from 50 m to 2000 m msl. The topography is undulating with mountains and valleys. Such wide variations present problems because of the necessity to identify varieties and technologies suited to the respective situations.

The region is a high rainfall area with the precipitation ranging from 1200 mm to 12000 mm. Most of the rainfall is concentrated during the months of June to September. Pod decay in short duration pulses like mung, urd, rajmash is very common. Delayed sowing of these pulses results in water stress and cold injury. Rainfall pattern and intensity may also vary at short distances at some places making the choice of varieties difficult. High humidity is conducive for the development and spread of diseases in an epidemic form, leading to great losses in crop. High humidity also affects seed drying and storage adversely. During the monsoon period small periods of droughts are also not uncommon. The winter months are almost dry in majority of the places.

Socio-economic factors : Jhuming on hill slopes, dry or wet terrace and valley land cultivation are very common; such situations have specific requirement of crops and varieties. The present day high yielding varieties may not be suitable for low input-low risk primitive technology. In this region, the land generally belongs to the village chief or owned by the community. As a result, the individual farmer does not own the land and this affects the development and maintenance of cultivated land.

Non-vegetarian food habit : The farmers of NEH region give low priority to the pulses because of lack of general consumer acceptance as the food habit of NEH region is non-vegetarian.

Quality seed : Availability of quality seed of improved varieties of pulses to farmers may ensure bumper harvest. Most of the time quality seeds are not available to farmers and they are compelled to use their own seed stock of traditional varieties. Sound seed production programme does not exist in this region and varied agro-climatic conditions of NEH region require the availability of quality seed of location specific varieties. Presently, the seed replacement rate is very low (2 – 5%) which has to be increased manifolds. High humidity in the region throughout the year does not let the seed moisture come down to storage level and this causes infection by various storage pathogens. The intensity of insect damage during storage is also increased.

Irrigation for rabi crop : This region faces severe drought during winter season which limits the production of *rabi* crops. Effective rainwater harvesting during rainy season and utilization of harvested water for irrigation of *rabi* pulses would increase the area and production of pulses in this region.

Integrated nutrient management : Most of the pulse growing areas in this region have low to medium population of native rhizobium. Seed inoculation with bio-fertilizer (Rhizobium and PSB) can increase pulse productivity by 10 – 12%. Lack of quality culture in adequate quantity is one of the major constraints in popularization of bio-fertilizers. Application of fertilizer and manure improves pulse productivity substantially.

About 20 kg N/ha is required as starter dose for most of the pulses except frenchbean which requires about 60 kg N/ha.

Perspectives of increasing pulse production in NEH Region

India is facing deficit of pulses, which can be met by horizontal as well as vertical expansion. North Eastern Region is one of the potential areas where pulse production can be increased horizontally by utilizing a part of 1.67 million ha area under *jhum* cultivation (Bujarbaruah, 2004). Considering that pulses could be grown mixed with other crops and their feasibility for relay cropping; the area and production of pulses could be increased vertically. Intercropping system not only covers the risk but also gives a bonus yield of pulse crop in addition to normal return of the main crop.

Among the various possible approaches for increasing pulse production, one approach is to increase production per unit area and time. Usually spring and (before *kharif* rice sowing) autumn (after *kharif* rice harvesting) seasons in this region go as fallow. Short duration pulses like mung, urd and rajmash offer the opportunity for utilizing these seasons. Incorporation of pulses in rice and maize based cropping systems of this region would also increase the soil fertility in a sustainable way.

From the pulse area, production and productivity analysis, it is apparent that productivity of pulses in this region is better than that of whole India. The main concern is to increase their area. Rice being the staple food of this region is favoured against pulse as *kharif* crop. Non-vegetarian food habit of the people of this region also places pulses as the crops of secondary importance. Availability of pulse varieties that may compete with rice in terms of economic return and their assured purchase by government may increase the area of pulses in this region. Extension activities to demonstrate the potential of high-yielding varieties may also augment the pulse area of this region.

Seed is the basic input for bumper crop harvest and must be supplied to farmers. Quality seed production and distribution is a major constraint for improving production of pulses in this region. A sound and co-coordinated seed production programme is urgently required in the region. Production of breeder and foundation seed has to be taken up by institutions like ICAR Research Complex for NEH Region, Barapani, AAU, NSC, etc. For the production of certified seed, active participation of farmers is required. Progressive farmers are required to be identified, trained in various seed production activities and assisted through seed producing agencies during seed production programme.

Integrated nutrient management is necessary for increasing pulse production in this region. Pulses are capable of meeting out their nitrogen requirement through symbiotic nitrogen fixation by Rhizobium. As stated earlier, these crops require initial starter dose of nitrogen. e.g. seed inoculation with biofertilizer (*Rhizobium* and PSB) - low cost inputs - can increase pulse productivity by 10-12%. Lack of quality cultures in adequate quantity is one of the major constraints in popularization of biofertilizers. *Rhizobium* inoculation to blackgram was found to increase seed yield by 23.5%, when the biofertilizers was coupled with FYM @ 10t/ha, the response was overwhelming (Raychaudhury and Nagachan, 2001). The potential of established/local strains of biofertilizers especially in problem soils of north east has been successfully applied. The survival of *Rhizobium* is greatly affected by soil reaction. In acid soils, *Rhizobium* population declines but increases with the application of lime. Liming increased the efficiency of *Rhizobium* by increasing cell number under agroclimatic conditions of Manipur. Raychaudhury and Nagachan (2001) reported from Manipur, that the *Rhizobium* colonies increase by 31.7 folds with liming. In acid soils of NE Region, the inoculated seed should be treated with 1.5 kg finally powdered lime and kept for five minutes after thorough mixing and uniform pellets of seeds are made (Sarkar *et,al.*, 1989).

Prevalent favourable conditions for the growth of microbes and insect pests result in significant yield losses due to diseases and insect-pests. Breeding of varieties resistant for diseases and insect- is complicated by the presence of various strains of pathogens. Characterization of pathotypes, screening of individual genotypes with individual pathotypes, identification of resistant genotypes and using these genotypes for breeding are required for development of resistant varieties. There is need to supplement the resistant breeding work with molecular markers for effective incorporation of multiple genes for resistance (gene-pyramiding).

Utilization of rabi season for growing crops may boost pulse production in this region. Lowland rice fallow with ample moisture and bright sunshine offers great potential for the production of rabi pulses like pea, lentil and chickpea. For effective utilization of *rabi* season it, is essential to check the open grazing and develop suitable water harvesting strategies since there is little or no rain during rabi season.

To avoid the post harvest losses, post harvest management is very necessary. Most of the post harvest losses are because of lack of proper storage structures. This causes rapid deterioration in seed quality. High moisture conditions throughout the year result in heavy disease and insect incidences during storage. Lack of small processing units compels farmers to sell their produce as such without any value addition. Low cost *dal* mills (Gupta *et al.*, 2006) and metal storage bins are available which can be popularized for improving storage condition and increasing value of pulses.

ICAR Research Complex for NEH Region, Barapani has been making sincere efforts to tap the potential of NE states for organic production. Under organic conditions, use of pulses becomes inevitable and research efforts directed towards the development of pulse varieties for organic conditions are required.

The common issues for pulse based systems

1. Resource conservation

- Since pulses are generally grown in rainfed areas, greater attention should be given for rain water harvesting in farm ponds and community reservoirs for providing at least one or two live saving irrigation at critical stages.

- In rainfed areas mainly conventional cultivation practices are followed due to lack of improved implements. In areas like NEH specific small implements/ tools needs to be popularized.
- Promoting ridge-furrow and raised bed planting would help diversify pulse based systems.
- Promoting IPM technology including weed management is vital for this region.
- Ensuring the availability of critical inputs like appropriate seeds, biofertilizers, superphosphate, sulphur, zinc, biopesticides *etc.* can never be understated.

2. Expansion of area

- Area expansion in potential areas such as rice fallows of North eastern India has a vast potential for the horizontal expansion of *rabi* pulses like lentil, lathyrus, chickpea and fieldpea.
- The major constraints of pulse production in rice fallow is the excess soil moisture due to land locked condition and release of water from hills during winters. To overcome this, a simple land configuration of raised and sunken beds with provision of drainage at appropriate intervals was found effective. Rice fields needs to be drained completely at physiological maturity so that the field will be dried by the time rice crop is harvested. Growing pulses like pea, lentil, lathyrus etc. are feasible as soil condition improves and at the same time the field remains at field capacity. This soil moisture is adequate for a good crop growth and productivity. On the other hand growing a pulse crop is hardly possible in upland during dry season due to lack of soil moisture and irrigation facilities.
- Similarly urdbean and mungbean have the potential in uplands and can be grown as spring/summer crops before *kharif* season rice. PD 4 black gram and TMV 37 green gram have been identified from mid altitudes of Meghalaya having good productivity.
- Promoting pulses in intercropping system would sustain soil health and generate extra income to farmers. Experimental results at ICAR Research Complex for NEH Region, Umiam indicated that pulses like pea can be grown profitably even with cabbage/broccoli etc as intercrop.

3. Issues related to transfer of technology

- Promoting pulses in intercropping system would sustain soil health and generate extra income to farmers.
- Sensitization of farmers about the potential and importance of pulse production in the region

- Feasibility of intensive mono crop pulse production under upland ecosystem needs to be tested.
- Popularization of pulses as pre *kharif* and pre *rabi* crops.
- Testing and development of suitable varieties for the region. Varieties tolerant to moisture stress will be suitable for upland ecosystem.
- Agronomic package of practices for different pulses under NEH condition need to be standardized.
- Different crop rotations need to be tested to find the most suitable ones.
- Package of practices for mixed cropping of pulses need to be developed.
- Effective management of Leaf spot and Powdery mildew diseases, pod borer in pigeon pea and blister beetle in green gram.
- Standardization of post harvest management and value addition practices.
- Establishment of mini '*Dal* Mills'

Cropping systems

Pulses are mainly grown in uplands in NEH region. *Jhum* cultivation is also common in NEH in which usually mixed cropping is practiced. The farmers get a piece of land for 1 to 3 years on lease and as soon as the soil fertility declines, weed and pest infestations increases and they abandon the land and return to previously abandoned farmland or start *Jhuming* again on another leased land. The cropping system can more precisely be described as relay mixed cropping as crops are sown in the same field before the harvest of previous sown crops in the same field. Depending on the size of farm land and their resources, pulses are grown by the farmers. The predominant cropping systems prevailing in the region are given below :

1. Upland rice + Pulses
2. Lowland rice – Pulses (Pea/lentil/green gram/black gram/frenchbean etc.)
3. Maize (green cob) – Pulses in pre *rabi* season
4. Maize (green cob) + Pulses
5. Rice/maize- French bean/pea/lentil

Besides the above combinations, there are possibilities of growing pulses like green gram and black gram during spring season in the region as this season starts receiving rains from the month of March onwards, when the *jhum* cultivation begins. Similarly, after harvest of maize crop, pulses can be grown during pre *rabi* season without any moisture stress to the crop to increase cropping intensity.

Zero tillage – a viable option for pulse production in rice fallow

Conventionally after *kharif* rice, fields remain fallow in lowland areas, mainly due to excess moisture caused by seepage from surrounding hillocks in mid altitude regions. Proper drainage, like draining water from rice field completely at physiological maturity creates favourable condition for cultivation of a successful *rabi* pulses like pea, lentil etc. A simple drainage system around the rice fields/plots with appropriate outlets creates the desirable situations. To study the performance of pulses like pea and lentil, 4 varieties each of pea and lentil were obtained from IIPR, Kanpur, UP during 2009. Pea and lentils were grown under zero tillage in lowland rice fallow using recommended dose of NPK (20:60:40kg/ha). One weeding cum hoeing was given manually at 30 DAS.

Among various pea varieties tried, IPFD 99-13 recorded maximum green pod yield (41 q/ha) followed by IPFD 1-10- (32.9 q/ha), IPFD -99-25 (30.32 q/ha) and HUDP (17 q/ha).Among the lentil varieties tried, DPL-15 recorded maximum seed yield (10.9 q/ha) followed by DPL 62 (8.71 q/ha) and IPL 406 (4.8 q/ha). Therefore, pea and lentil increased the system productivity and farm income. With appropriate agronomic interventions and varietal screening, pea and lentil could be popularized at mid altitude for food and nutritional security of small and marginal farmers, especially the tribal farmers of the region.

Managing soil acidity and health

The soils of the region are mostly clay loam to sandy clay loam and are acidic in reaction (pH 4.5 - 5.5). Soil acidity affects nutrient availability, biological activity and ultimately the crop productivity. The upland *jhum* soils are moderately to highly acidic in nature, and accordingly lime application is very essential to ameliorate the soil pH to improve nutrient availability and also the solubility of other plant nutrients like phosphorus. Aluminum and manganese which are toxic due to more uptake, are also neutralized.

Some species of Rhizobia tolerate acidity better than others. Reduction of soil acidity and associated production of legume nodulation and N_2 fixation can be achieved by liming. Phosphorus deficiency is a factor that commonly restricts the realization of the potential of nitrogen fixation by pulses. It has been found that dual inoculation of pulse crop with Rhizobia and phosphate solubilizing bacteria (*Pseudomonas striata, Bacillus polymixa, Aspergillus awamori*) enhances phosphorus uptake by plants and consequently gives better nodulation, plant growth and yield in various pulses crops. These phosphorous solubilizing bacteria (PSB) have the capability to solubilize inorganic phosphorus from insoluble sources.

Liming also improves biological nitrogen fixation in legumes apart from improving the physical and microbial activities of soil. Lime application in general increases calcium and magnesium availability to the crops/ plants. Application of about 500 kg lime/ha in furrow is sufficient for achieving optimum yield of most of the crops.

Early sowing of crops favours early crop root growth and thereby, improves the efficiency and uptake of nitrogen from organic nitrogen reserves of the soil. Growing of legume/ pulse crops either as mixed cropping or in rotation with non legume crops, helps in maintaining soil fertility due to biological nitrogen fixation. Ploughing plant residues in the soil will add nitrogen by the decomposition of such residues besides killing of hibernating insects. Seed treatment with bio fertilizers preferably by Rhizobium and Phosphorus solubulizing mycrolyze (PSM) will augment Nitrogen and Phosphorus availability in the soil. Application of organic manures is very essential for increasing pulse productivity. Application of FYM @ 5 -10 t/ha either by broadcasting or application in the furrows improves physico-chemical properties.

Conclusion

The fact that productivity of pulses in NE states is better than that of whole India suggests far bright prospects of increasing pulse production from this region. The area under shifting cultivation may slowly be diverted for production of pulses and other crops, which would help in conserving soil and improving its health. Even if 20 % of this area becomes available, the region may become self-reliant in pulse production. Development of short-term varieties of mungbean, urdbean and rajmash for spring and autumn season holds the prospect of avoiding competition of pulses with rice and increasing area and production of pulses in this region. There is vast potential for introduction of pulses in rice fallow which remain idle during rabi season. Zero tillage has got potential for pulses cultivation that would save time, labour and other resources. With the identification of NE region for organic crop production, pulses area and production are likely to increase many folds. Pulses like rice bean, urd bean etc, not only serve as multiple sources of food, feed, cover crops but also build soil fertility. Large scale demonstrations and training programmes need to be undertaken for popularizing pulse cultivation disseminated.

References

Anonymous 1997 Annual Report of ICAR Research Complex for NEH Region, Barapani.

Anonymous 2011 Publications of Various State Governments and NEDFI, Guwahati 2010-11.

Bolland, M.D.A., Siddique, K.H.M., Loss, S.P. and Baker, M.J. 1999. Comparing responses of grain legumes, wheat and canola to applications of superphosphate. Nutrient Cycling in Agroecosystems, 53 : 157-175.

Bujarbaruah, K.M. 2004 Organic Farming: Opportunities and challenges in North Eastern Region of India. In: Nature 2004, Souvenir of International conference on organic food 14-17 Feb at ICAR Research Complex for NEH Region, Barapani. pp 13-23.

Gupta. S., Bujarbaruah, K.M., Dhiman, K.R., Pathak, K.A., Das S.P., Das, Alpana, Kumar, Arvind and Vijaypal 2006 Pulse Production Technology for North East India. Technical Bulletin No. 23, ICAR Research Complex for NEH Region, Barapani (Umiam), Meghalaya.

Raychaudhury, M. and Nagachan, S.V. 2001. Selection and application of biofertilizers in north east region. In Souvenir, IInd NER conference on biofertilizers held at AAU, Jorhat w.e.f. 22-24 Jan, 2001 pp 59-62.

Kannaiyan S., 2000 Perspective of increasing Pulse Productivity in Tamil Nadu. In Pulses Production Strategies In Tamil Nadu, Centre for Plant Breeding and Genetics, Tamil Nadu Agricultural University, Coimbatore.

Sarkar, B.B., Bhattacharya, N.R. and Roy, S. 1989 Rhizobium culture helps increase the yield of black gram. Indian Farming, 38: 13-14.

Sarma, B.K., Singh, M., Gupta H.S., Singh, G. and Srivastava L.S. 1995. Studies in rice bean germplasm. Research Bulletin 34. ICAR Research Complex for North East Hill Region, Barapani (Umiam), India pp70.

Singh, S.P, Mishra, B.K, Chandel, K.P.S. and Pant, K.C. 1980 Major food constituents of rice bean (Vigna umbellata). J. Food Sci. Techno. 17: 238-240.

Singhal, V. (1999) Indian Economic Data Research Centre, New Delhi. pp. 105-110.

Vavilov, N.I. 1951 Origin, variation, immunity and breeding of cultivated plants. Chron. Bot 13: 1-364.

❑❑❑

Chapter - 4

Crop Diversification Through Pulses

B. Gangwar and V. K. Singh
Project Directorate for Farming Systems Research, Modipuram, Meerut- 250 110 (U.P.)

Indian agriculture is characterized by small farm holdings. The average farm size was only 1.21 hectares during 2005-10 and it would be 0.68 ha by 2030. Around 85% of farmers have land holdings smaller than 2 ha and they cultivate nearly 60% of the available land. The contribution of agriculture sector to the total (GDP) has decline from 38% in 1980 to 29% in 2000 and 15% in 2010-11. The per capita availability of land was 0.15 ha in 1999-2000, which will reduce to 0.11 ha by 2020. As per estimates, India would require about 345 million tones of food grains for feeding about 1.5 billion populations by the year 2020. Approximately 90% of total farm holdings are categorized as small and marginal. There is hardly any scope of horizontal area expansion. The situation is, further, aggravated due to globalization of market and trade relaxations, which call for competitiveness and efficient agricultural production system to face never challenges on the ecological, climatic, equity, social justice and employment front. In areas, endowed with adequate irrigation resources, short duration pulses, oilseeds and other high value crops will find their definite niche as sequential or intercrops, rather than replacing the major cereal crops having higher yield potential combined with high stability. Hence, an increased cropping intensity will contribute substantially to additional demands of food and cash crops.

To enhance land use efficiency augment crop productivity and maintain soil fertility, use of short duration legumes are considered as an important option to meet the future needs of food and partial protein requirement of the growing population. The PDFSR has identified 30 major cropping systems in India (Yadav *et al,* 1996). Majority of predominant systems are cereal based double cropping systems and there is considerable scope to intensify the existing systems using high value short duration crops specially pulses (Grain legumes). Any cropping sequence can be compatible and complete only when a legume is included in it. Pulses occupy a strategic position in intensive as well as subsistence agriculture, as they are an excellent source of dietary protein for millions of people, nutritional feed for livestock and mini-nitrogen plant having profound ameliorative effect on the soil. Legumes are excellent crops for natural resource management, environmental security, crop diversification and consequently for viable agriculture. Availability of short duration varieties coupled with matching agro technologies has resulted in identification of several remunerative, and more productive cropping systems which have either already shown their promise or have tremendous potential for expansion in new niches and diversification in the existing cropping systems. A vast potential exists for intercropping of quick maturing pulse crops with widley spaced crops like sugarcane, banana, cotton, sorghum etc., which is least utilized. With the introduction of drip irrigation, scope to grow short duration pulses with horticultural crops is also enormous.

Inclusion of pulses as intercrops in sequential as well as intercropping systems would not only make the country self reliant but also help in improving the soil health. The cropping systems involving legumes in particular and having considerable potential through intensification are discussed here for ready reference of researchers, planners and extension workers.

Pre- dominant cropping systems of India

The country is endowed with diverse climate and soil types to support diversified agriculture, which is the major strength to meet food, fodder, fibre and finance requirements of ever increasing population. The prevalent cropping systems are the outcome of the technological innovations, household needs, reflection of government policies, and availability of production inputs, market forces and socio-economic compulsions. Yadav and Rao (2001) enlisted 250 double cropping systems of primary and secondary importance in terms of their spread in the country out of which 30 are of primary importance. These are : rice-wheat, rice-rice, rice-gram, rice-mustard, rice-groundnut, rice-sorghum, pearl millet-wheat, pearl millet-gram, pearl millet-mustard, pearl millet-sorghum, cotton-wheat, cotton-gram, cotton-sorghum, cotton-safflower, cotton-groundnut, maize-wheat, maize-gram, sugarcane-wheat, soybean-wheat, soybean-gram, sorghum-sorghum, groundnut-wheat, sorghum-groundnut, groundnut-rice, sorghum-wheat, sorghum-gram, pigeon pea-sorghum, groundnut-groundnut, sorghum-rice and groundnut-sorghum. However, among the top ten popular cropping systems in the country, cereal-cereal with major share of rice-wheat (9.85 m ha) and rice-rice (5.89 m ha) are the most dominant systems contributing about 52.3 % to the national food basket (Table 1). There is a very common observation that a large diversity of cropping system exists in rainfed areas with an overriding practice of intercropping while in areas with assured irrigation facilities, the major cropping systems have tilted in favour of cereals mainly because of availability of high input responsive and photo insensitive varieties of rice and wheat. In central India (Madhya Pradesh and Maharashtra), pulses are intercropped with millets and oilseeds crops during *kharif* season. Maize + pigeon pea, maize + urd bean or mung bean, sorghum + pigeon pea, sorghum +mung bean or urd bean, groundnut + pigeon pea, cotton + mung bean or urd bean, cotton + pigeon pea, etc. are some of the popular cropping systems. In *rabi* season, chickpea crop is grown under rainfed condition as mono-crop as well as intercropped with linseed, safflower, coriander, etc.

Table 1. Predominant cropping systems of india

Cropping Systems	Area (m. ha)
Rice-wheat	9.85
Rice-rice	5.89
Pearlmillet-wheat	2.26
Soybean-wheat	2.23
Maize-wheat	1.86
Rice-vegetables	1.24
Cotton-wheat	1.09
Rice-groundnut	1.02
Sugarcane-wheat	0.97
Rice-fallow	4.42

Source : Yadav & Rao (2001)

Need of crop diversification through pulses

Although with the adoption of modern agricultural technologies, the country has become self sufficient in food grain production, some ecological imbalances have set in during the recent past . For example, at present, rice and wheat share 46% of cultivated area and area under other arable crops, especially pulses, is declining continuously (Timsina and Connor, 2001). This area expansion under rice and wheat crop has resulted in several new serious problems. They are as follows:

a) Excessive use of ground water leading to poor water use efficiency and depletion of ground water.

b) Deterioration in soil health.

c) Multiple infestations of weed flora, insect pests and diseases.

d) Indiscriminate use of energy (chemical, electricity or diesel).

e) Reduction in the availability of other protective food and high value crops.

f) Pollution of the ecosystem.

A changed scenario with respect to determinants of food requirement and natural resource management is foreseeable in next two decades. It is imperative that future agricultural systems be guided not only by the compulsion of improving food security, but also by the concerns for quality of food, environmental protection, resource sustainability and system profitability. Therefore, we may expect a major shift in cereal based agricultural production systems towards systems involving selected pulses, oilseeds, vegetables, fruits and flowers in view of emerging needs and global trade opportunities arising from open economy. Therefore, new potential crops with higher and stable yields and/or profits will continue to find a place in to the existing cropping systems of different agro-ecological regions. Further, because of integration of world markets, urbanization and rising personal incomes, significant shifts in cropping systems and other farm enterprises may also take place. In fact, there would be more growth in the demand for pulses and milk, relatively slow growth for cereals and decline in per capita demand for root and tuber crops. Therefore, intensification with pulses would be essential to cater the emerging needs.

Adoption of crop diversification helps in rotational use of basic resources like land, water and vegetation in such a way that it serves the objective of accelerated growth, employment and eco-system protection. For example, introduction of pulses in rice-wheat system has the ability to fix atmospheric N, its roots posses soil binding capacity, reduce sub-soil compaction by decreasing bulk density and help sustaining soil fertility (Dwivedi *et al*., 2003). Due to all these factors, the succeeding cereal crop have higher yield and productivity. Results of field experiments conducted at Project Directorate for Farming Systems Research (earlier PDCSR), Modipuram, Meerut, have revealed that by substitution of rice with pigeon pea in rice–wheat system an amount of 78- 89 ha-cm water can be saved.

Present scenario of pulses in cropping systems

Traditionally, pulses have been considered as important elements of cropping systems. They were popular because of their importance as a source of protein and ability to fix atmospheric nitrogen (N) and thus improve soil fertility. With the introduction of irrigation and due to high profitability of alternative sources of soil nutrients in the form of inorganic fertilizers in the mid–1960s, pulses were replaced or relegated to marginal lands. During the late 1960s and early 1970s, a large area under pulses in the Indo-Gangetic Plain (IGP) was substituted by high–yielding varieties of rice and wheat. The new technology of rice and wheat substantially changed the agricultural scenario and contributed largely to increase in agriculture production. With the passage of time, excessive use of chemical fertilizers and irrigation in rice and wheat to maintain their productivity has created an imbalance in soil fertility and threatened the sustainability of the most productive food grain belt in India. Pulses are an effective source of reversing the process, and can contribute significantly towards achieving the twin objectives of increasing productivity and improving the sustainability of the rice and wheat–based cropping systems (Ahlawat *et al.* 1998; Yadav *et al.* 1998). It is, therefore, important to diagnose and characterize the dynamics of the performance of pulses. Such an analysis will provide information to prescribe appropriate polices and strategies for inclusion of pulses in the existing cropping system.

The important grain pulses are black gram, mung bean, chickpea, lentil, field pea and pigeon pea. Production of grain pulses is relatively more risky. During the past decade, about 1 m ha area under pulses was substituted by other crops. The decline in area was attributed to relatively higher profitability of rice and wheat. Most of the pulses are typically low yielders with an average yield of 550-600 kg/ha. The reasons for low productivity are many, important among them are susceptibility to major abiotic or biotic or bote stresses and physiologically inefficient plant types. Most of the pulses have a combination of traits, which enable them to survive under adverse conditions but fail to yield commensurately under high input conditions. The price and risks of pulses are much higher than rice and wheat. Markets for pulses are thin and fragmented while price spread for pulses is much higher. Pulses are more prone to post harvest losses. Farmers lack knowledge of recently developed and released varieties. A technological breakthrough in pulses is required to substantially increase yield, minimize risk, and enhance profitability. The area under several pulses may be increased through availability of improved cultivars and high support prices. Their inclusion in cereal based cropping system results in increase in the yield of subsequent cereal crops is well known. Pulses contribute in saving of N fertilizers and thus, reduce the cost of production.

Most grain legumes are also grown during the post–rice phase, after which residues from the grain legumes are often used as fodder or green manure for rice. Out of 44 m ha rice area in our country, 17.4 m ha constitute rainfed low land ecosystem which is mostly mono-cropped with rice in *kharif* with low productivity ranging between 0.8 to 1.2 t/ha. Obviously, this gloomy production scenario keeps farmers away from rice farming. However, a paradigm shift in the system of cropping can convert it into an attractive and worthwhile remunerative enterprise. Consequently, crop intensification needs a new vision encouraging the possibilities of 'round the year farming.' Growing short duration pulses may be the right option due to availability of sufficient time after harvest of rice. It is also very obvious that the mere choice of a cropping system without viewing its specific considerations and profitability, would be a completely irrational decision. So, the socio-economic perspective and ecological feasibility has to be examined thoroughly for designing an improved cropping system. It has been reported that cropping system approches focusing on resource management have large potential towards enhancing resource use efficiencies and effective land utilization. New potential crops with higher and stable yields and/or profits will easily find a place in the existing cropping systems of different agro-ecological regions. A historical retrospection reveals that new crops have always made a dent as they spread faster and adapt well due to proper adoption of package of production technology. Some of the best examples are: rice in Punjab and Haryana, wheat in West Bengal, groundnut in Gujarat, soybean in Madhya Pradesh, extra-short duration pigeonpea in Western Uttar Pradesh. The adjustment of pulses in rice-fallow system, rajmash (French bean), mungbean and black gram have the potential in the spring season in the northern belt and as a winter crop in eastern and southern India.

Opportunities for diversification through pulses

The diversification strategies vary with available resource, local needs, social economic interactions and market. A large number of experiment carried out under All

India Coordinated Research Project on Cropping Systems (AICRP_CS) and other programmes in different agro-climates have indicated the beneficial effect of legumes (grown as intercrops, catch crop, substitute crop, green fodder crop or green manure) on the production profitability and soil fertility. The important avenues for inclusion of pulses are as follows.

Pulses as summer/catch crop

The most feasible way of including legumes in cereal based systems without decreasing land area of the cereal crops is to grow legume as catch crop. In RWCS, medium duration rice varieties followed by wheat varieties suited for normal sowing or late sowing in north-eastern plain did not leave any scope to grow legume between rice harvest and wheat sowing. However, the period between wheat harvesting and rice planting can be utilized for growing short duration (60-70 days) summer legumes. Experiments at Masodha, Varanasi and Ludhiana indicated that inclusion of mung bean in Rice-wheat cropping system (RWCS) enhances the total productivity and net returns as compared to (RWCS) alone All India coordinated Research project on cropping systems (AICRP__CS yen report). In other studies, the yield of rice in rice-wheat system generally remained higher when preceded by mung bean (Singh and Sharma, 2002). Results obtained under All India coordinated Research project on cropping systems (AICRP__CS) have shown promising significance of inclusion of pulses as catch or summer crop (Table 2). Singh *et al.* (1993) studied the effect of pulses on yield, economics and sustainability of rice-wheat system in eastern Uttar Pradesh for several years and found a significant improvement when summer mung bean was included in the system. However, a significant reduction was noted when wheat was substituted with chickpea (Table 3). Soil incorporated summer mungbean, either as a whole or its residues after pod picking, lead to increase in grain yield of rice as well as wheat grown in suquence. It also reduced bulk density and pH and enhanced organic carbon content in the soil after harvesting of rice (Saxena and Yadav 1998) (Table 4).

Moong bean, urd bean are the important crops of the summer season. Pulses are also grown as a spring crop, which is sown in March after the harvest of crops like mustard, potato, and sugarcane. Summer crop is sown after the harvest of wheat in mid April. The development of short duration varieties of grain pulses, in general, and mungbean, in particular, has provided a good opportunity to raise a successful crop in between harvest of wheat and planting of rice. Short duration (60-65 days) varieties of mungbean with synchronized maturity (PDM11, PDM 54, ML267, PDM139, Pusa vishal, SML 668, Pant mung 5, HUM2, Pant M-2 have helped in crop diversification through their introduction them as a catch crop during summer season in a wide range of intensive cropping sequences. The cultivation of summer mung bean in summer under rice-wheat cropping system is gaining ground in irrigated regions of Punjab, Haryana, Rajasthan, Uttar Pradesh, Bihar and West Bengal. This crop is also grown as a short duration *rabi* crop in rice fallow as catch crop or in rice-rice systems in southern and eastern regions.

At Modipuram raising mung bean for grain and residue incorporation in summer was found to be markedly better in sustaining productivity of rice-wheat system and improving soil fertility. This practice not only provided about 0.5 to 1 t/ha of pulse grain but also benefited the succeeding cereal crops through an improvement in N status of the soil (Gangwar *et al.* 2005).

Table 2: Effect of inclusion of pulses in rice-based crop sequences on yield and economic sustainability (Average of 7 years).

Crop sequences	Grain yield (t/ha)			Rice equivalent yield (t/ha)	Net return (Rs /ha/ year)	Sustainable value index
	Kharif	*Rabi*	Summer			
Rice-wheat	4.11	3.76	-	12.52	1495	0.35
Rice-wheat - mung bean	4.04	3.88	0.87	16.92	2063	0.64
Rice-mustard - mung bean	4.05	1.58	0.72	17.09	2257	0.70
Rice-chickpea	4.16	1.32	0.88	-	924	0.33

Source: Singh *et al.*, (1993)

Crop diversification can also be done by introduction of lentil in crop sequences and inter-cropping systems. Field pea can successfully be grown in upland irrigated ecosystem in sequence with maize, sorghum or pearl millet. In irrigated ecosystem under low lands, the field pea can also be grown in sequence with early rice. Development of high yielding genotypes such as Aparna, Malviya matar 2, KPMR 400, KFPD 103, Malviya matar 15, JP 885, DDR 23 and Pant matar 5 are highly productive and also resistant to powdery mildew. The adoption of these varieties improves the productivity and net returns.

The availability of early and extra early varieties of pigeon pea (UPAS 120, Manak, Paras, AH 201, Pusa 992, Pant A3, Pusa Ageti etc.) having tolerance to variations in photoperiod and temperature; have paved the way for its inclusion in sequential cropping systems during summer and *rabi* seasons. The pigeonpea–cereal cropping pattern is successful in states of Punjab, Haryana, Rajasthan, Gujrat, U.P. and M.P. Generally, wheat follows early pigeonpea in the sequence. Expanding irrigation facilities, availability of short duration varieties and necessity to raise cropping intensity has resulted in evolving successful sequential or relay cropping systems involving chickpea. Rajmash has been successfully introduced as *rabi* crop in irrigated areas in eastern UP, North Bihar, Vidharbha region of Maharastra and Orissa. The suitable varieties are Uday, HUR 15, HUR137, Ambar, Utkarsh and Arun.

Table 3 : Promising intensified cropping systems involving pulses as catch crop

Cropping Systems K R S	RYE (t/ha/yr)	System Productivity (Kg/ha/day)	RPE %	Net Return (Rs/ha/year)	System Profitability (Rs/ha/day)	REE %
Western Zone of U.P., Haryana (Hisar)						
P.Millet-Wheat- (E)	7.40	20.3	-	5505	15.1	-
P.Millet-Potato-G.Gram	13.75	37.7	85.8	15247	41.8	177.0
South Western Semi-Arid Zone of U.P. (Bichpuri)						
Pearlmillet-Wheat (E)	7.83	21.4	-	7405	20.3	-
Pearlmillet-Wheat-Greengram	9.51	26.1	21.5	10238	28.0	38.3
East & South Eastern coastal Zone of Orissa (Bhubneswar)						
Rice-G.Nut (E)	9.05	24.8	-	22675	62.1	-
Rice-G.Nut-Cowpea	14.12	38.7	56.0	42521	116.5	87.5
Rice-Tomato-Cowpea	18.77	51.4	107.4	51653	141.5	127.8
Central Plain Zone of Punjab (Ludhiana)						
Rice-wheat -(E)	10.95	30.0	-	12044	33.0	-
Rice-Gobisarson -Greengram	12.21	33.4	11.5	28438	77.9	136.1
Bhabar & *Tarai* Zone of Uttarakhand (Pantnagar)						
Rice-Wheat-(E)	11.92	32.7	-	23390	64.1	-
Rice-wheat-G.Gram	13.84	37.9	16.1	34709	95.1	48.4
Vindhyan Plain Zone of U.P (Varanasi)						
Rice-Wheat-(E)	8.48	23.2	-	11781	32.3	-
Rice-Potato-G.Gram	11.99	32.8	41.4	22735	62.3	93.0
South Bihar Alluvial Plain Zone of Bihar (Sabour)						
Rice-Wheat- (E)	7.70	21.1	-	14075	38.6	-
Rice-Linseed-G.Gram	11.16	30.6	44.9	27887	76.4	98.1
Rice-Potato-G.Gram	15.92	43.6	106.7	28557	78.2	102.9
Rice-Maize-G.Gram	11.96	32.8	55.3	29929	82.0	112.7
Upper Bhramputra Valley Zone of Assam (Jorhat)						
Rice-Rice (E)	8.05	22.1	-	29517	80.9	-
Rice-frenchbean-Rice	18.11	49.6	124.9	73813	202.2	150.1
North Kankan Coastal Zone of Maharastra (Karjat)						
Rice-Rice (E)	7.17	19.6	-	6422	17.6	-
Rice-Mustard-Cowpea	10.19	27.9	42.1	26137	71.6	307.0
Rice-Sunflower-Cowpea	8.59	23.5	19.8	13889	38.1	116.3
Rice-Watermilon -Cowpea	10.48	28.7	46.1	24165	66.2	276.3
Mid-hills sub-humid zone of H.P. (Palampur)						
Maize-Wheat	6.61	18.11	-	13069	35.81	-
Maize-Pea-Potato	11.92	32.67	80.39	14484	39.68	10.81
Maize-Pea-French bean	11.38	31.19	72.22	16998	46.57	30.05
Rice-Veg Pea-Potato	9.26	25.37	40.09	19073	52.25	45.92

Table 4 : Effect of summer green gram on grain yields of rice- wheat (under rice wheat) system and properties change in soil properties after 03 cycle

Crop sequences	Grain yield (t/ha)		Bulk density	pH	Soil organic
	Kharif	*Rabi*	(g.cm^{-3})		carbon (%)
Rice-wheat	4.20	4.10	1.40	7.83	0.36
Rice-wheat- mung bean*	5.09	4.46	1.32	7.70	0.44

*Residue incorporated after picking;
Source: Saxena and Yadav, (1998)

The intensified crop rotations involving pulses for higher net returns (Rs.54,000 to 61,000/ha) and production efficiency (6-18%) are rice – mustard Furrow Irrigated Raised Bed (FIRB)- mungbean Furrow Irrigated Raised Bed (FIRB), pigeonpea Furrow Irrigated Raised Bed (FIRB)- wheat Furrow Irrigated Raised Bed (FIRB), rice-vegetable pea Furrow Irrigated Raised Bed (FIRB)-wheat Furrow Irrigated Raised Bed (FIRB and rice- wheat Furrow Irrigated Raised Bed (FIRB)–mungbean Furrow Irrigated Raised Bed (FIRB) are compared to conventional rice-wheat (Rs. 49, 500/ha) system in trans-gangetic plain region of India. Similarly, instead of continuously practicing rice-wheat system, it is possible to improve productivity, profitability (8-25%) and resources use productivity with inclusion of legumes within the system using intensification or interruption with other crops once in three years in Upper-Gangetic Plain region of India (Gangwar *et.al* . 2005).

Inclusion of fast growing pulses during summer is another promising agro-technique, though the advantage of such legumes in terms of nutrient recycling, and improvement in soil physical and chemical properties are relatively less explored. Recent work at PDFSR, Modipuram revealed an appreciable increase in the use-efficiency of N and P fertilizers in rice-wheat system with the inclusion of summer cowpea (Table 5). Study conducted on NO_3-N leaching pattern at Modipuram showed that the inclusion of cowpea as fodder crop during summer under rice-wheat system not only helped in trapping the NO_3-N from deeper soil layers, but also made it available in upper profile layer for succeeding crops (Dwivedi *et al.* 2003).

Table 5: Recovery efficiency of N and P fertilizers in rice-wheat system as influenced by inclusion of summer forage cowpea.

Fertilizer NP	Rice		Wheat	
	Summer Fallow	Summer Cowpea	Summer Fallow	Summer Cowpea
	Recovery efficiency of N (%)			
$N_{120}P_0$	34.8	35.3	42.3	38.3
$N_{120}P_{60}$	36.4	41.2	54.5	61.7
	Recovery efficiency of P (%)			
N_0P_{26}	11.6	15.6	11.2	12.6
$N_{120}P_{26}$	22.7	25.0	27.9	30.4

Source: Dwivedi *et al.* (2003)

Pulses as intercrop

Inter-cropping is one of the important ways to increase the productivity and provide income stability under limited soil moisture conditions. Pulses such as pigeon pea, black gram, mungbean, and soybean are ideal intercrops in upland direct seeded rice. The scope of pulses intercrop under cereal dominated system further increases under water scarce conditions or aberrant weather situations, as it not only increase total productivity of the system but also play an important role in economizing the use of resources, particularity N fertilizer. Extensive studies carried out under AICRP_CS have helped in identification and standardization of several highly profitable inter-cropping systems. Some of the promising inter-cropping system are maize + black gram at Palampur, Ranchi and Banswara, maize + cowpea at Karjat, sorghum + pigeon pea at Indore, pigeon pea + mung bean at Bichpuri and Hanumangarh, and rice + soybean at Kalyani and Jabalpur. Most of these inter-cropping systems have been evaluated on farmers' fields and found to be highly remunerative (15-200%) over the sole crops (Table 6). It was generally observed that both the component crops should be fertilized at recommended rate, to achieve maximum benefit from a diversified inter-cropping system.

Table 6: Proven inter-cropping systems on farmers' fields

Inter-cropping system	District	Per cent additional benefit over sole crop
Pigeonpea+groundnut	Singhbhum	199
	Sangrur	50
	Mayurbhanj	69
Pigeonpea+blackgram	Aligarh	42
Pigeonpea+maize	Ghazipur	11
Pigeonpea+pearlmillet	Ghazipur	20
Sorghum+pigeonpea	Wardha	100
	Bidar and Bellary	103
Groundnut+blackgram	Tiruchirapalli, Pudukkottai and South Arcot	35
	Chengalpattu	25
Groundnut+greengram	Salem	10
Maize+blackgram	Bilaspur and Hamirpur	53
	Ghaziabad	30
	Dungarpur	38
Chickpea+mustard	Dungarpur and Jhunjhunu	12
	Rohtak and Jind	35

Source: AICRP-CS Reports

The success of intercropping depends upon choice of crops, varieties, planting method, crop geometry, nutrient and water management etc. Several pulses such as pigeon pea, black gram, mung bean and groundnut are ideal intercrops in upland direct seeded rice. The potential of legumes as intercrops in cereal based cropping system has not yet been fully exploited. Although, due to contrasting nature of rice/wheat and pulses with respect to requirement of soil moisture regimes, soil aeration and growth habit their cultivation as one of the associate crops in intercropping systems, has limited scope in intensively cultivated rice-wheat production systems of IGP. However, in upland direct seeded rice, several pulse crops like pigeon pea, black gram and mung bean are ideally suited for their introduction as intercrops (Lal and Singh 1990). The well-designed intercropping systems, besides increasing system productivity or profitability or both, play an important role in enhancing resource-use-efficiencies, particularly land use efficiency. For new alluvial zone of West Bengal, for example, inclusion of pulse crops in rice-wheat cropping system has been found more remunerative than rice-wheat cropping system (Mandal and Sinha 1993). Mandal *et al.* (1994) reported that intercropping of wheat with legumes gave higher yield and economic benefits than sole cropping of wheat.

In northern part covering the state of Uttar Pradesh, Rajasthan, Haryana, Punjab and Bihar, early varieties of pigeon pea are mostly mixed cropped with sorghum and maize. In this group, Bahar, Amar, Sarad, Pusa 9, MA 3 and Narendra Arhar are most suitable for cultivation. Mungbean and urdbean are grown as mixed crop with sorghum and maize during *kharif*. After harvest, early maturing fieldpea, linseed and barley etc may be grown alone or in association during *rabi*. Intercropping of *rajmash* with potato, moongbean, urdbean, chickpea and lentil with sugarcane are the best options for intensification in irrigated areas.

Field pea is also grown mixed with linseed and mustard in rainfed ecosystem but total productivity is very low in such cropping systems. The lentil crop in rainfed ecosystem is grown as mixed crop with mustard, field pea and linseed. The newly developed genotypes such as Pant L406, DPL 15, DPL 62, IPL 81, K75, L 4076 are highly productive and have the ability to increase the cropping intensity, system productivity and net return per unit area. Availability of pigeon pea varieties having highly diverse growth habits has made it one of the successful crops for inter cropping. This helps in reducing the adverse effect of competition for space and light and efficient utilization of resources. The important intercrops grown with pigeon pea are sorghum, maize, pearl millet, finger millet, rainfed rice and minor millets. Among legumes, groundnut, cowpea, mung bean, urd bean, soybean and beans are also grown. Pigeon pea intercropped with short duration pulses (mungbean, urdbean, cowpea etc.) is the most popular combination in the pigeon pea growing areas, in general, and in northern areas, in particular. The special feature of this intercropping system is that productivity of base crop of pigeon pea remains unaffected and an additional 4-5 q/ha of pulse grain is obtained without any additional input. Pigeon pea + groundnut, pigeon pea + soybean and pigeon pea + sesame are also common intercropping systems in India. While pigeonpea + groundnut is common in western and southern India, pigeon pea + soybean is popular in central and northern India. Growing of groundnut + pigeon pea in 6:1 row proportion gave groundnut yield at par with sole crop and 90% of sole crop of pigeon pea. In north Bihar, short duration varieties of moong bean and urd bean are grown in mixed cropping system in autumn season. In Rajasthan and Haryana states, short duration pulses like mung bean are cultivated as mixed crop with pearl millet and sesame during *kharif* season. In rabi season, lentil, field pea and lathyrus are grown as monoculture. In Orissa and coastal parts of West Bengal, moong bean and urd bean are cultivated during *rabi* season.

In central parts of India viz., Maharashtra, Madhya Pradesh, part of Rajasthan and Bundelkhand region of Uttar Pradesh, pigeon pea is cultivated in mixed and intercropping system with sorghum, cotton, and groundnut. For this purpose, medium duration varieties viz., ICPL 87, AKT 8811, Asha (ICPL 87119), BSMR 736, BSMR 853, Maruti and JKM 7, are most suitable. Similarly, moong bean and urd bean are grown in mixed and intercropping system with sorghum, maize, cotton and pearlmillet. Mungbean varieties PDM 11 (spring season), BM 4, HUM 1 (spring) and Pusa 9531, and urdbean varieties TPU 4, RBU 38, and KU 96-3, are available for general cultivation. Intercropping of blackgram in pigeonpea further improves the profitability of the system. There is a large scope for intensification of pigeonpea-wheat cropping system through intercropping with greengram, blackgram, cowpea etc. in pigeon pea. Intercropping legumes with short duration pigeon pea is more remunerative. Some of the promising cropping systems have been identified for different agro-eco regions (Table 7) as reported by Singh *et al* (2000).

However, for successful intercropping, the selection of suitable varieties and row ratio are important considerations. With the advent of modern tools like bed planting, multi-crop seed drills, it may emerge one of the promising options for rice-wheat system sustainability. Co- culture of pulses along with direct seeded/ bed planted rice reveals further scope for inclusion of legumes as intercrop in rice-wheat cropping system.

Pulses as substitute crop

The substitution of pre dominant crops like rice or wheat largely depends on nature of stress increased in different agro-ecologies. For instance, in Trans-Gangetic Plain (TGP), where water table depletion is a serious concern, there is scope for substituting rice with pigeon pea. Similarity in eastern part of India, where wheat productivity is generally low because of climatic constraints, particularity higher thermal regimes, wheat needs to be substituted with chickpea, lentil, pea or groundnut. Extensive studies on comparative performance of rice-wheat vs. rice–pulses or pulses-wheat at different locations of AICRP_CS reveal that growing legumes in the system with recommended rate of NPK, helped to improve system productivity and profitability than non-legume based system. In *Tarai* region, the annual productivity in terms of rice equivalent yields, energy production and net return increased tremendously in rice-wheat system, when maize+cowpea fodder crop was introduced during post-wheat summer season (Table 8). Diversifying wheat with chickpea was further advantageous. In double crop sequences involving cereal crops only, diversifying *kharif* or *rabi* cereal with a grain legume has shown promise.

Table 7. Efficient intercropping systems and row ratios recommended for rainfed agriculture in various Agro-eco-region (AER).

AER	Location	Intercropping system	Row Ratio
6.3	Akola	Sorghum + pigeonpea, Cotton + greengram/cowpea Cotton + pigeonpea Pigeonpea + Greengram	2:12:12:11:1
5.2	Indore	Maize + soybean, Soybean + pigeonpea	2:22:2
4.2	Arjia	Maize + blackgram, Groundnut + sesame	1/2:16:2
2.4	Rajkot	Groundnut + castor, Groundnut + pgeonpea	3:12/4:1
4.4	Jhansi	Sorghum + pigeonpea, Pearlimllet + fodder legume	2:11:1
10.3	Rewa	Sorghum + pigeonpea, Soybean + pigeonpea	2:14:2
3	Bijapur	Pearlimllet + pigeonpea, Sunflower + pigeonpea Chickpea + safflower	1:12:22/ 3: 1
6.1	Solapur	Pearlimllet + pigeonpea, Sunflower + pigeonpea Pearlmillet + mothbean, Chickpea + safflower	2:12:1/22/ 3:13:1
8.1	Kovilpati	Sorghum + blackgram, Sorghum + cowpea Cotton + blackgram	2:12:12:2
3.0	Anantpur	Groundnut + Pigonpea	1.5:1
7.3	Hyderabad	Sorghum + pigeonpea, Castor+ clusterbean	3:12:2
8.2	Bangalore	Fingermillet + pigeonpea, Groundnut + pigeonpea	8:18:2
12.1	Phulbani	Rice + pigeonpea, Maize + pigeonpea Pigeonpea + raddish	3/ 4: 12:12:1
12.3	Ranchi	Maize + pigeonpea, Rice + pigeonpea Groundnut + pigeonpea	1:13:13:1
23.	Hisar	Pearlmillet + blackgram/ Greengram	2:11:1
2.1	Jodhpur	Pearlmillet + greengram/ Cowpea	1:21:2
2.3	Dantiwada	Pearlmillet + greengram, Clusterbean + greengram Pearlmillet + clusterbean	1:32:13:1
4.3/ 9.2	Varanasi	Maize + blackgram, Pigeonpea + blackgram Pigeonpea + sesame, Rice + pigeonpea	1:11:11:13:1

Source : Singh *et al.* (2000)

Substitution of rice with pigeon pea, and that of wheat with field pea in a sequence in lighter soils of Ludhiana (Punjab) showed a fertilizer (NPK) economy of 25% of recommeneded dose both for rice and wheat (Singh *et al.* 1996) (Table 7). In a study involving rice-wheat and rice-pulse systems-lentil at Pantnagar, chickpea at Varanasi and Kanpur, and urd bean at Bhubneswar with recommended fertilizer application for both the crops in the sequence, rice productivity was consistently higher in rice-pulse system than that in rice-wheat system at all the locations (Hegde 1992) (Table 8). Singh and Verma (1999) have also reported similar benefits of preceding pulses on yield and economics of rice and proved the superiority of chickpea over lentil and field pea, as a substitute crop for wheat, in terms of rice equivalent yields and economic returns, under *Tarai* conditions of Pantnagar conditions. Many a times appreciable beneficial effects of pulses are recorded on soil and productivity of succeeding cereal crops but due to their low yield level, system productivity (equivalent yields in monetary terms) becomes questionable. Therefore, an efficient high yielder pulse having potential to adjust to the situation needs to be included in the system.

Table 8: Effect of legume inclusion on system productivity, energy production and economics at Pantnagar, India

Crop sequence	System productivity		Chemical energy equivalent (k cal* 10^{-4} ha)	Economics (Rs/ha)	
	Rice equivalent yield (t/ha)	Production efficiency		Gross return	Return over variable cost
Wheat-rice	10.3	28.28	3533	48302	32188
Chickpea-rice	10.8	29.79	3132	49203	35467
Wheat-maize+ cowpea (fodder) -rice	13.4	36.74	5051	61682	40793
Chickpea-maize+ cowpea (fodder) -rice	13.9	38.09	4865	62503	43992
CD 5%	0.5	6.64	333	3803	2955

Source: Singh and Sharma (2002)

The beneficial effects of substituting wheat or rice with a pulses on soil organic carbon, mineral-N and Olsen-P content was also reported by Yadav *et al.*, (2003); Singh *et al.*, (2005). Hegde and Dewivedi, (1992) reported that substitution of wheat with pea at Ludhiana, lentil at Pantnagar, chickpea at Kanpur, blackgram at Bhubaneswar, and groundnut at Chiplima resulted in marked increase in organic carbon and available NPK, when compared with RWCS. Raising a pulse crop in rice-wheat system can also minimize the adverse effect of continuous puddling on soil compaction. Data illustrated in Fig. 1 clearly show a rise in soil bulk density in sub-surface layers with rice-wheat cropping for three years. Taking a sort duration pigeon pea, on the other hand, mitigated such an effect of puddling. In other words, inclusion of legume may help in better crop establishment and root growth of wheat following rice, by way of reducing soil compaction.

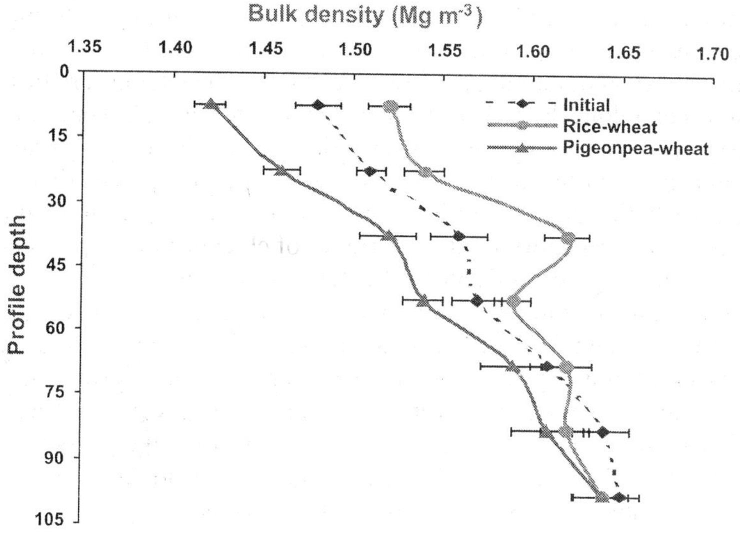

Fig. 1: Changes in bulk density of soil at different profile-depths after three crop cycles as influenced by substitution of rice with pigeon pea in RWCS (Singh *et al.*, 2005).

Pulses in fallow lands and new niche areas

There is a scope for introduction of pulses in new niches such as wastelands, reclaimed soils and rice-fallows efficient watershed management and as a replacement of less remunerative crops or intensification. To sustain and improve rice productivity, pulses have to be introduced in rice based cropping systems in different rice growing areas in Peninsular India including the deltaic areas of important rivers *viz.*, Krishna, Godavari, Kaveri, etc. To fit into different new niches there is an immediate need of short duration pulse crops. Rice fallows occupy about 15 m ha. Appropriate exploitation of these areas is possible through introduction of pulses like urd bean, moong bean. Availability of powdery mildew resistant urd bean variety (LBG 17) with matching agro-technology has expanded its acreage many folds in Andhra Pradesh. The new niches with reference to pulses have been discussed in detail by Ahlawat *et al* (2004).

An innovative approach of crop intensification in the Bundlkhand and other region, with respect to rice-pulse system, is cropping on rice bunds. Presently most of the bunds are left fallow. There is a great scope to boost production of *kharif* pulses by utilizing vast area under these bunds as peripheral cropping. Both old and new bunds may be used for growing pulse crops. Intercropping of pulses with rice is another possible way of crop intensification in the system. Research in the Bastar plateau zone has shown that rice + blackgram in the ratio of 2:1 can give yield of 2.4 t ha^{-1} of rice and around 0.25 t ha^{-1} of black gram. Similarly rice + pigeonpea in the ratio of 3:1 can yield 1.0 t ha^{-1} of rice and around 0.6 t ha^{-1} of pigeonpea. Further, by proper water harvesting techniques in rainfed zone, water for supplementary irrigation may be made available in *rabi* season and the cropping intensity may be increased from 100 to 200% by growing a pulse crop.

The advent of short duration varieties and irrigation facilities have provided an opportunity of double cropping system with pigeonpea either with planting early to facilitate early harvest or planting after the harvest of preceding crop. In the recent past, development of short duration, disease resistant and high yielding varieties with matching production technology have lead to introduction and popularization of several new cropping systems such as pigeonpea-wheat, rice/maize-wheat–mungbean, maize-potato/ mustard-mungbean/urdbean in North West Plains, rice chickpea/lentil, maize-pigeonpea in North-East Plains and rice-urdbean and sorghum-chickpea in southern peninsula. Development of extra-short duration varieties of chickpea may open a new avenue for their cultivation for green pods during September-December without sacrificing major cereal crops. Some of the possible new niches for pulses have been described by Ali (2004). An estimated 2.5 million ha additional area can be brought under different pulses through cropping system manipulation, crop diversification and multiple cropping systems. It can lift the national average of cropping intensity significantly. Based on series of experiments conducted at various locations over the years the most efficient systems involving legumes and other high value crops as identified for different zones in India for irrigated eco systems are presented in Table 9.

Table 9. Possible new niches for pulses.

Cropping System	Possible niches	Expected	Suitable varieties Area pulse crops
Pigeonpea-Wheat	Haryana, Punjab, North -West U.P. and North Rajasthan	1.0	UPAS 120, Manak, Pusa 33, AL 15, AL 201
Maize-Rabi Pigeonpa	Central and Eastern U. P., North Bihar, West Bengal, Assam	0.3	Pusa 9, Sharad
Maize-Potato/ Mustard+Mungbean /urdbean	Punjab, Haryana and West U. P.	1.0	**Mungbean:** Pant Mung 2, PDM 11, Hum 2, SML 668, Pusa Vishal **Urdbean:** PDU 1, Narendra Urd 1, Uttara
Spring Sugrcane + Mungbean /urdbean	East U. P., Bihar, West Bengal	0.15	**Mungbean**: Pant Mung 2, PDM 11, Narendra Mung 1 **Urdbean:** PDU 1, Pant U 19 TARM 1, Pusa 9072
Rice-Mungbean	Orissa, parts of Karnataka, Tamil Nadu, A.P.	0.35	TARM 1, Pusa 9072
Rice-Urdbean	Coastal areas of A.P., Karnataka, Tamil Nadu	0.35	LBG 17, LBG 402
Rice-Wheat-Mungbean	Western U. P., Haryana, Punjab	0.10	Pant Mung 2, Narendra Mung 1, PDM 139, HUM 2
Maize-Rajmash -Mungbean	Central and Eastern U. P., North Bihar	0.07	**Mungbean:** Pant Mung 2, PDM 11, Him 2 **Rajmash:** Hur 137, Hur 15, PDR 14, Amber
Rajmash + Potato	Eastern and Central U. P.	0.03	PDR 14, Amber

Source : Ali (2004)

Future thrust

The following aspects deserve focused attention:

1. There is need to generate diversified packages for diverse farming situations through characterization of resources available with the farmers of different holding size categories. The resource characterization should not be confined to documentation of land and water resources, but should necessarily include other factors *viz.*, socio-economic conditions, domestic needs, market infrastructure, input supply etc.
2. No fixed system is expected to sustain in the years to come. Therefore, diverse systems, dynamic in nature and capable to absorb change may be desirable. However, pulses should essentially be included in the dynamic systems.
3. The complimentary intensive intercropping systems deserve to be studied both under irrigated and rain fed situations with proper water management strategies. Under these situations, pulse based cropping systems are to be developed which can tap the full potential of different soil and environmental factors available throughout the country.
4. Intensification approaches are highly location specific depending upon the existing resources and driving forces. Therefore, location specific strategies need to be worked out for obtaining beneficial advantages of diversification using new short duration and suitable varieties of pulses.
5. The research on crop diversification should essentially be taken up in a farmer-participatory mode and it should be preferably conducted under an actual on-farm environment rather than on the research farms. A multi-disciplinary team of scientist's should involve the farmer right from planning till arriving at conclusions.
6. Suitable arrangements for price forecasting of pulses well before a crop is sown and simultaneous arrangements for procurement of the produce of these crops may provide extra opportunity of pulse inclusion in different system.
7. Generation and synthesis of pragmatic information on the area, productivity and limitations of pulses in cereal based production systems.
8. Extensive long-term studies are also required to study the influence of inclusion of pulses in rice-wheat system on chemical (nutrient dynamics), physical and biological health of soil, allelopathic effects and weed dynamics.
9. Short duration and disease resistant varieties of mung bean with better synchrony for fruiting and maturity need to be developed and popularized for summer cultivation, particularly in rice-wheat system.
10. To enhance the land use efficiency and augment its productivity, multiple cropping using short duration pulses or extra-short duration pulses are to be

focused upon. Development of such extra early new crop varieties with more efficient photosynthetic apparatus, shorter in duration and high yielding would be of immense help in increasing cropping Index intensity.

11. Modern Resource Conservation Technologies like furrow irrigated raised beds need to be widely evaluated as it has potential for bringing pulses under various cropping systems.

References

Ahlawat, IPS, Ali, M, Yadav, R.L., Rao, J.V.D. KK, Rego, T.J. and Singh, R.P. 1998. Biological Nitrogen Fixation and Residual effect of summer and rainy season Grain Legumes in Rice and Wheat Cropping Systems of the Indo-Gangetic Plain In: Residual effects of legumes in rice and wheat cropping systems of the Indo-Gangetic Plain (Edited: Rao J. V. D.K. K., C. Johansen C. & Rego T.J.), Oxford & IBH Publishing Co. Pvt. Ltd. PP 31-54

Ahlawat, IPS, Gangwar B. and Gangaiah B. 2004 Potential Pulse Based Cropping Systems for New Niches. In: Pulses in New Perspective (Edits: Ali M. Singh B. B. Kumar S. and Dhar V) ISPRD , Kanpur pp 260-270.

AICRP-CS Reports. Project Directorate for Cropping Systems Research Modipuram, Meerut, India (1984-2000).

Ali, M 2004. Role of Pulses in Crop Diversification In: Pulses in New Perspective (Editor: Ali M. Singh B. B. Kumar S. and Dhar V) ISPRD , Kanpur pp 245-259.

Dwivedi, BS. Shukla, K., Singh, VK and Yadav, RL. 2003. Improving nitrogen and phosphorus use efficiencies through inclusion of forage cowpea in the rice-wheat system in the Indo-Gangetic Plains of India. Field Crops Research 80 : 167-193.

Gangwar, B., Tripathi, S.C., Singh, J.P., Kumar, Ravi, Singh, R.M. and Samui, R.C. 2005. Diversification and Resource management of rice – wheat system. Research Bulletin No. 05/1. PDCSR, Modipuram: pp. 1-68.

Hegde, DM and Dwivedi BS. 1992. Nutrient management in rice-wheat cropping system in India. Fertilizer News 37:27-41

Hegde, D.M. 1992. Cropping systems research highlights. In: Coordinators report, 20th Workshop, Tamil Nadu Agricultural University Coimbatore, 1-4 June 1992. Modipuram, Meerut 250 110, India: Project Directorate of Cropping Systems Research pp. 21-22.

Lal, S and Singh, MP. 1990. Pulses in rice based sequences. Indian Farming 39 (4): 10-12.

Mandal, GK and Sinha, AC 1993. Sustainability of rice –wheat sequential cropping through introduction of legume crops and green manure crops in the system. Oryza 20 : 260-262

Saxena, A and Yadav DS. 1998 Effect of summer green gram on rice –wheat system. Indian Journal of Agricultural Sciences 68 : 583-586.

Singh, HP., Venkateswarlu, B.,Vittal, KPR., and Ramachanderan, K. 2000. Management of Rainfed Agro-ecosystem.In:Natural Resource Management for Agricultural Production in India (Ed.Yadav J.S.P. and Singh G.B.) Indian Society of Soil Science, New Delhi pp. 669-774.

Singh, MP, Verma SC and Singh RP. 1993. Effect of rice -based sequences on yield and economic sustainability under irrigated conditions of Eastern Uttar Pradesh. Annals of Agricultural Research 14 : 237-239.

Singh, N.B. and Verma KK. 1999. Impact of selected Rabi crops on productivity and nitrogen economy in rice -based cropping systems. Oryza 36 : 89-91

Singh, R, Rana, DS, Bhandari, AL and Brar, RS. 1996. Influence of legume based cropping systems on crop yield and nutrient uptake in Typic Ustochrept. Indian Journal of Agricultural Sciences 66 : 7-10.

Singh, V.K. and Dwivedi, BS. 2006.Yield and N use-efficiency in wheat, and soil fertility status as influenced by substitution of rice with pigeon pea in a rice-wheat cropping system. Australian Journal of Experimental Agriculture.46 : 1185-1194.

Singh, V.K. and Sharma, B.B. 2002. Economic evaluation of rice based cropping systems in foothills of Himalayas. Indian Journal of Agronomy 47 (1) : 12-19.

Singh, V.K. Dwivedi, B.S., Shukla K, Chauhan, YS and Yadav, RL. 2005. Diversification of rice with pigeonpea in a rice-wheat cropping system on a Typic Ustochrept: effect on soil fertility, yield and nutrient use efficiency. Field Crops Research 92, 85-105.

Timsina, J and Connor, DJ 2001. Productivity and management of rice-wheat cropping systems: issues and challenges. Field Crops Research 69, 93-132.

Yadav, R.L 1996. Intensive cropping system involving pulses. (In) special group meeting on constraints and opportunities in pulses production. Management options and Researchable Issues (12-13 March 1996) IARI, New Delhi, India.

Yadav, RL, Singh, VK, Dwivedi, BS and Shukla, AK. 2003. Wheat productivity and N use efficiency as influenced by inclusion of cowpea as a grain legume in rice–wheat system. Journal of Agricultural Sciences (Cambridge) 141 : 213-220

Yadav, R.L., and Rao. AVMS. 2001. Atlas of cropping systems in India. PDCSR Bulletin No. 2001-2, PDCSR, Modipuram, pp.96.

Yadav, R.L., Dwivedi, BS, Gangwar, K S. and Prasad, K. 1998. Overview and Prospects for Enhancing Residual Benefits of Legumes in Rice and Wheat Cropping Systems in India. In: Residual effects of legumes in rice and wheat cropping systems of the Indo-Gangetic Plain (Edit: Rao J. V. D.K. K., C. Johansen C. & Rego T.J.) , Oxford & IBH Publishing Co. Pvt. Ltd. pp. 207-226.

❑❑❑

Chapter - 5

Pulses Production in Organic System

S.S. Pal
Project Directorate for Farming Systems Research, Modipuram, Meerut, U.P - 250 110

The relevance and need for an eco-friendly alternative farming system arose from the ill effects of the chemical farming practices adopted worldwide during the second half of the last century. The methods of farming evolved and adopted by our forefathers for centuries were less injurious to the environment. People began to think of various alternative farming systems based on the protection of environment which in turn would increase the welfare of the humankind in various ways like clean and healthy foods, an ecology which is conducive to the survival of all the living and non-living things, low use of the non-renewable energy sources, *etc*. Many systems of farming evolved as an outcome of the efforts of many experts and laymen. However, organic farming is considered to be the best among all of them because of its scientific approach and wider acceptance all over the world.

Organic farming means farming in the spirit of organic relationship. In nature, organic relationship is a pervasive phenomenon. Everything is connected with everything else. Hence, organic farming is not mere non-chemicalism in agriculture. Since, organic farming means placing farming on integral relationship, we have to know the relationship between soil-water and plant; between soil and soil microbes and waste products; between the plant and animal kingdom of which the apex animal is the human being; between agriculture and forestry; between soil, water and atmosphere *etc*. It is the totality of the relationship that is the bedrock of organic farming. However, the understanding of this totality of relationship is not sufficient for success in organic farming. Understanding the local specificity in respect of each of these aspects of these multiple relationship is also important (Pal, 2006).

Indian agriculture has been successful in achieving increased food grains production. While the mission of increasing food grains production stands somehow achieved without a major jump in pulses productivity and production in the country.

These were accompanied by widespread problems of resource degradation. Though recently, production of pulses has reached 18.0 million tonnes after being stagnant for a period of more than 10 years, there is greater need to increase productivity using suitable management practices to improve livelihood and nutritional security of the large number of Indian vegetarians. Therefore, the conservation of natural resources becomes a necessity to achieve sustainable and profitable pulse production system and subsequently aims to improve livelihoods of the farmers.

Pulses are an essential source of protein in the diet of the predominantly vegetarian Indian population and the cultivation of these legumes has been a long-standing tradition. According to the statistics of the Department of Agriculture and Co-operation, 23 million ha land area is used for pulse production, which accounts for 35% of the world-wide pulse cultivation area. With average annual production of 12 to 14 million tonnes over the years, India ranks among the major pulse production countries even today. This is notable also that in spite of grave importance of pulses, attempts to increase area under pulses have not been very successful.

Since pulses are cultivated in rain-fed areas, the development and spread of irrigation has pushed them to relatively less productive lands resulting in reduced production and productivity. The importance of pulses as a source of protein and their inclusion in the diet is well understood but their consumption in rural India is much below the desired level. As per international standard dietary requirement is 55 g/person/day but this has stooped to level of 30 g/person/day. Four deades ago, it was 67 g/person/day in India. In rural India, pulses have been the major protein source which has unfortunately declined causing several abnormalities in growth. In mountain areas, Pulses are crops traditionally grown in the moutainous regions and the varying ecological situation have bestowed a huge diversity in them. There are about 16 types of pulses with their innumerable number of varieties within. They are also different in taste due to varied climate and soil quality. The pulses grown in the mountains have always been in high demand because of their unique taste. Since mountain agricultural land cannot be placed as quality land, a local traditional sculpture of farming practiced since ages can only be the answer. In this context, organic farming is the best option.

Present scenario of organic agriculture with special reference to pulse production in India

The growth of organic agriculture in India has three dimensions and is being adopted by farmers for different reasons. The first category of organic farmers are those which are situated in no-input or low-input use zones, for them organic is a way of life and they are doing it as a tradition (may be under compulsion in the absence of resources needed for conventional high input intensive agriculture). The second category of farmers are those which have recently adopted the organic in the wake of ill effects of conventional agriculture, may be in the form of reduced soil fertility, food toxicity or increasing cost and diminishing returns. The third category comprised of farmers and enterprises which have systematically adopted the commercial organic agriculture to capture emerging market opportunities and premium prices. While majority of farmers in first category are traditional (or by default) organic, they are not certified, second category farmers comprised of both certified and un-certified but majority of third category farmers are certified. These are the third category commercial farmers which are attracting most attention. The entire data available on organic agriculture today, relates to these commercial organic farmers.

Starting from 42,000 ha under certified organic farming during 2003-04, organic agriculture has grown almost 29 fold during the last 5 years. By March 2010, more than 4.48 million ha area has been brought under organic certification process in India. Out of this, cultivated area accounts for 1.08 million ha, while remaining 3.4 million ha is wild forest harvest collection area. Year wise growth of cultivated area under organic management is shown in Table 1. Overall status of organic production projects, processors, quantity produced, quantity exported and the value of export is given in Table 2. Details in respect of important commodities produced during 2008-09 are given in Table 3.

Table 1. Growth of area under organic management

S.No.	Years	Area under organic management (ha).
1.	2003-04	42,000
2.	2004-05	76,000
3.	2005-06	1,73,000
4.	2006-07	5,38,000
5.	2007-08	8,65,000
6.	2008-09	12,07,000
7.	2009-10	10,85,648

Source: Yadav (2009)

For quality assurance, the country has an internationally acclaimed certification process in place for export, import and domestic markets. National Programme on Organic Production (NPOP) defines the regulatory mechanism and is regulated under two different acts for export and domestic markets. NPOP notified under Foreign Trade Development and Regulation Act (FTDR) looks after the export requirement. The NPOP notified under this act has already been granted equivalence by European Union and Sweden. United States Department of Agricultural (USDA) has also accepted the conformity assessment system of NPOP. Due to this, the product certified by any Indian accredited certification agency under NPOP can be exported to Europe, Sweden and USA without the requirement of re-certification. To look after the requirement of import and domestic market, the same NPOP has been notified under Agriculture Produce Grading, Marking and Certification Act (APGMC). Regulatory body of NPOP under

FTDR act is Agricultural and Processed Foods Export Development Authority (APEDA) under Ministry of Commerce and of NPOP under APGMC act is Agricultural Marketing Advisor (AMA) under Ministry of Agriculture. Accreditation of Certification and Inspection Agencies is being granted by a common National Accreditation Body (NAB). Eighteen accredited certification agencies are looking after the requirement of certification process. Out of these 4 agencies are under public sector while remaining 14 are under private management.

Out of total 2099 operators, while processors account for 427 and individual farmers 753; majority of farmers *i.e.* 597,873 are small and marginal farmers covered by 919 grower groups. Out of the total organic producers in the world, approximately half of them are in India (Table 4). This is mainly because of small holdings with each producer.

Table 2. Overall of organic production projects, processors, quantity produced, quantity exported and the value of export (2008-09)

S.No.	Component	Quantum
1.	Area under organic certification process (Ha)	
	Full organic.	757978.71
	In-conversion	327669.74
	Total	**10.85648.45**
2.	Farmers under organic certification process (Nos)	
	Full organic	351297
	In-conversion	246576
	Total	**597873**
3.	Number of operators	2099
4.	Number of processors	427
5.	Number of grower groups	919
6.	Number of exporters	253
7.	Total production (MT)	1811111
8.	Total quantity exported (MT)	53918
9.	Value of export in US $	116.09 million
10.	Value of export in INR	591 crores

Source: Yadav (2009)

Some potential organic systems involving pulses

With the advent of high yielding and short duration varieties, new cropping systems with high productivity and monetary returns have been developed for rainfall and irrigated areas and have become popular among farmers. These cropping systems can be brought under organic farming with judicious choice of varieties which suits local conditions and choice of suitable combinations of organic sources of nutrients which includes on-farm produced organics and externally brought bio-fertilizers and manuring soils with naturally occurring minerals like rock phosphate, pyrites, gypsum *etc.* to have a balanced source of nutrients.

Among the important cropping systems, rice-wheat-mongbean, and rice-late sown chickpea for the irrigated areas of Punjab, Haryana, Delhi, Central and Western U.P., Bihar, West Bengal and Madhya Pradesh have been found efficient and remunerative.

Table 3. Production of important commodities under organic management (2008-09)

S.No.	Commodities	Quantity produced (MT)		
		Organic	In-conversion	Total
1.	Rice	44335	32354	76690
2.	Wheat	6892	15364	22560
3.	Other cereals/ millets	67333	63985	131318
4.	Pulses	17560	16785	34345
5.	Oil seeds including soybean	163185	59647	222832
6.	Cotton (raw seed cotton)	284832	86906	371740
7.	Spices	17419	20084	37504
8.	Tea/ coffee	16506	10838	27344
9.	Fruits and vegetables	194505	538073	732579
10.	Herbal/ medicinal plants	129543	58767	188310
11.	Other miscellaneous crops	8001	25235	33236

Source: Yadav (2009)

Table 4. Estimates of area covered by different crops under organic management (2008-09)

S.No.	Commodities	Area in ha		
		Organic	In-conversion	Total
1.	Paddy	18134	9766	27900
2.	Wheat	4056	7192	11284
3.	Other cereals/ millets	26184	37678	63862
4.	Pulses	12023	17617	29640
5.	Oil seeds including soybean	91849	87307	179156
6.	Cotton	259699	93299	352998
7.	Spices	6507	23291	29798
8.	Tea/ coffee	12711	12465	25716
9.	Fruits and vegetables	128879	41176	170055
10.	Herbal/ medicinal plants	32313	10690	43003
11.	Other miscellaneous crops	27995	28306	56301
12.	Crop details not available	19812	198110	217922
	Total area	640162	566897	1207059

Source: Yadav (2009)

Pulse crops also have vast scope for their introduction as intercrops with coarse cereals (*e.g.* sorghum, pearl millet, maize *etc*), oil seeds (castor, groundnut, safflower, linseed and mustard), fiber crops (cotton) and commercial crops (sugarcane, sunflower). Promising intercropping systems involving pulses are pigeon pea + soybean for Madhya Pradesh and Bundelkhand region (2:4/2:6 row ratio), pigeon pea + groundnut for Saurashtra region of Gujarat and Karnataka (1:2), pigeon pea + urdbean / mungbean for Uttar Pradesh, Bihar, Maharashtra, Gujarat and Tamil Nadu (1:1), pigeon pea + sorghum for Bundelkhand region, Madhya Pradesh and Maharashtra (1:1), chickpea + mustard for Rajasthan, Uttar Pradesh, Haryana and Madhya Pradesh (4/6:2), chickpea + linseed for Bundelkhand region, Madhya Pradesh and Maharashtra (2:1), chickpea + safflower/ coriander for Maharashtra, Karnataka, Tamil Nadu (2/4:1), lentil + mustard/linseed for Bundelkhand region of Uttar Pradesh and Madhya Pradesh (4:2/ 4:1), spring sugarcane + urdbean/mungbean for Uttar Pradesh and Bihar (1:2), spring sunflower + urdbean / mungbean for Punjab, Haryana and Bihar (2:6) and potato + rajma for central and eastern Uttar Pradesh and Bihar (3:2). Promoting these intercrops will help in bringing an additional area of 1-1.5 million ha under pulses.

Nutrient management in pulses under organic system

Farm yard manure and pulse production: Organic agriculture system advocates recycling of on -farm produced organic resource as nutrients by converting them to manure by suitable bio-conversion techniques. India has a vast potential of organic resources and if utilized in proper manner can produce 7.25 million tonnes of nutrients by 2025 (Table 5). They are also an excellent source of micro nutrients (Table 6).The farmyard manure is an excellent source of soil ameliorant and improves soil physio-chemical and biological properties and posseses disease suppressive actions. In a study on mitigating the adverse effect of drought in rain fed black gram, (Muthuvel *et.al.* 1985) reported that farm yard manure application significantly increased the soil moisture status besides addition of nutrients. Dry lands are usually low in organic matter status due to rapid decomposition and lesser addition of crop residues. In such lands, addition of organic manures or waste become necessary for increasing water holding capacity and improving the fertility status (Venkateswaralu,1987). Several authors have established the important role of farm yard manure in promoting the soil fertility status, improving soil physical properties, thus promoting soil moisture storage and its beneficial influence on yield of several crops including pulses. Application of Farm Yard Manure (FYM) at 5 t ha^{-1} resulted in significantly higher chickpea grain yield over no Farm Yard Manure (FYM) application.(Powar *et al.* 2009)

Enriched farmyard manure: Enriched farmyard manure is also an excellent source of nutrients for pulses.The term "enriched farmyard manure" implies the well decomposed farm yard manure pre incubated with the recommended dose of rock phosphate under aerobic conditions. The rock phosphate is a restricted item and can be used as an alternative source of phosphorus in organic agriculture system. Application of enriched farm yard manure increased the P availability in the soil which ultimately lead to increased solubilization of insoluble P fraction during humification and reduced P fixation in the soil due to the protective action of manure (Harisudan and Latha, 2007). Horse gram grown as a follow up crop of groundnut utilized the residual nutrient of enriched farmyard manure resulting in relatively higher grain yields. (Parasuran and Mani, 2003).

Table 5. Estimated trappable nutrients from different organic sources in India

Resources	Year	
	2010	**2025**
Generators		
Human population (million)	1120	1300
Livestock population (million)	537	596
Resources (considered trappable)		
Human excreta (dry) (million tonnes)	15	17
Livestock dung (dry) (million tonnes)	119	128
Crop residues (million tonnes)	112	162
Nutrient (genetic potential)		
(Million tones $N+P_2O_5+K_2O$)		
Human excreta	2.24	2.60
Livestock dung	7.00	7.54
Crop residues	7.10	20.27
Nutrient (considered trappable)		
Human excreta	1.80	2.10
Livestock dung	2.10	2.26
Crop residues	2.34	3.39
Total	**6.24**	**7.25**

Trappable = 30% of dung, 80% of excreta, 33% of crop residues

Source: Tandon (1997)

Residue management: It relies on the decomposition of crop residues to return organic carbon back to the soil. Residue decomposition can vary with depth of placement in the soil, crop type and residue quantity, allelopathic interactions between existing soil-biota and time. Subsequent benefits from partial disease control can be attained using residue management methods, such as crop rotation and tillage that lower the pathogen's inoculum density in the soil, reduce its ability to survive, deprive the pathogen of its host, and create conditions that favor the growth of other microorganisms at the expense of the pathogen. The soil-borne pathogens,*Cochliobolus sativus Drechs. Ex Dastur* etc, causes common root rot of cereals and can be applied for residue management. Agricultural practices that recycle organic residues by the process of decomposition for utilization by the pathogen and other microbial residents should be practiced for growing pulses organically.

Table 6. Potential of organic and biological resources and plant micronutrients

Name of resource	Annual potential Potential dung/ biomass (M t)		Zn	Fe	Cu	Mn	B	Total
Cattle	744.57	89.35	1824.18	10.42	215.92	16.38	14.15	2170.40
Buffalo	258.02	28.38	647.64	4.13	79.99	7.74	5.42	773.30
Goat and sheep	12.23	1.77	39.74	0.26	4.59	0.51	0.31	47.18
Pigs	4.60	0.69	12.26	0.10	4.60	0.26	0.16	18.07
Poultry	3.40	0.54	11.58	0.09	1.32	0.23	0.15	13.91
Other livestock	6.02	0.69	15.74	0.11	1.87	0.14	0.15	18.74
Human beings	30.38	3.65	79.60	1.70	7.93	3.04	0.82	96.74
Crop residues	100.00	3.60	20.30	2.30	8.80	0.21	0.15	35.36
Forest litter	15.00	0.30	1.55	0.18	0.68	0.02	0.01	2.74
Water hyacinth	3.00	0.29	11.69	0.26	0.45	0.15	0.07	12.91
Rural compost	226.00	27.73	12.60	11.30	54.47	10.40	4.75	127.25
Urban compost	6.00	0.69	15.25	0.33	1.54	0.29	0.15	18.25
Sewage sludge	0.30	0.18	2.37	0.08	0.11	0.01	0.01	2.76
Total	1409.5	153.86	2704.54	31.26	382.27	39.38	26.30	3337.61

Source: Paroda *et.al.*(1997)

Tillage : Tillage practices can be categorized according to the level of soil disturbance involved. Conventional tillage uses primary and secondary cultivation to prepare a seed-bed for planting resulting in considerable disturbance of the soil, while reduced tillage uses a single cultivation or no cultivation, thus resulting in a minimal amount of disturbance to the soil. Reduced tillage systems accumulate organic matter by sequestering carbon in the soil (Lupwayi *et al.* 1999) and by increasing the rate at which soil micro floral and micro faunal decomposition progresses (Doran, 1980; Kennedy and Smith, 1995). Tillage practices that remove crop residues (*i.e.* a primary source of inoculum) should lower the potential for the disease. Under semi-arid environmental conditions, wheat straw decomposition is greater when it is partially incorporated or buried in the soil than when it is left on the surface. When intensive tillage was a common practice, the severity of common root rot was higher with the use of phosphate fertilizers. Under reduced tillage, total soil nitrogen, organic matter and de-nitrification processes increased (Russell and Sallans 1940).

Composts : Disease suppressive conditions have been obtained in soils following addition of certain composts in greenhouse production systems (Hoitink *et al.*, 1993; Hoitink and Fahy, 1986). Composts have also been found effective as bio-control agents under field conditions, however, production issues need further research, such as the optimal time of application to maximize disease suppression (*i.e.* influence of seasonal applications) and the number of repeated applications. Concentration of chemical components could play a significant role in compost efficacy. High concentrations of acetic, propionic, isobutyric, butyric and isovaleric acids were shown to be present in fresh compost (De Vleeschauwer *et al.*, 1981). The levels of volatile fatty acids were above or at inhibitory levels for upto 4 months following application. Young compost was similarly inhibitory to growth of *Brassica rapa L.* and the activity was related to the content of volatile fatty acids, including acetic, propionic and n-butyric acids (Chanyasaket *et al.*, 1983).

Green manures also generate by-products of anaerobic fermentation. (Lynch, 1978). Phyto-toxicity following anaerobic decomposition of wheat straw was related to the production of acetic acid from microbial degradation (Lynech 1978). Similarly, additions of large quantities of organic matter were found to create anaerobic conditions that contributed to the reduction of inoculum of *Fusarium oxysporum, F. sp. asparagi Cohen, Rhizoctonia solani kühn* and *V. dahlia.*The vast majority of studies examining the benefits of using organic amendments have attempted to link beneficial effects to those arising from changes associated with soil microbial activity and biological control with disease suppressive composts. In the biological disease suppression the antibiotic,4-diacetylphloroglucinol turned out to be the essential active component produced by the fluorescent strain (Raaijamakers and Weller,1988). It is quite possible to miss sub-segments of microbial soil populations that may play key roles in disease suppression. There has been no definitive work linking biological control in soil to applications of organic amendments. Soils from organically managed farms, experienced lower incidence of root rot disease due to unspecified biological soil characteristics, whereas reduced root infection by *Phytophthora parasitica.*(Van Bruggen, 1995).

Plant root system is always in close association with multitude of microorganisms and other nutrients. The microbes in root zone are maintained by a variety of secretions from the roots and constitute what is often described as 'rhizodeposition'. These microbes in their turn supply nutrients to the soil system through their heterotrophic activity. Maintenance of these microbes in the rhizosphere, therefore, is also necessary for soil health, crop productivity and nutrient cycling. However, integral parts of the exploitation of soil health care and have led to soil degradation through nutrient depletion and erosion, so that long term strategies are needed to avoid the use of chemical fertilizers without adversely affecting crop productivity. The use of organic manures, composts, bio-fertilizers should be judiciously used in organic farming systems on location specific basis. As stated earlier, the estimated trappable nutrients from various organic sources of nutrients for organic agriculture in India for 2025 has been estimated about 7.25 M tons but at present it is about 6.24 M tons (Table 5).

Vermiculture and vermicompost: Organic matter when subjected to decomposition with the help of earthworms, the results in the production of vermicompost and the process is known as vermicomposting. The product is the result of organic waste consumed by earthworm, digested and excreted in the form of granules. Vermicompost is a nutritive 'organic fertilizer' rich in NKP (nitrogen 2-3%, potassium

1.85-2.25% and phosphorus 1.55-2.25%), micronutrients, beneficial soil microbes like 'nitrogen-fixing bacteria' and 'mycorrhizal fungi' and plant growth hormones. Nutrient content in vermicompost and other composts has been depicted in Tables 6 and 7. Kale and Bano, (1986) reported as high as 7.37% nitrogen (N) and 19.58% phosphorus as P_2O_5 in worm's vermicast. Earthworms host millions of decomposer (bio- degrader) microbes in their gut (as they devour them) and excrete them in soil along with nutrients nitrogen (N) and phosphorus (P) in their excreta. Bacterial flora associated with the intestine and vermicasts of the earthworms and species like *Pseudomonas, Mucor, Paenibacillus, Azoarcus, Burkholderia, Spiroplasm, Acaligenes and Acidobacterium* have the potential to degrade several categories of organics. (Singleton, 2003).

The earthworms release coelomic fluids that have antibacterial properties and destroy all pathogens in the media which they inhabit (Pierre *et. al.* 1982). Earthworms are found in a wide range of soils representing 60-80% of the total soil biomass. Significantly, the worms lead to total improvement in the quality of soil and land which they inhabit and also enhance total plant growth and crop productivity (Bonkowski and Schaefer, 1997). One acre of fertile land may contain more than 50, 000 earthworms of diverse species. Earthworm activity is so prolific that, on a average, 12 tonnes/ha/year soil or organic matter is ingested by this population; leading to upturning of 18 tons of soil/year and world over at this rate it may create a 2 inch humus layer over the globe (Bhawalkar and Bhawalkar, 1993). Earthworms can contribute between 20 to 40 kg nitrogen/ha/year in soil, in addition to other mineral nutrients and plant growth regulators and increase soil fertility and plant growth by 30-200%. About one-fourth of the organic matter is converted into humus. The humic acid has very good impact on plant growth (Atiyeh *et al.* 2002). Earthworms excretion (vermicastings) in soil carry ammonia, nitrates, nitrogen, phosphorus, magnesium and other micronutrients and nitrogen fixing microbes. It is well established that worm casts are richer in 'inorganic phosphorus compounds' extractable in water than the surface soil ingested. Graff, (1970) and Sharpley and Syers, (1976) found that exchangeable phosphorus (P), measured isotopically, was three (3) times greater in worms vermicasts than in the underlying soils. This is done partly by worm's gut enzyme 'phosphatases' and partly by the release of phosphate solubilizing microorganisms in the worm cast. Neilson, (1965) and Tomati, (1987) reported the growth promoting hormone 'auxins' and flowering hormone 'gibberlins' are secreted by earthworms.

Table 7. Average secondary and micro-nutrient contents of vermicompost and FYM

Nutrients	Vermicompost	FYM
Ca (%)	0.44	0.91
Mg (%)	0.15	0.91
Fe (ppm)	175.2	146.5
Mn (ppm)	96.51	69.0
Zn (ppm)	24.43	14.5
Cu (ppm)	4.89	2.6

Source: Chhonkkar *et.al.* (2004)

Table 8. Average nutrient content of vermicompost and other composts

	Compost Nutrient content (% of dry matter)		
	N	P_2O_5	K_2O
Vermicompost	1.6	2.2	0.7
Rural compost	1.2	1.1	1.5
Urban compost	1.2	1.9	1.5
Paddy straw compost	0.9	2.1	0.9
Maize stalk compost	1.1	1.3	1.0
Cotton wastes compost	1.6	1.1	1.5
Water hyacinth	2.0	1.0	2.3
Poultry manure	2.9	2.9	2.4
Castor	5.8	1.8	1.0
Cotton seed	3.9	1.8	1.6
Neem	5.2	1.0	1.4
Niger	4.8	1.8	1.3
Rapeseed	5.1	1.8	1.0
Linseed	5.5	1.4	1.2
Sunflower	4.8	1.4	1.2

Source: Chhonkkar *et.al.* (2004)

Meeting the phosphorus requirement in organic pulse production systems

Sustainable agricultural production depends on an adequate P supply. Nutrient inputs for organic production are typically focused on-farm carbon-based nutrient sources (*e.g.*, crop residue, compost, manure) and non-processed mineral sources (*e.g.*, rock phosphate, lime, and gypsum). This nutrient export after crop harvest creates P defecit, necessitating regular P additions to replace the harvested P. Several studies investigating whole-farm P budgets have found nutrient P deficits in many organic farms and illustrate the need for nutrient additions.

Studies have shown that soil organic matter increases on organically managed farms, while other long-term studies do not show such a buildup due to differences largely on management practices such as tillage intensity, heavy manure additions, return of crop residues, the extent of cover cropping and climatic factors. Soil organic matter contains a variety of organic P compounds, such as inositol phosphate, nucleic acid, and phosphor-lipid. These compounds must be first converted to inorganic phosphate by soil enzymes before being used for plant growth. Phosphatase enzymes are produced by soil micro-organisms, *mycorrhizal* fungi, or excreted by the plant root. Some organic P compounds are stable for many years in the soil, while others are converted to inorganic P within a few days or weeks.

Phosphate solubilizing bacteria play a major role in the solubilization and uptake of native and applied soil phosphorus. Phosphorus, the key element, is an essential plant nutrient required for early establishment and better plant growth. It accelerates tillering, flower initiation, good pod and seed setting. Most of the Indian soils are deficient in available form of P and its requirement is met by the addition of phosphate fertilizers but the use efficiency of applied phosphorus rarely exceeds 30 % due to its fixation as Fe and Al phosphates in acid soil and Ca and Mg phosphates in alkaline soils. In this context, phosphate solubilizing microorganisms play an important role in the utilization of unavailable native phosphates as well as added phosphates.

The bacteria, *Bacillus megatrium, B. polymyxa, Pseudomonas striata, Micrococcus, Streptomyces,* are the important phospho-bacteria commonly present in the soil. The population of these bacteria is more in the rhizosphere compared to non-rhizosphere. The experiments conducted on soybean (Natarajan and Gunasekaran, 1991), pea, French-bean and ground-nut revealed the synergistic effect of phospho-bacteria. The trials conducted at National Pulses Research Centre, Vamban reported that the basal application of phosphorus as rock phosphate along with phospho-bacteria inoculation recorded significant yield increase in the blackgram and greengram.

Nitrogen and phosphorus are the two major plant nutrients required for higher productivity and combined inoculation of nitrogen fixers and phosphate solubilizing microorganisms may benefit the plant better than individual inoculation. Natarajan and Gunasekaran, (1991) reported the beneficial effect of combined inoculation of *Rhizobium* and phosphate solubilizing bacteria on soybean. Trials conducted at Coimbatore reported that combined inoculation of *Rhizobium* and phosphobacteria with 50% of N and P fertilizer recorded the equal yield of 100% N and P alone. The same results are obtained at National Pulses Research Centre, Vamban when tried with blackgram var. Vamban 1 and in horsegram also (Prabakaran *et al.*, 1999). The above results clearly suggested that combined inoculation of *Rhizobium* and Phosphobacteria will help the macro symbiont pulses to get maximum quantity of N and P nutrients so as to get higher yield in organic pulse production systems.

Cover crops : Cover cropping has many advantages in cropping systems, and weed suppression is certainly an important one. Cover crops grown with the main crop are sometimes called smother crops or living mulches. It can also be grown before or after the crop season, providing weed suppression and other benefits during the off-season. Cover crops can provide many benefits to cropping systems whenever the main crops are not covering the ground. This may be before main crop seeding and after main crop harvest, or as a substitute for fallow in other years. Cover crops can include forage legumes and grasses, as well cereals. As a fallow substitute, cover crops effectively suppress weeds while providing cover for the soil.

Cover crops are one of the primary fertility and soil management tools available to the organic farmer, and are an important strategy for preventing nutrient and soil loss from a field. There are crops grown primarily for soil or ecosystem improvement rather than cash, and can provide a variety of services, from increased nitrogen (N) input to soil protection, to weed and disease suppression. However, they can also have negative consequences if they are not managed properly or the wrong species are chosen. There are number of benefits the cover crops can provide. These include : nitrogen; add organic matter; improve soil structure; reduce soil erosion; control; manage nutrients; provide food for pollinators; and furnish moisture-conserving mulch provide habitat for beneficial organisms etc.

Generally, cover crop selection is based on each farmer's situation and production goals. For example, If the purpose of a cover is to provide readily available, biologically fixed nitrogen for cash crops, then the farmer should choose a legume, If the cover crop will be managed as a surface mulch for weed suppression or incorporated to improve soil quality, then the farmer should choose a grass cover crop, such as cereal rye or a sorghum-sudan grass mix. Both of these grass cover crops produce large amounts of biomass with high C : N ratios at maturity and both are reported to suppress some weeds.

The advantage of cover crops for P nutrition involves the accumulation of soil P by the cover crop. This P is subsequently released when the cover crop is incorporated. This is attributed to the ability of some species to drawdown soil P concentrations below what some cash crops can and also to their extensive root system. This P drawdown may also be the result of root exudates and the efficient P uptake by the cover crop roots. Some cover crops can be excellent hosts for *mycorrhizal* fungi, which may allow a greater exploitation of the soil P reserves (Nelson and Mikkelsen, 2008). There are considerable differences in the ability of various cover crops to provide additional P for the subsequent crop. Research has generally shown a greater P benefit from legume cover crops than from grass cover crops, but the effects of cover crops on P nutrition can be highly variable. In many cases, supplemental P is still required after the cover crop to eliminate P deficiency. In some circumstances, P uptake by the cash crop following the cover crop has actually reduced due to low residual soil P caused by uptake by the cover crop and poorly synchronized P release. Cover crops offer some P nutritional benefits in some circumstances. However, it must be remembered that cover crops do not provide any new P to the soil, but only allow the existing soil P reserve to be used more efficiently. With removal of P from the field in harvested products, the nutrient supply must be ultimately replaced with an additional supply to maintain sustainability.

Mycorrhizal Fungi : Enhanced P uptake is frequently cited as a primary benefit of *mycorrhizal* fungi colonization. In this symbiotic relationship, the plant root provides the energy (carbohydrate) for the fungi in exchange for improved nutrient uptake and other plant root benefits. Almost all crop plants form this relationship with *mycorrhizal* fungi, which is present in the root zone of most soils. Many organic growers encourage the associations of *mycorrhizal* fungi with crop roots through the use of cover crops and rotations. However, frequent tillage commonly used for weed control, causes a disruption of the soil fungal network and may reduce its effectiveness for providing nutrients to the plant. The value of *mycorrhizal* fungi for supplying P for crops is most apparent in low-P soils. In most cases, plants growing in soils with medium to high concentrations of P have less *mycorrhizal* association than plants in low-P conditions. Therefore, the value of *mycorrhizal* fungi is greatest in soils without an adequate supply of P. Similar to cover crops, *mycorrhizal* fungi do not provide any additional P to the soil, but can allow better utilization of the existing soil resource. Commercial sources of *mycorrhizal* fungi are available and may be used in specialized conditions (Goslinga *et al.* 2006).

Rock Phosphate: Rock phosphate (apatite) is a general term used to describe a variety of globally distributed P-rich minerals. Of the two main types (sedimentary or igneous), sedimentary rock deposits are the source of over 80% of the total world production of phosphate rock. Depending on its geologic origin, rock phosphate has widely varying mineralogy, texture, and chemical properties. Some rock P is found in hard-rock deposits, while other rock P is found as soft colloidal (soil-like) material. This great variation in properties and the accompanying elements present in the rock (such as carbonate and fluoride) has a large effect on its value as a source of plant nutrient. This range in properties makes some rock P sources excellent nutrient sources and cover crops can improve soil properties, reduce erosion, and increase the nutrient supply to the following crop.

Manure and Composts: These materials are generally excellent sources of P for plants. Though these materials are considered as organic products, over 75% of the total P is present as inorganic compounds. It is commonly recommended that the P in manure and compost be considered as 70% available for soils with low soil-test P, but 100% available for soils testing adequate or high for P. The ratio of nutrients in composts and manures does not closely match with required amount by plants or the harvested products. When manure and compost are used as a primary N source for crops, P is typically over applied by 3 to 5 times compared with the crop removal rate. Long-term use of manures and compost as the primary N source leads to an accumulation of P in the soil that can become an environmental concern for surface water quality.

Bone Meal: Bone meal, prepared by grinding animal bones, is one of the earliest P sources used in agriculture. Most commercially available bone meal is "steamed" to remove any raw animal tissue. The primary P mineral in bone material is "calcium-deca hydroxyapatite" $[Ca_{10}\text{–}x(HPO_4)x(PO_4)_6\text{–}x\,(OH)_2\text{–}x\,(0 < x < 1)]$, which is more soluble than rock phosphate, but much less soluble than conventional P fertilizers. Similar to rock P, bone meal is most effective in acidic soils and when the particle size is small. When used properly, it can be an effective P source. There are no restrictions on the use of bone meal and most commercial bone meal products are heat treated for use in organic production systems.

Meeting the nitrogen requirement in organic pulse production systems

Pulse crops have been an important component of agriculture since ancient times. These leguminous plants on symbiotic association with soil bacterium, the *Rhizobium* forms nitrogen fixing root nodules which are agronomically significant as they provide an alternative to the use of energy expensive nitrogenous chemical fertilizer. Despite the large and increasing use of nitrogenous fertilizers in agriculture, estimates suggest that biological nitrogen fixation contributes at least four times more nitrogen to the soil throughout the world. On a global basis, these symbiotic associations between legume and *Rhizobium* may reduce about 70 million tons of atmospheric nitrogen to ammonia per annum which amounts to about 40 % of all biologically fixed nitrogen per year (Burns and Hardy, 1975). The legumes improve soil fertility through their nitrogen fixing ability. Nitrogen fixing values estimated for various legumes are given in Table 9.

Table 9. Estimates of nitrogen fixed by legumes

Crop	Nitrogen fixed Kg /ha
Black gram	119-140
Chickpea	23-97
Cluster bean	378-196
Cowpea	9-125
Green gram	50-66
Pigeon pea	4-200
Soyabean	49-450

Source: Peoples *et al.* (1995)

Large scale screening should be carried out to identify ideal inoculant strains for different pulse crops. The criterion for selections may vary for particular soil types like acidic, sodic, sodic-saline, saline, nitrate, rich or heavy metal contaminated. Yeast extract mannitol agar (YEMA) medium is employed for culturing of *Rhizobia*. Large scale multiplication of *Rhizobia* can be carried out using rotary shaker or fermenter. In shake flask culture, liquid medium in flasks is agitated by circular motion of rotary shaker. Fermenters are used for industrial scale production of bio-fertilizers. Culture vessels ranging from 5 to 1000 liters can be used depending upon the requirement. The amount of inoculum culture to be added into the fermenter vessel depends on the size of the fermenter, but the ratio between the inoculum and the medium in the vessel should be maintained at 1:2 (5% inoculum rate). The broth is continuously aerated by forcing sterile air. Various fermentation requirements like aeration, agitation and fermentation time vary from strain to strain. The broth should be added to the carrier for preparation of carrier based inoculant when the number of *Rhizobia* in the liquid medium attains the required standard (10^8-10^9 cells/ml).

Considerable input of biologically fixed N can be achieved in almost all agricultural ecosystems through the activity of symbiotic associations. Strategies are available for pulse crops to manipulate N inputs through symbiotic nitrogen fixation. This can be achieved by changing the proportion of plant nitrogen derived from symbiotic nitrogen fixation. Management practices can be imposed which either increase the numbers of effective *Rhizobia* in soil, reduce the levels of or legumes sensitivity to soil nitrate or enhance the potential for legume growth. For increasing pulse crops yield through bio-fertilizers, the following strategy is suggested. The most important constraints to effective exploitation of symbiotic nitrogen fixation are:

- The quality of the inoculants.
- Lack of knowledge about inoculation technology for the extension personnel and the farmers.
- Effective inoculant delivery system.
- Formulation of the policy to exploit symbiotic nitrogen fixation successfully.

For success of bio-fertilizers, concerted efforts right from production, demonstration to distribution is required. The next step consists of convincing and educating the farmers regarding the benefits of these inoculants.

Future thrust

To increase area and production of pulse crops in organic agriculture system, we need crop specific and region specific approaches are needed, which should be adopted in the overall framework of systems approach. The major thrust areas to be addressed are as follows:

- Replacement of cereal crops in the prevailing rice-wheat cropping systems with high yielding varieties of pulses.
- Inclusion of short duration varieties of pulses as catch crop.
- Development of multiple disease and pest resistant varieties and effective organic modules for management of targeted pest and diseases.
- Reducing storage losses and improving market information and infrastructure.
- Linking minimum support price to market prices for bridging the gap between demand and supply.
- Developing high nitrogen fixing varieties, which will play a crucial role in organic systems
- Coordination of research, extension and farmers to encourage farmer's participatory reseach.
- Research on judicious use of secondary and micro nutrients on site specific basis should be initiated to mitigate the soil related micro nutrient disorder by applying suitable combination of organic manures and bio fertilizers.
- Native strains should be developed for seed inoculation with *Rizobium* and phosphate solubilizing bacteria.
- Research on site-specific water management based on soil and agro climatic conditions for specific pulse crops using pressurized irrigation systems like drip and sprinkler should be carried out.

References

Atiyeh, R.M., Lee Edward,C.A.,Arancon, N.Q. and Metzger, J.D. 2002. The influence of humic acids derived from earthworm-processed organic wastes on plant growth; Bioresource Technology, 84 : 7-14.

Bhawalkar, V.U. and Bhawalkar,U.S. 1993. Vermiculture: The Bio- nutrition System; National Seminar on Indigenous Technology for Sustainable Agriculture, I.A.R.I, New Delhi, March 23-24, pp: 1-8.

Bonkowski, M. and Schaefer, M. 1997. Interactions between earthworms and soil protozoa: A trophic component in the soil food web. J. of Soil Biology and Biochemistry, 291: 499-502.

Burns, R.C. and. Hardy, R.W.F. 1975. In: Nitrogen Fixation in Bacteria and Higher Plants. New York: Springer-Verlag.

Chanyasak, V., Katayama, A., Hirai, M.F., Mori, S.and Kubota, H.1983. Effects of compost maturity on growth of Komatsuna (Brassica rapa var,Pervidis) in Neubauer's pot. II. Growth inhibitory factors and assessment of degree of maturity by org.-C/org.-N ratio of water extract. Soil Sci. Plant Nutr, 29 : 251–259.

Chhonkkar, P. K. and Dwivedi B. S., 2004.Organic farming and its implications in Indian food security- Fert. News, 49 : 15-38.

De.Vleeschauwer, D., Verdonck, O. and Van Assche, P. 1981. Phytotoxicity of refuse compost.Bio Cycle,2: 44–46.

Doran, J.W. 1980. Soil microbial and biochemical changes associated with reduced tillage. Soil Sci. Am. J., 44 : 765–771.

Goslinga, P., Hodgeb, A.G., Goodlassc, and Bending, G.D. 2006.Arbuscular mycorrhizal fungi and organic farming.Agriculture, Ecosystems and Environment, 113 : 17–35

Graff, O., 1970. Phosphorus Contents of Earthworms Casts (In German), LandbauforchungVolkenrode, 20 : 33-36.

Harisudan, C., Latha, K.R., Subbian, P., Vaidyanathan, R. and Manivannan, V. 2009. Nutrient managment for rainfed pulses – A review, Agric. Rev., 30 : 224 - 228.

Hoitink, H.A.J., Boehm, M.J., and Hadar, Y. 1993. Mechanisms of suppression of soil borne plant pathogens in compost-amended substrates.In: Hoitink, H.A.J., Keener, H.M. (Eds.), Science.

Hoitink, H.A.J.andFahy, P.C. 1986. Basis for the control of soilborne plant pathogens with composts.Annu. Rev. Phytopathol. 24 : 93–114.

Kale, R.D. and Bano,K., 1986.Field trials with vermicompost.An Organic Fertilizer; In Proc. of National Seminar on 'Organic Waste Utilization by Vermicomposting'.GKVK Agricultural University, Bangalore, India.

Kennedy, A.C., and Smith, K.L.1995. Soil microbial diversity and the sustainability of agricultural soils. Plant Soil, 170 : 75–86.

Lupwayi, N.Z., Rice, W.A. and Clayton, G.W. 1999. Soil microbial biomass and carbon dioxide flux under wheat as influenced by tillage and crop rotation. Can. J. Soil Sci. 79 : 273–280.

Lynch, V.M. 1977. Phytotoxicity of acetic acid produced in the anaerobic decomposition of wheat straw. J. Appl. Bacteriol. 42 : 81–87.

Lynch, V.M., 1978. Production and phytotoxicity of acetic acid produced in anaerobic soils containing plant residues. Soil Biol. Biochem., 10 : 131–135.

Natarajan, N. and Gunasekaran, S. 1991. Coinoculants for soybean.Abst. XXXI. Annual Conf. AMI., TNAU, Coimbatore. p.127.

Neilson, R.L., 1965. Presence of plant growth substances in earthworms, demonstrated by the paper chromatography and Went Pea Test, Nature, 208: 1113-1114.

Nelson, N. and Mikkelsen, R. 2008. Meeting the phosphorus requirement on organic farms.Better Crops. 92 : 12-14

Pal, S. S. 2006. Scope of organic farming in India- A Review J. Farming System Research & Development, 12 : 1-12

Parasuraman, P. and Mani, A.K. 2003. Growth, yield and economics of Rice (Oryza sativea). Finger Millet (Eleusine coracana) crop sequence as influenced by Coir Pith and Farmyard manure with and withour Inorganic fertilizers. Indian Journal of Agronomy. 48 : 12-15

Paroda, R.S. 1997. Keynote address in FAO-IFFC) International Seminar on IPNS for sustainable development. November 25-27, Vigyan Bhawan, New Delhi.

Peoples, M.B., Herridge, D.F. and Ladha, J.K. 1995. Biological nitrogen ûxation: an efûcient source of nitrogen for sustainable agricultural production. Plant and Soil, 174 : 3–28.

Pierre, V., Phillip,R., Margnerite,L. and C. Pierrette, 1982.Anti-bacterial activity of the haemolytic system from the earthworms Eisiniafoetida Andrei.Invertebrate Pathology, 40 : 21-27.

Prabakaran., 1999. Rhizobial inoculants for problem soil. In: Reason Advances in Microbial inoculant. Dept. of Agrl.Microbiol., Tamil Nadu Agric. Univ. Coimbatore, India, p.18.

Raaijamakers, J.M. and Weller, D.M. 1988. Natural plant protection by 2,4-diacetylphloro-glucinol-producing Pseudomonas spp. in take-all decline soils. Mol. Plant–Microbe Interact, 11 : 144–152.

Russell, R.C. and Sallans, B.J. 1940. The effect of phosphatic fertilizers on common root rot.Sci. Agric., 21 : 44–51.

Sharpley, A.N. and Syers,J.K. 1976. Potential Role of Earthworms Casts for Phosphorus Enrichment of Run-Off Waters.Soil Biology and Biochemistry, 8 : 341-346.

Singleton, D.R., Hendrix,B.F.,Coleman,D.C. and Whitemann,W.B. 2003. Identification of uncultured bacteria tightly associated with the intestine of the earthworms Lumricusrubellus. Soil Biology and Biochemistry, 35 : 1547-1555.

Tandon, H.L.S. 1997. Organic resources:An assessement of potential supplies, their contribution to agricultural productivity and policy issues for Indian agriculture for 200 to 2025. In. Plant needs Supply, Efficiency and Policy, National Academy of Agricultural Sciences, New Delhi. P. 15-28.

Tomati, V., Grappelli,A. and Galli,E. 1987.The Presence of Growth Regulators in Earthworm-Worked Wastes. In Proceeding of International Symposium on 'Earthworms'; Italy; 31 March-5 April, 1985; pp: 423-436.

Van Bruggen, A.H.C. 1995. Plant disease severity in high-input compared to reduced-input and organic farming systems. Plant Dis. 79 : 976–984.

Venkateswaralu, B. 1997.Water stress and biological nitrogen fixation in legumes.Ind. J. Dryland Agric. Res. Dev., 12 : 51-53.

Yadav, A.K. 2009. Organic agriculture concept, scenario and practices. National Center of Organic Farming, department of Agriculture and Cooperation, Ministry of Agriculture, Govt. of India; pp 7-11.

❑❑❑

Chapter - 6

Legume Inclusion Substitution in Cropping Systems

V. K. Singh
Project Directorate for Farming Systems Research, Modipuram, Meerut, - 250 110 (U.P.)

Among thirty major cropping systems in the country, which contribute significantly to the national food basket, the majority are cereal-based. In fact, continuous cropping of cereal-based systems during last three decades has resulted in second generation problems and the sustainability of these systems is under threat. The major challenge is to produce more from each unit of land on the one hand and to sustain soil health on the other. The depletion of soil fertility by the intensive cereal-cereal production systems is considered to be a major cause of the yield decline. For example, a rice-wheat cropping system (RWCS) yielding 7 t/ha of paddy and 4 t/ha of wheat removes 315 kg nitrogen (N)/ ha, 28 kg phosphorus (P)/ ha and 333 kg potassium (K)/ ha, and significant amounts of micro-nutrients (Hegde and Dwivedi, 1992). These off-takes often exceed fertiliser input in many Indian states. Refinement of nutrient management strategies would help maintain the crop productivity and soil fertility, but other rotational strategies could also help especially in situations where greater fertiliser use may be uneconomic or environmentally unacceptable. Legume crops fix atmospheric N and enrich soil fertility, and could help to sustain the long-term productivity of cereal-based cropping systems. Depending on the soil and ecological stresses, a cropping system can be diversified using legumes under different mode. Results from the All-India Coordinated Project on Integrated Farming Systems (AICRP-IFS) of Indian Council of Agricultural Research show consistently better productivity with inclusion of legumes in cropping systems. The benefits of legumes in rotation are not solely due to biological nitrogen fixation but also because it increased nutrient availability, improved soil structure, reduced disease incidence and increased mycorhyizal colonization (Wani *et al.*, 1995). Inclusion of pulses as intercrops in sequential as well as intercropping systems would not only make the country self reliant but also help in improving the soil health. Series of experiments are being conducted under (AICRP-IFS) and elsewhere in the country to identify need based cropping systems for different agro-climatic zones of India. The pre-dominent cropping systems involving legumes as a strategy for nutrient management through different approaches are being discussed here for ready reference of researchers, planners and extension workers.

Predominant cropping systems and yield gaps

Cropping system of a given region depends on endowed natural resources, farming situations and socio- economic conditions of the region. Yadav and Rao (2001) listed 250 double cropping systems of primary and, secondary importance in term of their spread in the country. However, among the top ten popular cropping systems in the country, cereal-cereal with major share of rice-wheat (10 m ha) and rice-rice (5.9 m ha) are the most dominant systems contributing to about 52.3% to the national food basket. Notwithstanding the projections highlighting the need for increasing agricultural production, a vast untapped potential in respect of the yield for all the crops prevails in different agro-climatic zones of the country. It is worth noting here that the gaps between potential and realisable and between realisable and averaged realised yield in the country generally ranged around 50 to 100%, respectively. The agro-climatic zone wise yield potential computed in terms of Rise Equivalent Yield (REY) at Project Directorate for Farming Systems Research, Modipuram indicates that highest productivity ranging between 8.46-12.04 t/ha for rice-wheat system, 8.61-11.35 t/ha for maize-wheat system and 5.84-7.46 t ha for pearl millet-wheat system in Trans Gangetic Plains. While the lowest system productivity of rice-wheat system (4.72-8.73 t/ha) and maize-wheat system (4.35-9.34 t/ha) was in Western Himalayan region (Fig. 1).

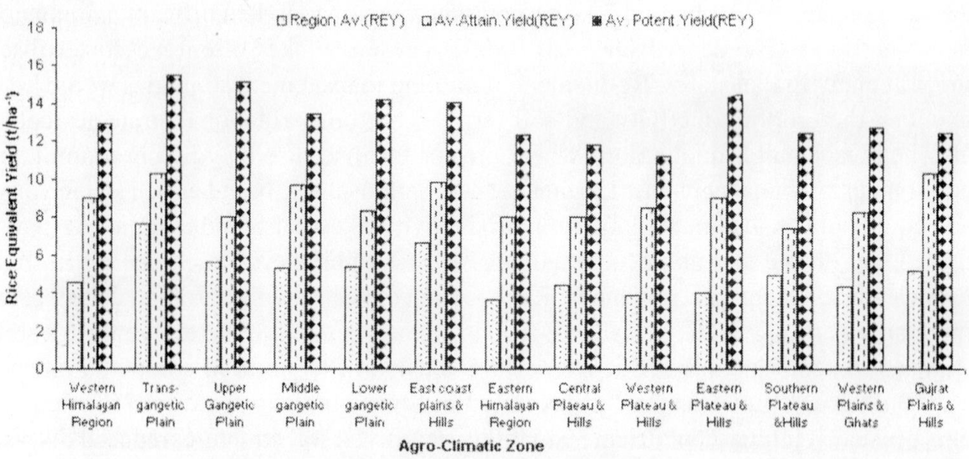

Fig.1: Yield gap between agro-climatic zones average yield, attainable yield and potential yield (Gill *et al.*, 2008)

Assessing yield gap in terms of Rise Equivalent Yield (REY) for various cropping systems varied from 0.88 to 7.54 t/ha in different agro-climatic zone. Further, the yield gap between state average and on-station research average varied from cropping system to cropping system and from region to region within a system (Gill *et al.*, 2008). The most important viable options for vertical productivity improvement is efficient agronomic management options. Among the different factors, fertilizer plays a key role in bridging such gaps as it alone can contribute to 50% in achieving the production target (Tiwari, 2002). These facts call for scientific use of fertilizers in a integrated and balance manner for enhanced productivity.

Role of legumes in addressing non- sustainability issues

During post green revolution era the non-sustainability issues causing production fatigue are declining soil organic matter and emergence of multiple nutrient deficiencies (Dwivedi *et al.*, 2006). Farmers' participatory surveys conducted in Indo-Gangetic Plain (IGP) region indicate that farmers' have resorted to higher doses of nutrients each year to obtain the same yield as obtained in the previous year (Dwivedi et al., 2001). The probable reasons behind these are continuous mining of nutrients, inadequate use of organic sources, imbalance fertilization and decline of soil organic matter. For example, rice and wheat shares 46% area of cultivated area at present and the area under other arable crops especially pulses are declining continuously (Timsina and Connor, 2001). Such area expansion under rice and wheat crop has posed several new generation problems *viz.*, depletion of ground water table, sub-soil surface compaction, poor crop establishment after rice, deterioration in soil health, multiple infestations of weed flora, insect pests and diseases, excessive energy use and ecosystem pollution.

Therefore, a major shift is expected in cereal based agricultural production systems towards systems involving legumes which have the potential attribute to overcome the second generation problems as they benefit the system through (i) Improved nutrient use efficiency, (ii) Increased soil organic matter content, (iii) Improved soil physical condition along with microbial activity and nutrient availability, (iv) Decreased NO_3-N leaching and (v) Suppression of weed population

Nutrient management strategies through inclusion of legumes in cropping systems

Legumes are known to fix atmospheric N and enrich soil fertility, and therefore, could help in sustaining the long–term productivity of cereal based cropping systems. Soil fertility restoring capacity of legumes has been known since historic times even when their capacity to fix atmospheric N was not known. Depending upon the crop and its growth period legumes can fix 26-200 kg N/ha (Table 1). From the computations made using area under pre-dominant pulses it is estimated that pulses fix 1.14 mt N annually in India. Apart from these, other legumes such as fodder legume and green manure crops also contribute significant amount of N through BNF.

Table 1: Estimates of nitrogen fixed by important pulse crops of India

Crops	Biological N fixed (kg/ha)[@]	N released into the soil (kg/ha)[@]	Country level N contribution to soil through BNF (mt)[@]
Chickpea	26-63	50.3	0.37
Lentil	35-100	34.5	0.10
Mung bean	50-55	-	0.20
Pigeonpea	68-200	59.4	0.27
Field pea	46	38.3	0.33
Urd bean	51	-	0.17

Source: *Brahmprakash *et al.*, (2004); [@]Singh *et al.*, (1981)

Pulses meet most of their N demands by atmospheric N fixation with the help of *Rhizobia* that grow on their roots and N released for succeeding crops ranging from 35-59 kg N/ ha (Table 1). However, some legumes such as soybean (*Glycine max*) are poor starters and need an application of 40-60 kg N/ha. Rajmash or French bean (*Phaseolus vulgare*) is an extreme example, which when grown in plains needs 100-120 kg N/ha, because these crops have very poor nodulation. From the IPNS viewpoint, more important is the amount of N that legumes leave for the succeeding crops which varies from 25-100 kg N/ha. Depending upon the situation and site-specificity, legumes can be included in the system as described below:

Legume as summer/catch crop : Legumes can easily be included in a cropping system as catch crop during fallow periods under cereal based systems without decreasing land area of the cereal crops. In rice-wheat cropping system, medium duration rice varieties followed by wheat varieties suited for normal sowing or late sowing in north-eastern plain, did not leave any scope to grow a legume between rice harvest and wheat sowing. However, the period between wheat harvest and rice planting can be utilized for growing short duration (60-70 days) summer legumes. Experiments conducted on *Tarai* soils indicated that yield of rice in rice-wheat system generally remained higher when preceded by mung bean (Singh and Sharma, 2002). Singh *et al.* (1993) studied the effect of pulses on yield, economics and sustainability of rice-wheat system in eastern Uttar Pradesh for several years and found a significant improvement when summer mung bean was included in the system. Mungbean and urdbean are the important crops of the summer season. At Modipuram, raising mungbean for grain and residue incorporation in summer was found to be markedly better in sustaining productivity of rice-wheat system (Fig. 2) and improving soil fertility. This practice not only provided about 0.5 to 1 t/ha of pulse grain but also benefited the succeeding cereal crops through an improvement in N status of the soil (Singh and Mishra, 2010).

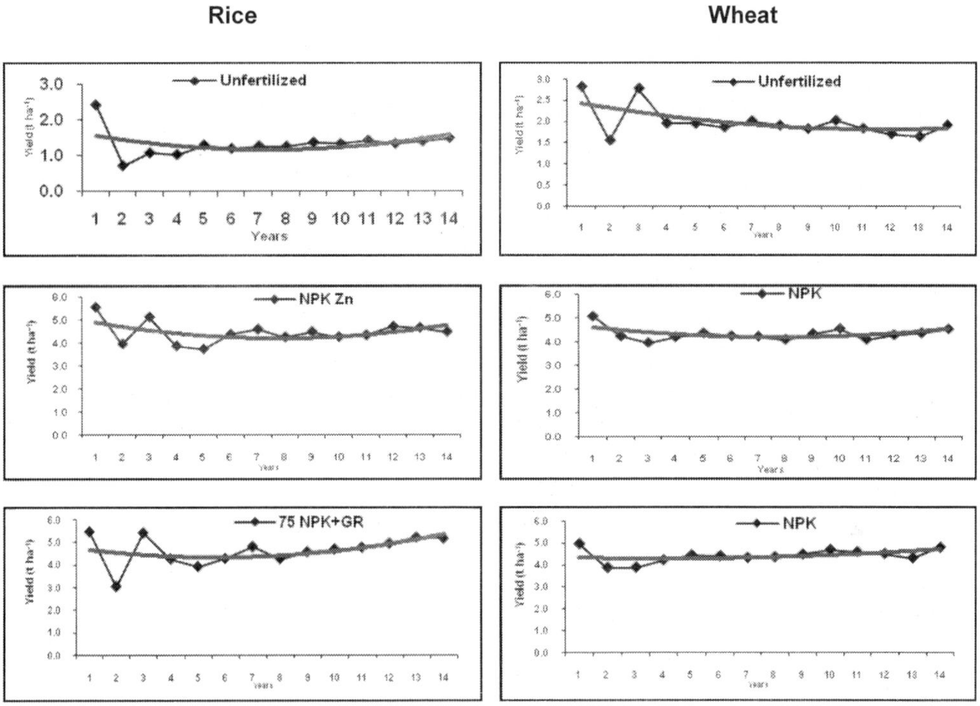

Fig.2: Yield trend of rice and wheat with continuous use of fertilizers vis a vis green gram residue (GR) incorporation under long-term experiment (LTE) at Modipuram. (Singh and Mishra, 2010)

The possibility of summer legume fodder was also explored at Modipuram and an appreciable increase in the use-efficiency of N and P fertilizers in rice-wheat system was noticed. Such strategy not only helped in trapping the NO_3-N from deeper soil layers, but also made it available in upper profile layer for succeeding crops (Fig. 3).

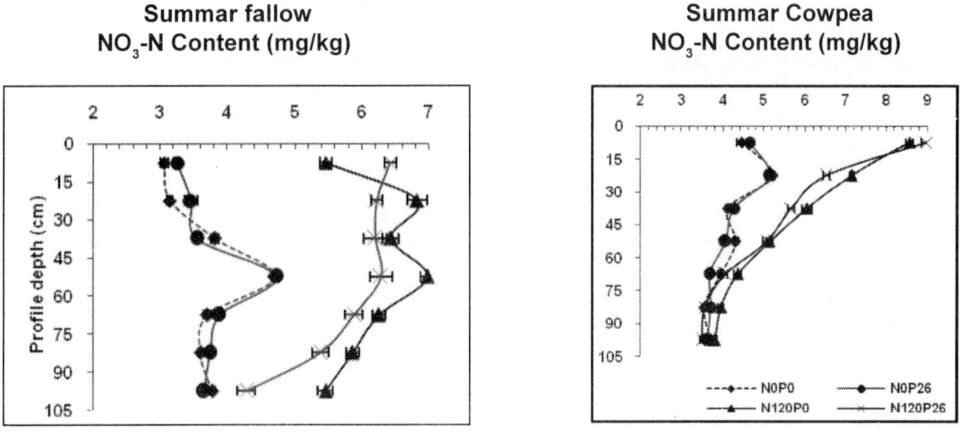

Fig. 3: Effect of fertilizer N and P, and summer cowpea on the distribution of nitrate-N in soil profile after wheat harvest under rice-wheat system. (Dwivedi *et al.*, 2003)

Legume as intercrop: Inter-cropping is one of the important ways to increase the productivity and provide income stability under limited soil moisture conditions. Legumes such as pigeon pea, black gram, mungbean, and soybean are ideal intercrops in upland direct seeded rice. The scope of pulses as intercrops in cereal dominated systems further increases under water scarce conditions or aberrant weather situations, as it not only increases total productivity of the system but also play an important role in economizing the use of resources, particularity N fertilizers. Extensive studies carried out under (AICRP-IFS) have identified promising intercropping systems such as maize + black gram at Palampur, Ranchi and Banswara, maize + cowpea at Karjat, sorghum+ pigeon pea at Indore, pigeonpea + mungbean at Bichpuri and Hanumangarh, and rice+soybean at Kalyani and Jabalpur. Most of these inter-cropping systems have been evaluated on farmers' fields and found to be highly remunerative (15-200%) over the sole crops. The success of intercropping depends upon choice of crops, varieties, planting method, crop geometry, nutrient and water management etc. It was generally observed that both the component crops should be fertilized at recommended rate, to achieve maximum benefit from a diversified inter-cropping system. There is a large scope for intensification of pigeonpea-wheat cropping system through intercropping with greengram, blackgram, cowpea etc. in pigeon pea crop. Intercropping of legumes with short duration pigeon pea is more remunerative. With the advent of modern tools like bed planting, multi-crop seed drills, pneumatic seeder, punch planter machine, intercropping of legumes may become a promising option. Co- culture of pulses along with direct seeded/bed planted rice reveals further scope for inclusion of legumes as intercrop in rice-wheat cropping system.

Legume as substitute crop : The substitution of rice or wheat largely depends on the nature of stress being experienced in different agro-ecologies. For instance, in Trans-Gangetic Plains (TGP), where water table depletion is of serious concern, there is scope for substituting rice with pigeon pea. Similarity in eastern part of India, where wheat productivity is generally low because of climatic constraints, particularity higher thermal regimes, wheat needs to be substituted with chickpea, lentil, pea or groundnut. In fact, legumes enriched the system by the means of nutrient recycling through leaf litter fall, root and stubble incorporation, BNF and felicita for proliferation of succeeding crop roots.

Extensive studies on comparative performance of rice-wheat vs. rice-legume or legume-wheat at different locations of AICRP_IFS reveal that growing legumes in system with recommended rate of NPK helped to improve system productivity and profitability compared to a non-legume based system. In double crop sequences involving cereal crops only, diversifying *kharif* (Singh *et al.*, 2005) or *rabi* cereal with a grain legume has shown promising results (Singh and Sharma, 2002). Reports from different parts of the country indicated that inclusion of legumes in rice-wheat cropping system (RWCS) as a substitute crop, increased the productivity of succeeding crop along with saving of 30-40/kg/ha fertilizer N (Fig. 4).The inclusion of legume as substitute crop at Pantnagar had positive effect on soil organic carbon and Olsen-P content. In contrast, available K was generally depleted under different crop sequences, but the greatest depletion (nearly 8%) was under the rice-wheat sequence (Table 3).

Table 3: Changes in organic carbon, and available P and K under different crop sequences after 2-years of cropping

Crop sequence	% Change over initial content		
	Organic carbon	Available P	Available K
Two crop sequence			
Wheat-rice	1.1	15.5	-7.8
Chickpea-rice	4.5	45.5	-1.3
Lentil-rice	4.5	32.8	-4.9
Pea-rice	2.2	34.5	-2.2
Three crop sequence			
Wheat-fodder-rice	1.1	28.7	-4.1
Chickpea-fodder-rice	0.0	41.9	-2.9
Lentil-fodder-rice	0.0	39.7	-4.0
Pea-fodder-rice	1.1	41.9	-3.7

Source: Singh *et al.*, (2002)

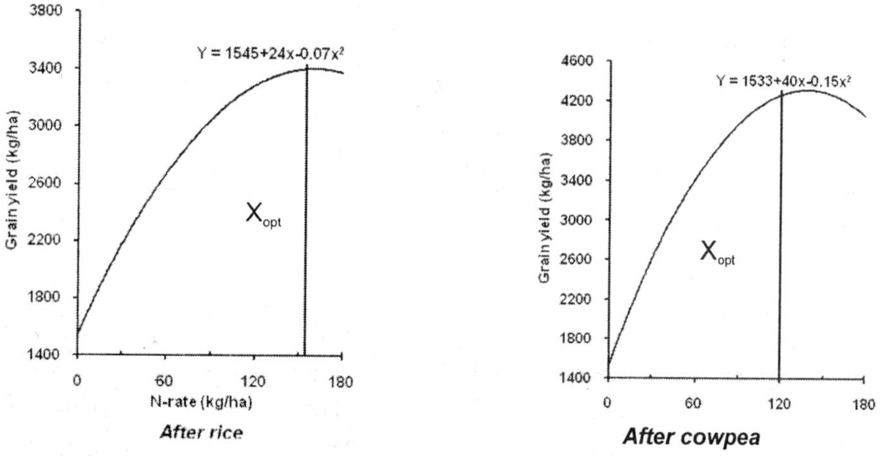

Fig.4: Response functions and economic optimum rate (Xopt) of N in wheat grown after rice or cowpea. (Yadav *et. al.*, 2003)

The benefits of legumes cannot be explained solely in terms of residual N but there are also other advantages. Improvements in soil structure such as decrease in soil bulk density following legumes (Table 4), breaking of cycles of cereal pests and diseases, ability to mine the nutrients from different soil layers, control of weeds, and the phototoxic and allelopathic effect of different crop residues can result in extra yield in a sequential cropping system.

Legume as fodder crop : Inclusion of fast growing forage legumes is another promising agro-technique, though the advantage of forage legumes in terms of nutrient recycling, improvement in soil physical and chemical properties are relatively less explored. Studies conducted at Project Directorate for farming systems research (PDFSR) Modipuram revealed an appreciable increase in the use-efficiency of N and P fertilizers in rice-wheat system with the inclusion of summer cowpea (forage) (Dwivedi *et al.*, 2003). Further, inclusion of cowpea as fodder crop during summer under rice-wheat system not only helped in trapping the NO_3-N from deeper soil layers, but also made it availability in upper profile layer for succeeding crops. A number of leguminous fodder crops have been evaluated for contribution which they made in meeting N demand of succeeding cereals and it was as much as 35-120 kg N/ha.

Table 4: Change in soil bulk density in different soil profile layers after 3 years of rice-wheat and pigeon pea-wheat cropping

Soil depth (cm)	Bulk density (Mg/m^3) after 3 crop cycles		% Change over initial	
	Rice-wheat system	Pigeon pea-wheat system	Rice-wheat system	Pigeon pea-wheat system
0-15	1.50 ±0.013@	1.42 ±0.009	+5.4	-4.2
15-30	1.58 ±0.011	1.46 ±0.010	+3.3	-4.8
30-45	1.62 ±0.012	1.49 ±0.016	+3.8	-4.7
45-60	1.60 ±0.010	1.52 ±0.011	+1.9	-3.2
60-75	1.62 ±0.014	1.59 ±0.018	+0.6	-0.1
75-90	1.62 ±0.013	1.61 ±0.020	-0.1	-0.2
90-105	1.64 ±0.016	1.64 ±0.015	-0.1	-0.1

@Mean and standard error of mean
Source: Singh and Dwivedi, (2006)

Legume as green manures : Since FYM has limitations because of its competitive use for domestic fuel, legume green manuring especially during fallow period like summer in rice-wheat system, is another viable option for sustained productivity. Results from AICRP-IFS indicated that a saving of 25 to 50% N can be made under rice-wheat, rice-rice, rice-maize, maize-wheat, pearl millet-wheat and sorghum-wheat systems (Table.5). Growing legumes as green manure can provide the equivalent of 40-60 kg of urea N/ha to the suceeding rice crop, with added improvement in rice equivalent yield (REY) of rice- wheat system. The increase in REY over rice-wheat system with inclusion of a green manure crop in post wheat summer season was ascribed to incorporation of additional amounts of plant nutrients through green biomass amounting to 23.8 to 27.4 t/ha in case of sesbania green manure and 25.4 to 28.5 t/ha in case of mungbean green manure (Singh *et al.*, 2001). In fact, green manure hastens the nutrient transformation (Wani *et al.*, 1995), increases root growth and N use efficiency of cereal crops, which ultimately results in greater productivity of cereal-based production systems (Buresh and De-Datta 1991; Yadav *et al.*, 1998).

Table 5: Effect of integrated use of green manuring and inorganic fertilizers on productivity of cereal based cropping system under AICRP-IFS (mean over 2005-06 to 2009-10)

Treatments		Grain yield (kg/ha)		
Kharif	***Rabi***	***Kharif***	***Rabi***	**Rice equivalent yield**
Rice-wheat	**Ludhiana**			
100% NPK	100% NPK	5361	4639	10315
50% NPK+ 50% N (GM)	100% NPK	6027	4780	11132
75% NPK+ 25%N (GM)	75% NPK	6096	4401	10796
	Varanasi			
100% NPK	100% NPK	4496	3360	8084
50% NPK+ 50% N (GM)	100% NPK	4759	3512	8510
75% NPK+ 25%N (GM)	75% NPK	4625.2	3275	8123
	Kalyani			
100% NPK	100% NPK	3904	3021	7131
50% NPK+ 50% N (GM)	100% NPK	4065	3527	7831
75% NPK+ 25%N (GM)	75% NPK	4230	3257	7708
Rice-rice	**Rajendra Nagar**			
100% NPK	100% NPK	4744	5166	9909
50% NPK+ 50% N (GM)	100% NPK	4865	5433	10299
75% NPK+ 25%N (GM)	75% NPK	4949	5292	10241
	Chiplima			
100% NPK	100% NPK	5335	5517	10852
50% NPK+ 50% N (GM)	100% NPK	6046	6061	12108
	Bhubaneswar			
100% NPK	100% NPK	5358	5919	11277
50% NPK+ 50% N (GM)	100% NPK	6081	6737	12818
75% NPK+ 25%N (GM)	75% NPK	5871	6129	12000
	Karmana			
100% NPK	100% NPK	4601	4432	9431
50% NPK+ 50% N (GM)	100% NPK	4993	4494	9488
75% NPK+ 25%N (GM)	75% NPK	4970	4441	9411
Rice-mustard	**Rudrur**			
100% NPK	100% NPK	4869	810	6308
50% NPK+ 50% N (GM)	100% NPK	5371	815	6818

Contd.

Maize -wheat	**Ranchi**			
100% NPK	100% NPK	4849	5157	9462
50% NPK+ 50% N (GM)	100% NPK	5517	5806	10700
75% NPK+ 25%N (GM)	75% NPK	5095	5257	9769
Rice-maize	**Kathalagere**			
100% NPK	100% NPK	6590	4372	10156
50% NPK+ 50% N (GM)	100% NPK	6459	4685	10280
Pearl millet-wheat	**S. K. Nager**			
100% NPK	100% NPK	1071	2305	2753
50% NPK+ 50% N (GM)	100% NPK	1453	2940	3583
75% NPK+ 25%N (GM)	75% NPK	1530	2994	3690
	Hisar			
100% NPK	100% NPK	3557	5852	7674
50% NPK+ 50% N (GM)	100% NPK	3715	5999	7922
	Junagadh			
100% NPK	100% NPK	1450	3325	3895
50% NPK+ 50% N (GM)	100% NPK	1585	3643	4264
75% NPK+ 25%N (GM)	75% NPK	1507	3351	3962
Sorghum-wheat	**Ranchi**			
100% NPK	100% NPK	2850	3437	5570
50% NPK+ 50% N (GM)	100% NPK	2854	3458	5591

GM= Green manure
Source : AICRP-IFS Reports

During the past two decade, extensive studies on different aspects of green-manuring in rice-wheat and maize-wheat cropping systems were undertaken on alluvial soils of Punjab with some practically important conclusions as (i) sesbania green-manuring was more promising than other legume crops, (ii) efficiency of green-manure N was comparable to that of urea N in rice, (iii) benefit accruing through green manure was higher in rice-wheat than in maize-wheat system, and (iv) incorporation of green-manure crop a day before rice transplanting gave better efficiency than incorporation 1 to 2 weeks before transplanting (Singh *et al.*, 1993). The major bottleneck for summer green-manuring is that it involve the loss of a crop season. To overcome this, alternatively, short-duration grain legumes like green gram can be grown in summer and after picking the pods, green biomass may be incorporated to the field. Singh and Mishra (2010) reported that this option has an advantage over Sesbania green manure as apart from green manure, it also produce an additional grain yield (0.8-1.2 t ha) of mung bean (Table 6).

Table 6: Comparative performance of green manuring and legume residue incorporation in rice-wheat system

Treatment	Grain yield (t/ha)		
	Rice	Wheat	Total
Faizabad (Mean of 3 years)			
Green manuring	5.14	4.91	10.05
Green gram residues	5.15	4.79	9.94
Fallow	4.70	4.55	9.25
Chiplima (Mean of 2 years)			
Green manuring	4.19	2.21	6.40
Green gram residues	4.16	2.12	6.29
Palampur (Mean of 2 years)			
Green manuring	4.82	2.76	7.58
Green gram residues	5.00	3.07	8.07
Kharagpur (Mean of 2 years)			
Green manuring	3.65	2.96	6.61
Green gram residue	3.86	2.94	6.80

Source: Hegde (1992)

Legume as break crop: The fodder legumes yielding 40-90 t/ha^{-1} green fodder offer ample scope for inclusion as break crop. Studies revealed an assured advantage of inclusion of legume in continuous RWCS and also in other predominant cropping systems. As both rice and wheat are staple food grains, substitution of either crop may not be advisable in the RWCS practicing areas. The idea of inclusion of legumes as summer forage or as break crop may be more saleable in most of the rice-wheat growing areas. Studies conducted at Modipuram indicated that noxious weed population like *Phalaris minor* in wheat and *Echinocloa* in rice were substantially reduced when legume was introduced as a break crop of cowpea or berseem at 2-3 years interval under RWCS. Apart from these, sustained productivity of the system was noticed on long term-basis (Table 7).

Table 7: Weed dynamics as influenced by break crop of legume under rice-wheat cropping system.

Break crop (every 3rd rice-wheat cycle)	% weed reduction		
	Before introduction of break crop	1-year after break crop	2-year after break crop
Berseem in place of wheat	-	94	62
Cowpea in place of rice	-	84	59

Source: Singh and Mishra (2010)

Multipurpose Legume Shrubs : In the farming situations where farmers are reluctant to spare their meager land resources and inputs for green manure crops, fresh lopping of perennial leguminous trees such as *Gliricidia, Casuarina, Leucaena* and *Sesbnia sesban* grown in hedge rows (alleys) and on field bunds may be used for incorporation into the soil as source of N (Kundu et al., 1991). Green leaf manure applied before transplanting rice releases more N then FYM (Chatterjee *et al.*, 1979). In addition to improving soil health, these multipurpose legume shrubs also provide nutritious fodder for animals. Despite all the benefits, the scope for large-scale adoption of legumes in alley cropping under RWCS is limited to upland marginal lands.

Harnessing advantages of legumes through modern technologies

Urea spray in extra short duration pigeon pea: In extra short duration pigeon pea crops foliage remain green even after reaching physiological maturity, which if allowed to defoliate, could augment the residual benefit of pigeon pea to the following wheat in a pigeon pea-wheat rotation. Studies at Modipuram indicated that the use of 10% urea spray on mature extra short duration pigeon pea cv. ICPL 88039 added more leaf litter (10%) to the soil, increased mean wheat yield by 29% over no-urea spray (Fig. 5) along with 21% additional net profit.

The addition of pigeon pea leaf litter to the soil resulted in a saving of 40-60 kg N for the following wheat crop indicating that such a practice, if adopted by farmers, may enhance sustainability of wheat production system in an environmentally friendly way by reducing amount of fertilizer N application to soil and enhancing wheat yield.

Inclusion of legumes in systems through RCTs : In the Indo-Gangetic Plain region, a major deterrent to large scale adoption of the otherwise advantageous pigeonpea-wheat cropping system is the poor crop stand of pigeonpea and other legumes due to temporary water logging during the rainy (monsoon) season, leading to its low productivity and a smaller residual effect on the succeeding wheat crop. The modern technologies like permanent raised bed (PRB) system of planting, as widely used particularly for wheat in different countries, has seldom been studied in Indian conditions. Studies conducted at Modipuram indicated that growing pigeonpea on PRB had plant lesser mortality due to Phythopthoral infestation (Fig. 6) and greater N and P recycling through its residue comprising of root, stubble and leaf litter and lower economic optimum doses of fertilizer N and P for succeeding wheat in the pigeonpea-wheat system as compared to conventional flat beds (FB) owing to increased N and P supply, greater P use efficiency and a better crop growth environment under PRB planting (Singh *et al.* 2009). The lower NO_3-N at deeper profile under PRB, suggesting that PRB planting might reduce NO_3-N leaching to deeper soil layers. The greater irrigation water use efficiency and systems net returns and relatively smaller specific energy requirement under PRB make the practice further attractive, particularly in view of dwindling ground and surface water resources due to over-exploitation in the high productivity transects of IGP.

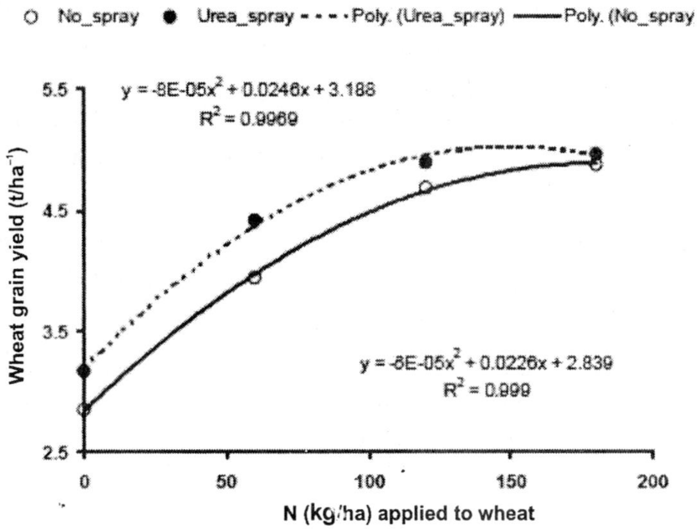

Fig. 5: Effect of foliar spray of urea in pigeonpea on the grain yield of wheat grown with varying fertilizer N rates

Source: Chauhan *et al.*, (2004)

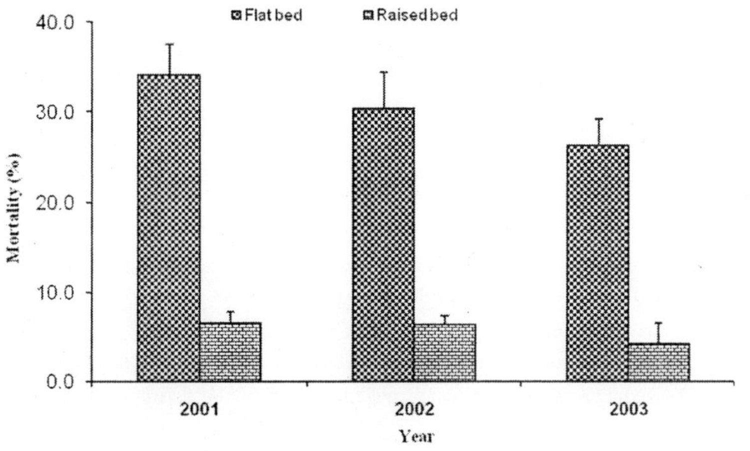

Fig. 6: Effect of planting technique on pigeon pea mortality in different years. (Singh *et. al.*, 2009)

Conclusion and future thrust

Legumes play a critical role in natural ecosystems and agriculture with their ability to fix N symbiotically, making them excellent colonizers of low N and fragile environments. Also legumes have the potential to be a major factor in an effective crop diversification programme particularly in breaking mono cropping situations or to recoup soil resources in degrading environments. Studies revealed an assured advantage of inclusion of legumes in continuous cereal-cereal based or other cropping systems. In the most pre-dominant rice-wheat system, both the crops are staple food grains, substitution of either of them may not be advisable in these areas. However, attempts should be made to harness the advantages of legumes in bi-exploiting one or the more avenues indicated in these studies. The idea of inclusion of legumes as summer forage or as break crop may be more saleable in most of the rice-wheat or cereal-cereal growing areas.

References

AICRP-IFS Reports. 2005-2010. Project Directorate for Farming Systems Research Modipuram, Meerut, India.

Brahmaprakash, G. P., Girisha, H. P., Navi, V. and Hegde, S. V. 2004. Biological nitrogen fixation in pulse crop. In: Pulses in new perspective (Eds. M. Ali, B. B. Singh, Shiv Kumar and Vishwa Dhar). India Society of Pulses Research and Development. Kanpur. pp.271-286.

Buresh, R. J. and De-Datta, S. K. 1991. Nitrogen dynamics and management in rice-legume cropping systems. Advances in Agronomy 45 : 1-59.

Chatterjee, B.N., Singh, K. I., Pal, A., and Maiti, S. 1979. Organic manure as substitution of for chemical fertilizers for high yielding rice varieties. Indian Journal of Agricultural Sciences 49 : 188-192.

Chauhan Y.S., Apphun A, Singh VK and Dwivedi BS. 2004. Foliar sprays of concentrated urea at maturity of pigeonpea to induce defoliation and increase its residual benefit to wheat. Field Crops Research 116, 127-139.

Dwivedi, B.S., Shukla, A. K., Singh, V. K., Yadav, R. L. 2001. Results of participatory diagnosis of constraints and opportunities (pdco) based trials from the state of Uttar Pradesh. In: A. Subba Rao, S. Srivastava (Eds), Development of farmers' resource-based integrated plant nutrient supply systems: experience of a FAO-ICAR-IFFCO collaborative project and AICRP on soil test crop response correlation Bhopal: Indian Institute of Soil Science. pp. 50-75.

Dwivedi, B.S., Singh, D., Chonkar, P.K., Sahoo, R.N. Sharma, S. K and Tiwari, K. N. 2006. Soil Fertility evaluation-a potential tool for balanced use of fertilizer, pp.1-66, IARI New Delhi-PPI-PPIC, India programme, Gurgaon, India.

Dwivedi, B.S. Shukla, Arvind K., Singh, VK and Yadav, RL. 2003. Improving nitrogen and phosphorus use efficiencies through inclusion of forage cowpea in the rice-wheat system in the Indo-Gangetic Plains of India. Field Crops Research 80 : 167-193.

Gill, M.S., Shukla, Arvind, K. and Pandey, P.S. 2008. Yield, nutrient responses and economic analysis of important cropping systems in India. Indian Journal of Fertilzers 4 (4) : 11-48.

Hegde D.M. and Dwivedi B.S., 1992. Nutrient management in rice-wheat cropping system in India. Fertilizer News 37 : 27-41

Hegde, D.M. (1992.} Cropping systems research highlights. In: Coordinators report, 20th Workshop, Tamil Nadu Agricultural University Coimbatore, 1-4 June 1992. Modipuram, Meerut 250 110, India: Project Directorate of Cropping Systems Research pp. 21-22.

Kundu, D. K., Rao, K.V., and Pillai, K.G. 1991. Agronomic efficiency of green manure and urea nitrogen in wet land rice as influenced by seasonal conditions. Indian Journal of Agricultural Sciences 61: 422-424.

Singh, H.P., Venkateswarlu, B.,Vittal, KPR., and Ramachanderan, K. 2000. Management of Rainfed Agro-ecosystem.In:Natural Resource Management for Agricultural Production in India (Ed.Yadav J.S.P. and Singh G.B.) Indian Society of Soil Science New Delhi pp. 69-774.

Singh, MP, Verma S.C. and Singh, RP. 1993. Effect of rice -based sequences on yield and economic sustainability under irrigated conditions of Eastern Uttar Pradesh. Annals of Agricultural Research 14 : 237-239

Singh, V.K. and Dwivedi B.S. 2006. Yield and N use-efficiency in wheat, and soil fertility status as influenced by substitution of rice with pigeon pea in a rice-wheat cropping system. Australian Journal of Experimental Agriculture. 46 : 1185-1194.

Singh, V.K. and Sharma, B.B. 2002. Economic evaluation of rice based cropping systems in foothills of Himalayas. Indian Journal of Agronomy 47 (1) : 12-19.

Singh, V.K., Dwivedi, B.S., Shukla, K, and Mishra, R.P. 2009. Permanent raised bed planting of the pigeonpea-wheat system on a Typic Ustochrept: Effects on soil fertility, yield, and water and nutrient use efficiencies. Field Crops Research 116 : 127-139.

Singh, V.K., Dwivedi, BS, Shukla, K, Chauhan Y.S. and Yadav, RL. 2005. Diversification of rice with pigeonpea in a rice-wheat cropping system on a Typic Ustochrept: effect on soil fertility, yield and nutrient use efficiency. Field Crops Research 92 : 85-105.

Singh, V.K., Sharma B.B. and Dwivedi BS. 2002. The impact of diversification of a rice-wheat cropping system on op productivity and soil fertility. Journal of Agricultural Science, Cambridge 139 : 405-412.

Singh, R. C., Faroda, A.S., and Lal, S. 1981. Pulses in aid of soil nitrogen economy. Indian Farming 30 : 3-4.

Singh, V. K. and Mishra, R. P. 2010. Integrated nutrient management in transplanted rice-wheat system. PDCSR Annual Report 2007-2008, pp. 44-50, PDCSR, Modipuram, Meerut, India

Timsina, J and Connor, DJ 2001. Productivity and management of rice-wheat cropping systems: issues and challenges. Field Crops Research 69 : 93-132.

Tiwari, K.N. 2002. Nutrient management issues and strategies. Fertilizer news, .47 (8) : 23-30, 33-40 & 43-49.

Wani, S.P., Rupela, O. P. and Lee, K.K. 1995. Sustainable agriculture in the semi-arid tropics through biological nitrogen fixation in grain legumes. Plant and Soil 174 : 29-49.

Yadav, R.L 1996. Intensive cropping system involving pulses. (In) special group meeting on constraints and opportunities in pulses production. Management options and Researchable Issues (12-13 March 1996) IARI, New Delhi, India.

Yadav, R.L, Singh, V.K., Dwivedi B.S. and Shukla A.K. 2003. Wheat productivity and N use efficiency as influenced by inclusion of cowpea as a grain legume in rice-wheat system. Journal of Agricultural Sciences (Cambridge) 141: 213-220

Yadav, R.L., and Rao, AVMS. 2001. Atlas of cropping systems in India. PDCSR Bulletin No. 2001-2, PDCSR, Modipuram, pp.96.

Yadav, R.L., Dwivedi, B.S., Gangwar, K S. and Prasad K. 1998. Overview and Prospects for Enhancing Residual Benefits of Legumes in Rice and Wheat Cropping Systems in India. In: Residual effects of legumes in rice and wheat cropping systems of the Indo-Gangetic Plain (editor: Rao J. V. D.K. K., C. Johansen C. & Rego T.J.) , Oxford & IBH Publishing Co. Pvt. Ltd. PP 207-226.

❑❑❑

Chapter - 7

Tillage Management and Crop Establishment Strategies

K. K. Singh
Principal Scientist (FMP), Project Directorate for Farming Systems Research, Modipuram, Uttar Pradesh.

Tillage management and seedbed preparation is crucial for crop establishment, growth and ultimately, yield. Typically, the aims of cultivation are to incorporate crop residue, bury weeds and loosen soil to allow appropriate soil–seed contact, easy flow of nutrients, air and water and unimpeded root penetration and crop growth (Hermawan and Bomke, 1997; Bengough *et al.*, 2006). The use of tillage to prepare seedbeds and the subsequent benefits are well documented. However, the amount of tillage and best practice in terms of optimum soil physical properties specifically for the establishment period has received less attention in comparison. The influence of a seedbed can vary greatly in terms of soil aggregation and subsequent porosity. This soil arrangement, therefore, has direct impacts on soil temperature, water content, oxygen availability and strength, all of which have the potential to affect the performance of the seedbed and its ability to provide an adequate environment for crop establishment. In a review, Braunack and Dexter (1989) stated as summary of previous works conducted on soil aggregation that aggregates will exhibit differing physical and chemical properties depending on the size of the aggregates, thus influencing the suitability of a seedbed for germination, emergence and root development, by influencing factors such as intra- and inter-aggregate aeration. Schjønning and Rasmussen (2000) found that cultivation methods can have large impacts on surface soil layers, altering both strength and pore dynamics (air and water filled), which can restrict plant growth. The relationship between soil physical properties and the plant root systems are vitally important in this process since they directly affect seedbed quality (Awadhwal and Thierstein, 1985; Aubertot *et al.*, 1999; Dexter, 2004).

Undoubtedly, crop establishment is the key to successful yields (Blake *et al.*, 2003). Different crops, however, require different soil physical properties for successful establishment. Excessive compaction and soil resistance can result in detrimental effects on root growth and development, which may lead to changes in root behaviour and characteristics to accommodate increased strength (Bingham and Bengough, 2003; Clark *et al.*, 2003). While modelling soil function for *Triticum aestivum* L., Jakobsen and Dexter (1987) found that crop development was not affected at soil strengths <3.0 MPa for germination, <2.3 MPa for root elongation, <1.7 MPa for coleoptile growth and <0.8 MPa for emergence. In a review by Bengough and Mullins (1990), it was stated that hindrance to growth starts at levels >1 MPa and becomes non-existent at 5 MPa. Plant growth is also affected by the soil physical environment created by cultivation. Dexter (1986), stated that root behaviour is dependent upon pore space, with large pore spaces resulting in deflection and contact with compacted sub-layers. *Kvasnikov (1928)* stated that maximum yields in wheat, barley and oats are achieved with seedbeds of 1–2 and 2–3 mm size aggregates. Ha°kansson *et al.* (2002) found rolling after drilling cereals resulted in improved final emergence by 4% and improved yields by 2%. The spatial distribution of roots and plant water uptake is strongly affected by soil physical conditions (Pardo *et al.*, 2000). Recent studies have shown a link between effective stress and soil strength in relation to reduced impedance of roots in unstructured soils at matric potentials ≥ 250 kPa (Whalley *et al.*, 2005).

Various resource conservation technologies are described here to attain the goal of increasing productivity and meeting food security needs while simultaneously using natural resources efficiently, including water, providing environmental benefits and improving the rural livelihoods of farmers. The major success in the last few years has been the development and deployment of resource conservation technologies (RCTs) with farmers in the rice-wheat system. RCTs range from simple surface seeding, where wheat seed is broadcast on the non-ploughed soil; no-till with a special opener for placing seed in the soil also without ploughing; reduced tillage and bed planting. This has benefited farmers through less cost, more yields and more income. RCTs are defined as any practice that will result in improvement of the efficiency of natural resources. These technologies are rapidly gaining popularity among farmers as they result in higher production at less cost with significant benefits to the environment and more efficient use of natural resources. This ultimately results in higher profits, cheaper food, and improved farmer livelihoods. Crop diversification is also easier as less land is needed to produce staple cereals, freeing up land for other crops.

Description of resource conservation technologies

A basket of technologies have been developed and made available to farmers. Some are based on reduced tillage including zero-tillage. Bed planting systems increase water productivity and when combined with reduced tillage in a permanent bed system provide even more savings. Laser leveling combined with these tillage systems provides additional benefits. Various system based technologies are described briefly below:

Zero tillage : This is RCT where the seed is placed into the soil by a seed drill without prior land preparation. This technology is more relevant in the higher yielding, more mechanised areas of north-western India, where most land preparation is now done with four-wheel tractors. However, in order to extend the technology in other parts, equipment for 2-wheel hand tractors and bullocks is being modified. The basis for this technology is the inverted-T openers. This coulter and seeding system places the seed into a narrow slot made by the inverted-T as it is drawn through the soil by the four-wheel tractor. The coulters can be rigid or spring-loaded depending on the design and cost of the machine. This type of seed drill also works well with little surface residue after manual harvesting. Where combine harvesting is becoming popular, loose straw and residue creates a problem for the inverted-T opener. Farmers presently burn residues to overcome this problem of loose stubble whether they use zero till or the traditional seed drill. This practice needs to be discouraged because of major environmental and air pollution issues. Future strategies will have to look at alternative machinery and techniques to overcome this problem. Leaving the straw as mulch on the soil surface has not been given much thought. However, results suggest that this may be very beneficial to early establishment and vigour of crops planted this way and for soil moisture conservation, water infiltration and erosion. Significantly fewer weeds were found under zero-tillage compared to conventional tillage. Fields with zero tillage and those with normal tillage were sprayed with weedicide, but significantly lower weed counts were found in fields with zero tillage. This difference can be explained by the nature of the weeds found in the system. Most of the weeds affecting the crop germinate during the crop season, and since the soil is disturbed less under zero-tillage, fewer weeds were exposed leading to reduced weed germination. Also, before the weeds are able to grow and compete, the main crop is able to cover up the surface and significantly reduce weed biomass. Weed problems typically were more severe under conventional tillage than under zero tillage, at least in the near term. Earlier planting is the main reason for the additional yields obtained under zero-tillage. The results of many trials suggest that the longer the farmer delays planting, the lower the yield. This finding has been confirmed in trials throughout the Indo-Gangetic plains in the past few years. In Haryana, surveys and crop cuts have shown that zero till produces 4-500 kg/ha more grain than traditional systems. This is attributed to earlier, timely planting, less weeds, better plant stands and improved fertilizer efficiency because of placement with the seed drill. Some experiments are now in their 14th year of continuous zero till and find no deleterious effects that would make them revert to the traditional system.

Reduced tillage : The strip and rotary till drills have been developed that prepare the soil and plant the seed in one operation. This system consists of a shallow rotavator followed by a seeding system. Soil moisture was found to be critical in reduced tillage system. The rotavator fluffs up the soil, which then dries out faster than with normal land preparation. The seeding coulter does not place the seed very deep, so soil moisture must be high during seeding to ensure germination before the soil dries appreciably. A tractor can also be used with a rotavator to quickly prepare the soil and incorporate the seed after a second pass. This speeds up the planting and results in better stands with less cost than traditional methods. However, the strip and rotary till drills do a better job because the seeds are placed at a uniform depth in the single pass.

Bed planting : In bed planting systems, crops are planted on raised beds. This practice has increased in the last decade. Farmers had their reasons for adopting the new system *viz*., management of irrigation water was improved, bed planting facilitated irrigation before seeding and thus provided an opportunity for weed control prior to planting, plant stands were better, weeds could be controlled mechanically between the beds early in the crop cycle, seed rates were lower, after the crop was harvested and straw burned, the beds were reshaped for planting the succeeding crops, burning can be eliminated, herbicide dependence was reduced and hand weeding and roguing was easier as well as less lodging occured. Two bed widths and two or three rows of crop planted per bed were compared with conventional flat bed planting. Two rows on 70 cm beds were best. Two of the major constraints in obtaining higher yields are weeds and lodging. Both can be reduced in bed planting. The major weed species affecting crop are normally controlled using the herbicides. Preliminary observations indicated that weeds were less prolific on dry tops of raised beds than on the wetter soil found in conventional flat bed planting. Cultivation between the beds can also reduce weeds. Thus, bed planting provides farmers with additional options for controlling weeds. Lodging is also less of a problem on raised beds. Additional light enters the canopy and strengthens the straw, and the soil around the base of the plant is drier. Reduced lodging can have a significant effect on yield. An additional advantage of bed planting becomes apparent when beds are "permanent" – that is, when they are maintained over the medium term and not broken down and re-formed for every crop. In this system, crop is harvested and straw is left or burnt. Passing a shovel down the furrows reshapes the beds. The next crop can then be planted into the stubble in the same bed. The use of beds also provides a way for improving fertilizer use efficiency. This is achieved by placing a band of fertilizer in the bed at planting or topdressing. Using slow release formulations and experimenting with urea super granules can make further improvements. Both can be applied in the bed at the time of planting with the seed cum fertilizer drill.

Broad bed planting: Planting of crops, both cereals and pulses, under intercropping systems under zero tillage have wide possibilities. The preliminary studies conducted under bio-intensive complimentary systems using land configuration have shown positive results which are under perfection.

Effect of resource conservation technologies on productivity

Intensification of rice based cropping systems with legumes/pulses under different tillage systems: To study the effect of inclusion of legumes and pulses on soil and cropping system productivity and profitability of rice based cropping systems under different tillage options, ten treatment combinations: T_1=RWG(Z): Rice-Wheat (ZT)-Green gram (ZT), T_2=RPG(Z): Rice-veg. Pea (ZT)-Green gram (ZT), T_3=RWC(Z): Rice-Wheat (ZT)-Cowpea (ZT), T_4=RWG(C) : Rice-Wheat (CT)-Green gram (CT),

T_5=RPG(C): Rice- veg. Pea (CT)- Green gram (CT), T_6: RWC(C):Rice-Wheat (CT)-Cowpea (CT), T_7=RWG(B): Rice- Wheat (BP)-Green gram (BP), T_8=RPG(B): Rice-veg. Pea (BP)- Green gram (BP), T_9=RWC(B): Rice-Wheat (BP)-Cowpea (BP), and T_{10}=RW(C): Rice-Wheat (CT), were evaluated.

Conventional tillage adopted in previous *rabi* and summer season crops produced 5.25 t/ha rice that was, respectively, 6 and 19% more than bed planting and zero tillage. Rice- vegetable Pea-Green gram (CT) cropping system produced maximum rice (5.38 t/ha) followed by 5.29 t/ha in Rice-Wheat-Cowpea (CT) whereas Rice-Wheat (CT) produced only 5.08 t/ha rice. Rice-vegetable Pea-Green gram system produced 4% more rice than Rice-Wheat-Green gram and Rice-Wheat-Cowpea systems.

The wheat yield obtained under bed planting, ZT (5.82/t/ha) was, 5.5 and 1.7 % higher than conventional tillage (CT) and bed planting (BP) respectively. Rice-Wheat-Green gram (ZT) cropping system produced maximum wheat (5.88 t ha^{-1}) followed by 5.77 t/ha in Rice-Wheat-Cowpea (ZT) whereas Rice-Wheat (CT) produced only 5.38 t/ha wheat. Rice- Wheat-Green gram system produced 7 and 1.5% more wheat than Rice-Wheat and Rice-Wheat-Cowpea systems.

The maximum vegetable pea yield (5.12 t/ha) was obtained under Rice-vegetable Pea-Green gram (BP) cropping system that was, 108 and 75 % higher than Rice-vegetable Pea-Green gram (ZT) and Rice-vegetable Pea-Green gram (CT) system respectively. The green gram yield obtained under bed planting (0.73 t/ha) was, 41 and 58 % higher than conventional till sowing and zero till drilling respectively. The green gram grown after vegetable pea yielded (0.64 t/ha) that was 48 % higher than that grown after wheat. The cowpea yield obtained under bed planting (2.35 t/ha) was, 78 and 27 % higher than zero till drilling and conventional till sowing respectively.

Rice-Vegetable Pea (BP)- Green gram (BP) cropping system provided maximum rice equivalent system yield of 16.4 t/ha followed by 16.3 t/ha in Rice-Wheat (BP)-Cowpea (BP) and 15.9 t/ha in Rice-Wheat (CT)- Cowpea (CT) compared to 11.9 t/ha in rice – wheat (CT) system (Fig. 1). Rice-Vegetable Pea (BP)- Green gram (BP) cropping system provided maximum net returns (Rs 95354 ha) & B:C ratio (3.15) followed by Rs 81948 ha^{-1} & 2.46 in Rice-Wheat (BP)- Cow pea (BP) and Rs 76909 ha^{-1} & 2.32 in Rice-Wheat (CT)- Cow pea (CT) compared to Rs 28900 ha & 1.61 in rice – wheat (CT) system (Fig. 2). Rice-Vegetable Pea (BP)-Green gram (BP) cropping system also provided maximum soil organic carbon (OC), mean weight diameter (MWD) of aggregates and available N, P, K after three years, compared to other treatments as well as respective initial values (Fig. 3 and 4).

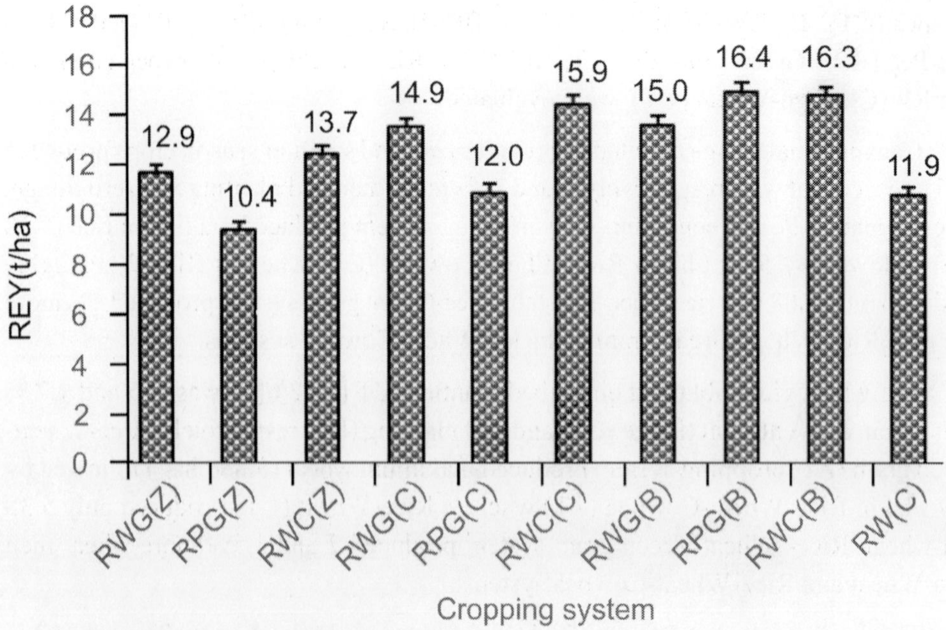

Fig. 1. Effect of tillage and cropping systems on rice equivalent system yield (REY)

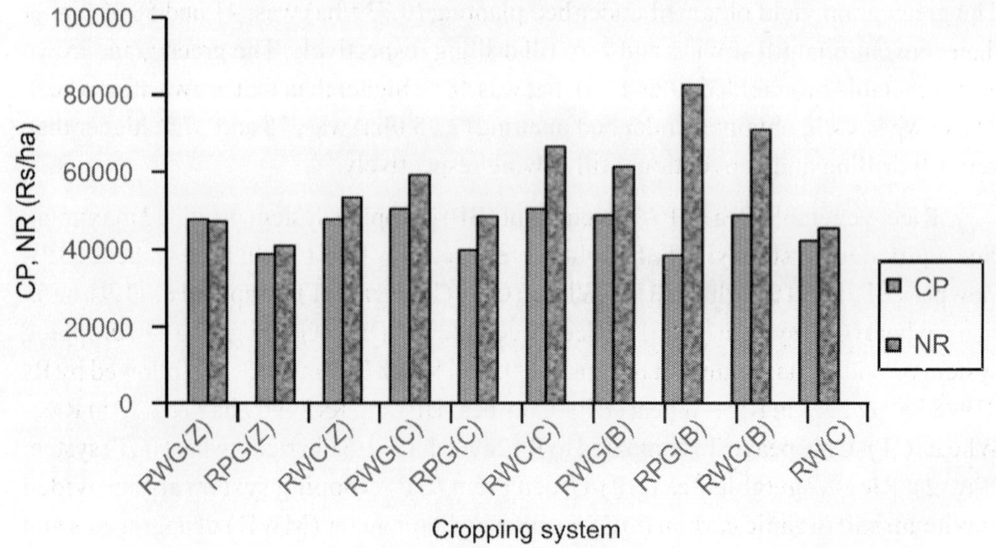

Fig. 2. Effect of tillage and cropping systems on cost of production (CP) and net returns (NR)

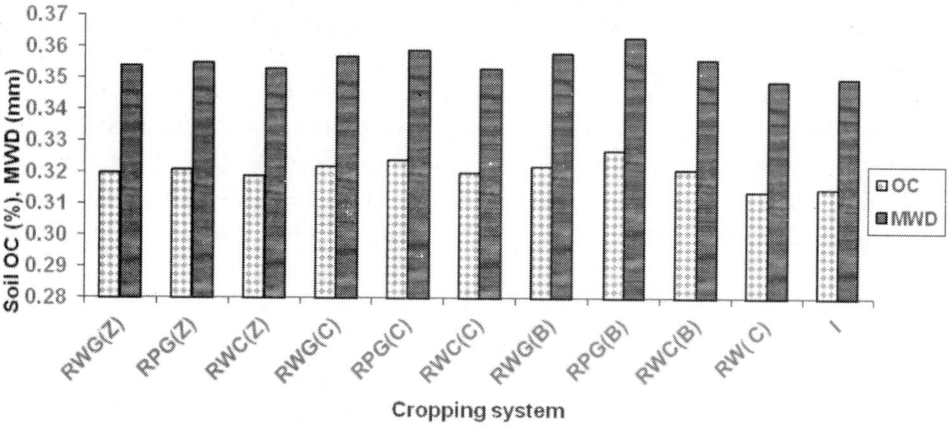

Fig. 3. Effect of tillage and cropping systems on soil organic carbon (OC) and mean weight diameter (MWD) of aggregates (*I = Initial values*)

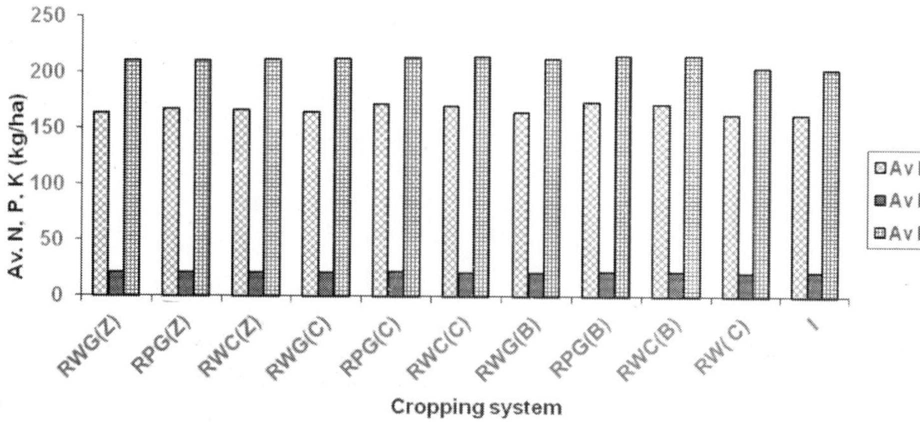

Fig. 4. Effect of tillage and cropping systems on available N, P and K of soil (*I = Initial values*)

Resource conservation through cropping system management and land configuration: During *kharif*, [Pigeon pea + Black gram] (BB) + sesbania (F) system produced maximum rice equivalent yield, (REY) (6.88 t/ha) followed by Pigeon pea (RB) + sesbania (F) system (6.38 t/ha) and Pigeon pea (FB) system (5.12 t/ha) while Maize (FB) produced lowest REY (2.33 t/ha). During rabi, Vegetable pea (RB) + Wheat (F) system produced maximum REY (10.12 t/ha) followed by vegetable Pea (BB) + Wheat (F) system (9.51 t/ha) and vegetable Pea (FB) (7.86 t/ha) while Wheat + Mustard produced lowest REY (4.72 t/ha). During summer, Green gram after Veg. pea (FB) produced REY of 3.40 t/ha.

The bio-intensive complementary cropping system Maize -Veg. Pea -Green gram produced maximum REY of 13.6 t/ha, which was followed by Maize (B) + Sesbania (F) – vegetable Pea (B) + Wheat (F) – Green gram (13.4 t/ha) and [Maize + vegetable Cowpea] (BB) + Sesbania (F) - Veg. Pea (BB) + Wheat (F) - Green gram system (13.2

t/ha) (Fig. 5). The system Maize -Veg. Pea - Green gram also provided maximum net return of Rs 92424 ha, which was followed by Maize (B) + Sesbania (F) – Vegetable Pea (B) + Wheat (F) – Green gram system (Rs 87806 ha) and [Maize + vegetable Cowpea] (BB) + Sesbania (F) - Veg. Pea (BB) + Wheat (F) - Green gram (Rs 83688 ha) (Fig. 6). The system [Pigeon pea + Black gram] (BB) + Sesbania (F) – Wheat + Mustard used least water (46 ha-cm) (Fig. 7). Averaging over cropping systems, about 29.6 and 43.5 % irrigation water was saved using raised bed and broad bed than flat bed configuration.

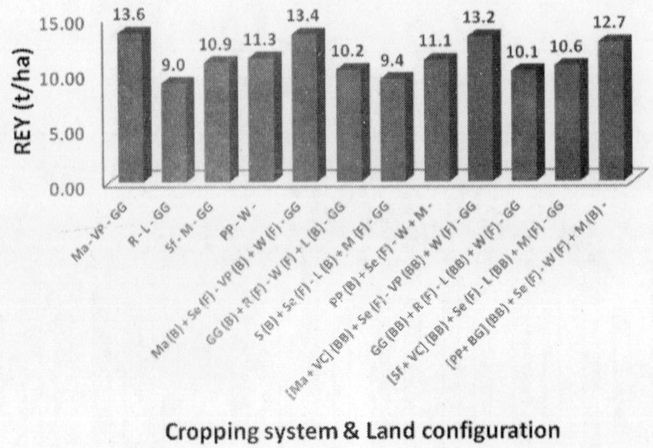

Fig. 5. Effect of land configuration and cropping systems on Rice Equivalent Yield *(Ma – Maize, VP – veg. Pea, GG – Green gram, R – Rice, L – Lentil, Sf – Sorghum fodder, M – Mustard, PP – Pigeon pea, W – Wheat, Se – Sesbania, VC – veg. Cowpea, BG – Black gram, B – Raised bed, BB – Broad bed, F – Furrow)*

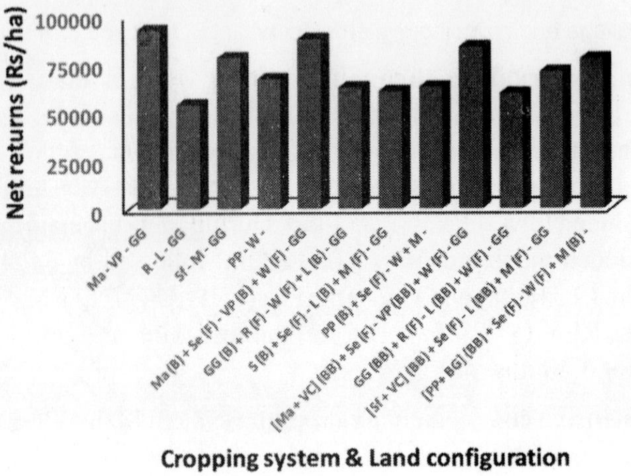

Fig. 6. Effect of land configuration and cropping systems on net returns

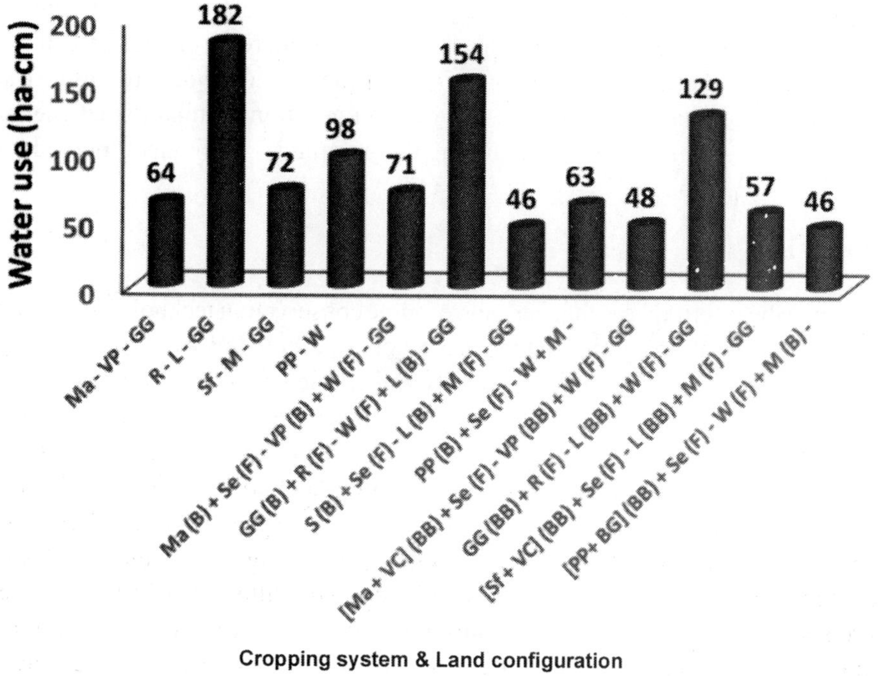

Fig. 7. Effect of land configuration and cropping systems on water use

Effect of resource conservation technologies on conservation of resources: Farmers are adopting these technologies quickly. The adoption could be even faster if it were possible to have sufficient machinery available from small-scale manufacturers. The RCTs of zero, strip and rotary till drilling, and bed planting saved 64 to 85 % resources (time, labour, cost, fuel and energy). Farmer feedback on water savings with these technologies essentially says that they save water. For zero-tillage, farmers reported about 25-30% savings because zero tillage was possible with the residual moisture of previous crop and untilled soil had less infiltration than tilled soil so farmers were able to apply irrigation much faster. Because zero tillage took immediate advantage of residual moisture from the previous crop, as well as cutting down on subsequent irrigation, water use was reduced by about 10 cm-hectares, or approximately 1 million liters per hectare. One additional benefit was less water logging and yellowing of the plants after the first irrigation that is a common occurrence on normal ploughed land. In zero till, less water is applied in the first irrigation and this yellowing is not seen. Averaging over cropping systems, about 29.6 and 43.5 % irrigation water was saved using raised bed and broad bed than flat bed configuration.

Environmental Impact of Conservation Agriculture Technologies: The widespread adoption of one or several reduced tillage methods would also bring significant environmental benefits. Use of zero till drill on one hectare of land would save 60-70 liters of diesel and approximately 1 million liters/ha of irrigation water.

Using a conversion factor of 2.6 kilograms of carbon dioxide per liter of diesel burned, this represents about a quarter ton less emissions per hectare of carbon dioxide, the principal contributor to global warming. These benefits increase dramatically if extended to larger areas. Adoption of zero-till on, say, 5 million hectares would represent a savings of 5 billion cubic meters of water each year. In addition, annual diesel fuel savings would come to 0.5 billion liters (including the pumping) - equivalent to a reduction of nearly 1.3 million tons in CO_2 emissions each year.

Conclusion

It may be inferred here that various resource conservation technologies are available for use in different cropping systems. The benefits included improved production at lower cost, enhanced the water productivity and efficiency of natural resources, benefits to the environment, and improved livelihoods of farmers. The success of these technologies is, however, dependent on rapid adoption by farmers. In fact, "conservation agriculture" which includes resource conservation technologies is a key for ensuring sustainable food production in the next decade. Overcoming mindsets that hold traditional beliefs about excessive tillage and providing the enabling factors that allow exposure of the technology to all those who are involved in agriculture will be the key factor for future success. This technology revolution is seen as one way to sustainably increase food production to meet the future demands on the one hand while conserving natural resources and improving farmer livelihoods on the other hand.

References

Aubertot, J.N., Du¨rr, C., Kie^u, K., Richard, G., 1999. Characterization of sugar beet seedbed structure. Soil Sci. Soc. Am. J. 63 : 1377–1384.

Awadhwal, N.K., Thierstein, G.E., 1985. Soil crust and its impact on crop establishment: a review. Soil Tillage Res. 5 : 289–302.

Bengough, A.G., Mullins, C.E., 1990. Mechanical impedance to root growth: a review of experimental techniques and root growth responses. J. Soil Sci. 41 : 341–358.

Bengough, A.G., Bransby, M.F., Hans, J., McKenna, S.J., Roberts, T.J., Valentine, T.A., 2006. Root responses to soil physical conditions; growth dynamics from field to cell. J. Exp. Bot. 57 (2) : 437–447.

Bingham, I.J., Bengough, A.G., 2003. Morphological plasticity of wheat and barley roots in response to spatial variation in soil strength. Plant and Soil 250 : 273–282.

Blake, J.J., Spink, J.H., Dyer, C., 2003. Factors affecting cereal establishment and its prediction. HGCA Research Review 51. HGCA.

Braunack, M.V., Dexter, A.R., 1989. Soil aggregation in the seedbed: a review. II. Effect of aggregate sizes on plant growth. Soil Tillage Res. 14 : 281–298.

Clark, L.J., Whalley, W.R., Barraclough, P.B., 2003. How do roots penetrate strong soil. Plant and Soil 255 : 93–104.

Dexter, A.R., 1986. Model experiments on the behaviour of roots at the interface between a tilled seedbed and a compacter subsoil. 1. Effects of seedbed aggregate size and subsoil strength on wheat roots. Plant and Soil 95 : 123–133.

Dexter, A.R., 2004. Soil Physical Quality. Part I. Theory, effects of soil texture, density and organic matter, and the effects on root growth. Geoderma 120 : 201–214.

Hermawan, B., Bomke, A.A., 1997. Effects of winter cover crops and successive spring tillage on soil aggregation. Soil Tillage Res. 44 : 109–120.

Ha°kansson, I., Myrbeck, A °., Etana, A., 2002. A review of research on seedbed preparation for small grains in Sweden. Soil Tillage Res. 64 : 23–40.

Jakobsen, B.F., Dexter, A.R., 1987. Effect of soil structure on wheat root growth, water uptake and grain yield. A computer simulated model. Soil Tillage Res. 10 : 331–345.

Kvasnikov, V.V., 1928. The structure of the soil and yields. Russ. J. Agric. Sci. 5 : 459–482.

Pardo, A., Amato, M., Chiaranda, F.Q., 2000. Relationship between soil structure, root distribution and water uptake of chickpea (Cicer arietinum L.). Plant growth and water distribution. Eur. J. Agron. 13 : 39–45.

Schjønning, P., Rasmussen, K.J., 2000. Soil strength and soil pore characteristics for direct drilled and ploughed soils. Soil Tillage Res. 57 : 69–82.

Whalley, W.R., Leeds-Harrison, P.B., Clark, L.J., Gowing, D.J.G., 2005. Use of effective stress to predict the penetrometer resistance of unsaturated agricultural soils. Soil Tillage Res. 84 : 18–27.

❑❑❑

Chapter - 8

Integrated Nutrient Management

N. Ravisankar, B. Gangwar and Anil Kumar Singh*
Project Directorate for Farming Systems Research Modipuram, Meerut - 250 110
*Indian Council of Agricultural Research, New Delhi - 110 012.

Integrated nutrient management is considered to be the key for achieving higher yield of crops and cropping systems. In general, the productivity of pulse crops has remained low mainly for two reasons: cultivation on agriculturally marginal soils and little if any crop inputs. Among production inputs, fertilizer plays a key role in pulses too like other crops for enhancing productivity levels. Most of the pulses fix atmospheric N, which is the predominant mechanism to meet their N requirement. It can enrich soil fertility by fixing nitrogen. It has been estimated that chickpea can fix (convert atmospheric nitrogen to organic nitrogen which would be available for subsequent crops) up to 140 kg nitrogen per hectare in a growing season, although measured values are usually in the range of about 20-60 kg nitrogen per hectare (Reddy 2004). It has been well established that long duration pigeonpea in northern India can fix nitrogen in the order of 200 kg/ha when grown over a 40 week period. Pigeonpea can also have substantial residual effects on subsequent crops. For example, medium-duration pigeonpea grown at the International Crop Research Institute for Semi-Arid Tropics (ICRISAT) (Asia Centre) could benefit a subsequent maize crop to an extent equivalent to 40 kg nitrogen per hectare. However, this capability is jeopardized through insufficient supply of other plant nutrients. General recommendations for phosphorus (P) fertilization are made in most states. However, K application is generally neglected, resulting in imbalanced nutrient supply and lower crop yields. Under intensive cropping systems, large amounts of K are removed, leading to serious depletion of soil K reserves. Pulses such as chickpea and pigeonpea remove about 60 and 52 kg K_2O/t of grain, respectively. Information compiled from various sources on integrated nutrient management in pulse/pulse based cropping systems in different agro-ecological regions of the country suggest that there is a wide scope for practicing integrated plant nutrient (IPN) practices in pulses and pulse based cropping systems.

Practicing of INM not only economize inorganic fertilizer use but also helps in balanced fertilization and improving soil health and productivity through complimentary effects of fixed N by pulses and applied phosphorus and potassium.

The major pulse growing regions in India are Uttar Pradesh, Rajasthan, Madhya Pradesh, Maharashtra, Bihar, Andhra Pradesh, Karnataka, Tamil Nadu, Odisha, Punjab, Jharkhand, Haryana and Chattisgarh. The soils in these regions invariably experience deficiency of single or multiple nutrients depending upon site characteristics crop/ cropping intensity and management practices adopted. As farmers have realized that pulses can restore the soil fertility, these crops are mostly grown in marginal and sub marginal lands. Despite release of high yielding and disease tolerant varieties, the yield level of pulses remained static which is mainly attributed to misconception that pulses can grow and produce better yield in poor fertility soils or without any fertilization. Protein rich pulses require more energy inputs per unit of production as compared to cereal crops. It is therefore, not surprising to note that nutrient deficiencies are emerging as the most important constraints in pulse growing regions of India and they can only be addressed through integrated crop management strategies including nutrient management.

Nutrient deficiencies : Release of high yielding varieties apart from increased cropping intensity led to extraction of more and more nutrients from soils. As the food production increased, the number of nutrients becoming deficient in soils has also increased (Fig 1). Deficiencies of N and P have become almost common along the length and breadth of the country irrespective of soils and agro-ecological regions. Many pulse growing soils are found to be N deficient (Table 1). Also, in general, it was felt that Indian soils are having adequate K supply and application of optimum K supply was never considered seriously more so in rainfed regions.

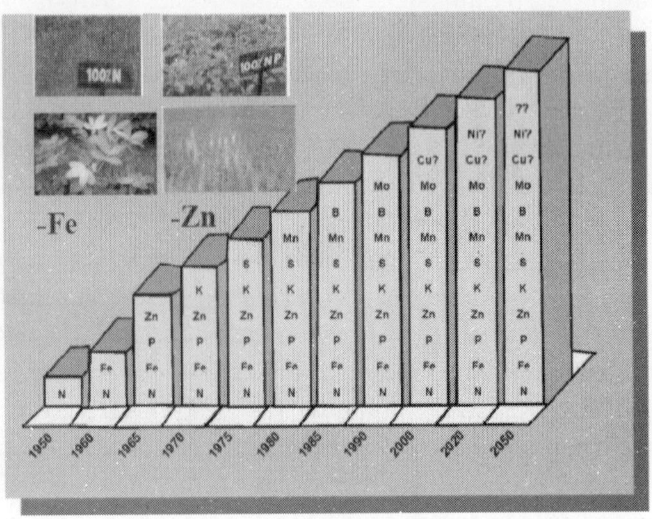

Fig 1. Emerging deficiencies of multi micro nutrients in Indian crops in Indian soils due to depletion in soil fertility during green revolution (Singh, 2008)

Table 1. Fertility status of N, P and K (% of samples) in major pulse growing states of India

State	Nitrogen			Phosphorus			Potassium		
	Low	Medium	High	Low	Medium	High	Low	Medium	High
Madhya Pardesh	40	41	19	39	38	23	10	32	58
Uttar Pardesh	80	15	5	71	26	3	12	55	33
Maharashtra	67	26	7	6	12	2	8	18	74
Andhra Pardesh	62	21	17	57	29	14	9	30	61
Karnataka	29	37	34	31	48	21	7	32	61
Odisha	0	23	17	59	28	13	33	41	26
Tamil Nadu	75	16	9	24	41	35	12	36	52
Panjab	41	51	8	29	49	22	6	48	46
India	63	26	11	42	38	20	13	37	50

Source : Ganeshamurthy *et al.*, (2003)

The Sulphur (S) requirement of pulses is relatively higher and its deficiency is also cropping up in major pulse growing regions. Under intensive cropping involving S responsive crops like pulses, S fertilization should form a part of the package of practices. Among the micro nutrients, Zn, B and Fe are deficient in Indian soils (Table 2). Integrated and balanced fertilization are essentially required which invariably includes micro nutrient application for realizing the yield potential of pulse varieties that are being released to enhance the productivity of pulses.

Table 2. Micro nutrient deficiencies in pulse growing.

State	No. of samples analyzed	% deficiency of available micronutrients			
		Zn	Fe	Cu	Mn
AP	5219-6563	51	2	1	2
Bihar	17802-19078	54	6	3	2
Gujarat	29532	24	8	4	4
Haryana	21598	61	20	2	4
MP	11204-12000	63	3	1	3
Punjab	15897	47	14	1	2
TN	19559-20580	53	15	3	8
UP	24425-25122	45	6	1	3
Karnataka	24411-25542	78	39	5	19

Source : Ganeshamurthy *et al.*, (2003)

Nutrient Balance : Sizeable quantities of nutrients are removed by pulses as they are energy rich crops. Generally they remove 30 to 50 kg of N, 2-7 kg of P, 12-30 kg of K, 3-10 kg Ca, 1-5 kg Mg, 1-3 kg S, 200-500 g Mn, 5g B, 1 g Cu and 0.5 g Mo to accumulate 1 t of biomass (Ahlawat and Ali, 1993). Pigeonpea removes 85 kg N, 8 kg P, 16 kg K and 9 kg S (Tandon, 1995) to produce 1.2 t seed/ha while chickpea requires 91, 6, 49 and 13 kg N, P, K and S ha^{-1}, respectively, to yield 1.5 t of seed/ha. The removal of nutrients by important pulse crops are presented in Table 3. Considerable variation in nutrient removal exists due to variation in length of growing season and yield potential of crops.

Table 3. Nutrient uptake by pulses

Crop	Yield (kg/ha)	Total uptake (kg/ha)			
		N	P	K	S
Pigeonpea	1200	85	8	16	9
Blackgram	890	70	6	50	5
Greengram	1000	106	21	90	6
Soybean	2500	125	43	101	22
Chickpea	1500	91	6	49	13
Lentil	2000	114	13	36	6

The nutrients added to the soil by way of fertilization are very meager resulting in a negative balance. It was found that a total of 197 thousand t of nitrogen, 156 thousand t of P_2O_5 and 6.6 thousand t of K_2O are being used in pulse crops. However, the removal of nutrients exceeded its application resulting in a negative balance of 710 and a total removal of 977.4 thousand tonnes (Table 4). There is a severe stress on production system and immediate attention is needed to increase production of pulses through adequate and balanced use of nutrients.

Table 4. Removal, addition and balance of nutrients ('000 mt) in pulses (All India).

Nutrient	Removal	Addition	Balance
N	592.4	197.4	-395.00
P	88.8	67.2	-21.60
K	296.2	2.73	-293.47
Total	977.4	267.33	-710.07

Source : Rao and Hansa Moodi, (2009)

N fixation : Biological N fixation by legumes has been studied with more focus on evaluation of efficient strains of Rhizobium inoculants. Results of experiments in the Indo-Gangetic Plains (IGP) in the winter pulses indicates that 19 to 63 % yield improvement in chickpea and 11 to 32 % in lentil was possible at farmers fields. The potential of N fixation was maximum in faba bean (300 kg N/ha) followed by pea (244 kg N/ha), lentil (192 kg/ha) and chickpea (144 kg/ha). Estimates of N_2 fixation by pulse crops are given in Table 5.

Table 5. Estimates of N_2 fixation by pulse crops

Crop	% Ndfa
Pigeonpea	88-96 (ID method)
	0-36 (NA method)
Blackgram	95-98 (NA method)
Greengram	8-90 (NA method)
Soybean	78-87 (NA method)
Cowpea	54-70 (ID method)

%Ndfa: Per cent of foliar N derived from atmosphere, NA: Natural Abundance, ID: Isotope dilution

Fertilizer consumption: The study conducted in Madhya Pradesh revealed that the fertilizer use in pulses was very low with chickpea receiving the highest priority and pigeonpea the least. Though the recommended dose was 15-25 kg of nitrogen and 20-50 kg of super phosphate/ha in pulses, the actual application of fertilizer on an average, was found to be 2.8 kg in pigeonpea, 6.4 kg in lentil, 8.13 kg in blackgram and 12.04 kg in chickpea (Rao and Hansa Modi, 2009). Phosphorus is absolutely necessary and application of 40 kg P_2O_5/ha has been recommended for most of the pulses but its application in the field was generally suboptimal.

Fertilizer requirement: Fluctuation in soil temperature and moisture during monsoon leads to variability in the proportion of N derived from fixation and it varies widely during kharif and summer pulses. In order to produce targeted pulses production of 32 mt in 2030 (Vision 2030, 2011), 2.24 mt of N, 0.25 mt of P, 1.18 mt of K, 0.36 mt of Ca, 0.17 mt of Mg, 0.11 mt of S, 0.02 mt of Mn, 280 t of B, 56 t of Cu and 28 t of Mo are required which has to come from different sources like farm yard manure, enriched compost, pressmud, and micro nutrient mixtures.

Integrated nutrient management

Pulses grown as companion or complimentary sequential crop under intensive cropping systems without application of sufficient nutrient addition or imbalanced fertilization, result in deterioration of native soil fertility and pose a serious threat to long-term sustainability. In India, short supply of indigenous fertilizers and exorbitant costs of nutrient based fertilizers have further aggravated the problem as pulses are always considered to be low nutrient requiring crops by farmers and extension functionaries. Integration of chemical, organic and biological sources and their efficient management have shown promise in not only sustaining the productivity and soil health, but also meeting a part of chemical fertilizer requirement of many crops and systems. Another dimension of the INM is balanced fertilization of pulse crops leading to increased productivity and efficiency of applied nutrients.

Balanced nutrient application : It is often felt that pulses will meet their nutrient requirement from the residual nutrients applied to preceding crop in the system. Many experiments have conclusively established that application of P and K with N enhances the yield of all the crops especially pulses with additional benefit of enhanced N fixation for the subsequent crops. On-Farm experiments conducted at farmers field have proved that application of NPK increases the yield by 47.7 % over N alone, 20 % over NP and 22.7 % over NK application in blackgram indicating the usefulness of balanced application of nutrients. The results of the experiments conducted at different locations of the country from 1999 to 2011 have been pooled and are presented in Table 6. The yield increase due to application of P with N was found to be 23 % in blackgram, 20 % in greengram, 27 % in chickpea and 20 % in soybean. Recommended dose of NPK to the pulse crops, namely, blackgram, greengram, chickpea and soybean, resulted in 582, 954, 1603 and 2170 kg of yield/ha which is 88, 74, 85 and 75 % higher than control, respectively. Similarly, the Partial Factor Productivity (PFP) which is a measure of the ability of unit quantity of applied and native nutrients to produce the yield of crops were also assessed under on-farm conditions for various pulses. The results reveal that PFP of N can be increased to 28.7 kg/kg of native + applied N by applying the recommended quantity of N with recommended quantity of P to blackgram (Table 7). Balanced application of NPK increased the efficiency of applied and native N by 43, 38, 43 and 42 % in blackgram, greengram, chickpea and soybean, respectively, compared to application of N alone. PFP of P increased by only 12.5 % when P was applied with N and K instead of N alone. In case of PFP of K, in all the pulses evaluated, it was found that efficiency of K was higher when it is applied with N and P instead of N alone. The effect was more pronounced in soybean (22.2 %) followed by chickpea (16.8 %), blackgram (16.6%) and greengram (16.4 %). Agronomic efficiency being a measure of response with respect to only applied nutrients, was also higher in pulses with combined application of NPK (Fig 2.). The effect of balanced application in terms of agronomic efficiency was more pronounced in soybean and chickpea compared to blackgram and greengram.

Balanced application of nutrients is necessary to enhance the yield of pulses as evidenced in the on-farm experiments. However, it will not sustain the productivity as occurrence of secondary and micro nutrient deficiency in pulse crop based systems is being noticed in several regions of the country. Hence, a combination of organic, inorganic and biological fertilizers is essential to enhance the soil fertility and sustain the growth rate of pulse production in the country. The interactive advantages of combining use of all possible sources of nutrients and their scientific management in INM have proved superior to the use of each component in isolation for optimum growth, yield and quality of different crops and cropping systems.

Organic manures: Since time immemorial, pulses have been cultivated on marginal and sub-marginal lands, which are characterized by poor soil fertility, moisture

Table 6. Yield of pulses under various macro nutrient combinations

State	District	No. of Trials	Fertilizer dose (kg/ha)			Yield (kg/ha)				
			N	P	K	Con-trol	N	NP	NK	NPK
				Blackgram						
Maha-rastra	Beed	21	25	50	25	310	394	485	474	582
				Greengram						
Orissa	Khurda	16	20	40	20	470	591	688	755	859
	Nayagarh	48	20	40	20	347	511	667	596	764
	Kendrapara	24	20	40	20	529	685	830	880	986
West Bengal	South 24 Parganas	24	20	40	40	501	577	673	669	779
Gujarat	Broodhra	20	20	40	40	894	1093	1265	1206	1381
	Mean	-	20	40	28	548	691	825	821	954
				Chickpea						
Maha-rastra	Bhandra	9	20	40	20	662	839	1064	967	1158
	Sindhewahi	24	20	40	20	996	1148	1440	1425	1636
	Gondia	12	20	40	20	690	1003	1346	1154	1571
	Aurangabad		25	50	25	1160	1481	1597	1685	1834
Karn-ataka	Bheemara - yangudi	42	25	50	10	800	1081	1472	1351	1569
	Gulberga	56	25	50	0	864	1284	1604	1506	1725
	Gadag	5	25	50	0	860	1020	1461	1200	1731
	Mean	-	23	46	14	862	1122	1426	1327	1603
				Soybean						
Madhya Pradesh	Chindwara	149	20	60	20	970	1168	1493	1324	1692
Uttar Pradesh	Bundhelkhand	67	20	80	20	1080	1240	1487	1374	1706
Maha-rastra	Wardha	26	30	75	30	1090	1435	1766	1741	2084
	Nagpur	13	30	75	30	1292	1586	1863	1922	2283
	Beed	93	30	60	30	1475	1693	1898	1915	2092
	Ahmednagar	60	50	75	25	1295	1703	1954	2249	2484
	Sindhwahi	24	30	75	30	918	1227	1471	1469	1760
Karn-ataka	Bheemara - yangudi	12	40	80	25	788	1027	1223	1183	1690
	Dharwad	22	40	80	25	1402	1632	2115	1764	2405
	Belgaum	15	40	80	25	2124	2504	3008	2935	3505
	Mean	-	33	74	26	1243	1522	1828	1788	2170

stress and consequently their yield potentials have not been realized. Further, more than 87 per cent area under pulses is rainfed (Amarender Reddy, 2009). Therefore, there is a great scope of increasing the production in rainfed areas through integrated nutrient management using organic manures as that will play a critical role as for as nutrient-water interactions in the soil are concerned. Organic manures are available in

Table 7: Partial factor productivity (Kg grain/kg of NPK) in pulses

State	District	PFPn				PFPp		PFPk	
		N alone	with P	with K	with PK	with N	with NK	with N	with NP
				Blackgram					
Maharastra	Beed	20.1	25.7	24.6	28.7	12.8	14.4	24.6	28.7
				Greengram					
Orissa	Khurda	29.6	34.4	37.8	43.0	17.2	21.5	37.8	43.0
	Nayagarh	25.5	33.3	29.8	38.2	16.7	19.1	29.8	38.2
	Kendrapara	34.3	41.5	44.0	49.3	20.8	24.7	44.0	49.3
West Bengal	South 24 Parganas	28.9	33.7	33.5	39.0	16.8	19.5	16.7	19.5
Gujarat	Broodhra	54.7	63.2	60.3	69.0	31.6	34.5	30.1	34.5
	Mean	34.6	41.2	41.1	47.7	20.6	23.9	31.7	36.9
				Chickpea					
Mah-arastra	Bhandra	42.0	53.2	48.4	57.9	26.6	29.0	48.4	57.9
	Sindhewahi	57.4	72.0	71.3	81.8	36.0	40.9	71.3	81.8
	Gondia	50.2	67.3	57.7	78.6	33.7	39.3	57.7	78.6
	Aurangabad	59.2	63.9	67.4	73.4	31.9	36.7	67.4	73.4
Kar-nataka	Bheem - arayangudi	43.2	58.9	54.1	62.7	29.4	31.4	140.2	158.3
	Gulberga	51.4	64.2	60.2	69.0	32.1	34.5	-	-
	Gadag	40.8	58.4	48.0	69.2	29.2	34.6	-	-
	Mean	49.2	62.6	58.2	70.4	31.3	35.2	77.0	90.0
				Soybean					
MP	Chindwara	58.4	74.6	66.2	84.6	20.4	23.3	66.2	84.6
UP	Bundhelkhand	62.7	74.3	69.2	85.4	18.6	21.4	69.2	85.4
Maha-rastra	Wardha	47.8	58.9	58.0	69.5	23.5	27.8	58.0	69.5
	Nagpur	52.9	62.1	64.1	76.1	24.8	30.4	64.1	76.1
	Beed	56.4	63.3	63.8	69.7	31.6	34.9	63.8	69.7
	Ahmednagar	34.1	39.1	45.0	49.7	26.1	33.1	90.0	99.4
	Sindhwahi	33.0	41.0	36.0	46.6	16.4	18.6	36.0	46.6
Karn-ataka	Bheema - rayangudi	25.7	30.6	29.6	42.3	15.3	21.1	47.3	67.6
	Dharwad	40.8	52.9	44.1	60.1	26.4	30.1	70.6	96.2
	Belgaum	62.6	75.2	73.4	87.6	37.6	43.8	117.4	140.2
	Mean	47.4	57.2	54.9	67.2	24.1	28.5	68.3	83.5

the form of green and dry plant residues, fresh animal wastes, decomposed materials of plant and animal origin and biologically active preparations. When added to soil they undergo microbial decomposition. In this process, the nutrients held in organic combinations are slowly released in available forms in addition to improving the availability of other secondary and micro nutrient elements present in the soil. They also promote the microbial and soil enzyme activities. As mentioned earlier, besides the NPK nutrients, organic matter also provides the most critical and essential micro nutrients which the chemical fertilizers are unable to provide. At the same time, the organic fertilizers are unable to provide all the required quantity of NPK nutrients. Therefore both organic manures and chemical fertilizers are used in integrated nutrient management. A minimum of 10-12 t/ha of organic manures should be applied. On-farm trials of INM for pulses (chickpea and lentil) conducted at 60 farmers' fields in Raisen, Bhopal,

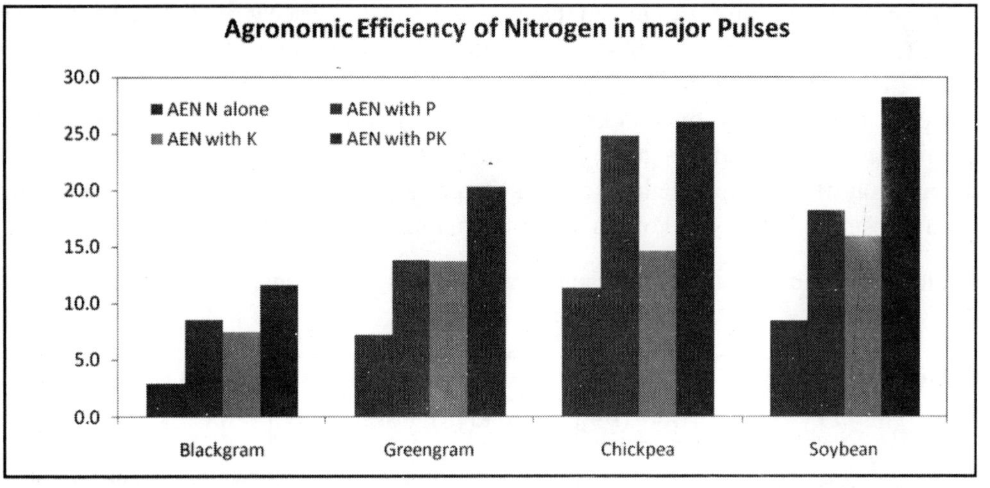

Fig 2. Efficiency of applied nitrogen in various pulses when N is applied with P and K

Rewa and Satna districts of Madhya Pradesh not only resulted in higher yields but also saved on fertilizer cost. Application of 75% NPK + 25t FYM + seed inoculation through Rhizobium + soil application of 3 kg PSB/ha to the soybean during kharif season and 50-75% of the recommended NPK dose to rabi crop (chickpea/lentil) (based on residual moisture availability) harvested 12-25% more chickpea and 15-28% more lentil as compared to traditional practice (Table 8).

Table 8. Yield (tha^{-1}) of pulse crops due to improved nutrient management and moisture conservation over farmer's practice in Madhya Pradesh

Treatment	Bhopal & Raisen	Rewa & Satna	
	Chickpea	Lentil	Chickpea
Integrated Nutrient Management	1.3 (13.5)	1.1(20.7)	3.3 (31.0)
Integrated Nutrient Management + Moisture Conservation	2.4(25.0)	1.3(28.3)	4.6 (42.7)

Data in parenthesis are % increase in seed yield

The average profit to the farmers with improved nutrient and soil moisture management practices was Rs 14000/ha for chickpea in Bhopal and Raisen, Rs 7500 ha^{-1} for lentil and Rs 15000/ha for chickpea in Rewa and Satna districts of Madhya Pradesh. Research evidences have also shown that the application of organic manure to dry lands not only improved the soil fertility status but also improved the water holding capacity of the soil. Judicious nutrient management encompassing organic manure and inorganic fertilizers could bring out a definite improvement in the cost of utilization of nutrient with increased nutrient use efficiency and will certainly improve the productivity and economic returns. The use of organic manures along with fertilizers has great significance for maintenance of soil fertility, physical condition and biological activity. The direct and residual effects of organic manures and chemical fertilizer on the grain yields of rice and chickpea showed that maximum yield of rice was obtained with neem cake (5.3 t/ha) followed by Sesbania rostrata (4.9 t/ha) which was on par with each other (Table 9) (Singh *et al.*, 1999). The residual effect of various organic manures differed significantly on chickpea yield. The residual effect of poultry manure significantly increased the grain yield of chickpea compared to all other treatments. Integration of recommended dose of fertilizer along with enriched farmyard manure would result in better yield of blackgram under rainfed conditions.

Table 9. Yield and economics of rice and chickpea as affected by organic manures and chemical fertilizers

Treatment	Yield (t/ha)		Economics (Average of both the crops)		
	Rice (D)	Chickpea (R)	Net return (Rs/ha)	B:C ratio	Rate of return Rs/Re on extra investment
Control	3.2	1.1	10790	0.75	-
Sesbania rostrata	4.9	1.5	19985	1.27	4.15
Sunhemp	3.9	1.4	14660	0.96	1.72
Poultry manure	4.5	1.7	20435	1.36	5.36
FYM	3.7	1.2	13050	0.80	0.90
Nadep compost	3.9	1.4	15210	0.94	1.26
Neem cake	5.3	1.4	2055	1.05	0.58
NPK @ 80:50:30 kg/ha	4.6	1.2	16055	0.97	3.63
CD at 5 %	0.4	0.12	-	-	-

Source : Singh *et al.*, (1999)
D : Direct, R : Residual

Gypsum / pyrites / liming agents: Gypsum, pyrites or rock phosphate are the cheapest sources of important secondary nutrients. Their application is known to increase both production and quality of pulses. Gypsum application requires special attention and thrust because farmers mostly use non-sulphur containing fertilisers like DAP and mixtures. High transport costs is one of the major constraints for farmers in using gypsum.

Bacterial agents : Rhizobium culture and phosphate solubilising bacteria (PSB) are bacteria that enable crops to fix crucial inputs from the soil. Rhizobium culture is one of the cheapest inputs in increasing production of pulses and other leguminous crops. The treatment of seed with this culture helps in fixation of atmospheric nitrogen through its symbiotic activity. The treatment is particularly beneficial in areas where groundnut and soybean are a new introduction. PSB has a capacity to release phosphorus and has been recommended as one of the low cost inputs for all crops. It helps to reduce nearly 20 per cent of phosphatic fertiliser input to crops. But though the treatment of seed with these cultures is highly recommended, it is necessary to follow certain precautions for usage: The culture should be obtained from a reliable source. It should not be outdated. The date of expiry should be checked before using the culture. Only cultures with ISI/BIS specifications / standards should be used. The culture should be well protected from heat and light during transportation and storage. It should not be stored in places contaminated with either pesticides or fertilisers. Before using the culture, ensure that the culture is meant for that specific crop. The seed treatment with culture should be done either on the same day or the previous night. After treatment, the treated seed should not be exposed to sunlight or high temperatures, as this kills the bacteria present in the culture. As far as possible, it should be stored at lower temperature. Rooms contaminated with either pesticides or with fertiliser materials should not be used for storing the culture. It is advisable that the sowing operation is undertaken either early in the morning or in the evening, so that mortality of the bacteria is reduced to minimum. If the culture does not possess stickiness, then either rice starch or jaggery solution should be used to moisten the seed, so that the culture sticks to the seed coat. Most pesticides are toxic to culture. Thiram is the least toxic fungicide and dieldrin least toxic insecticide. If these chemicals are used, culture treatment (slurry-inoculate) should be given to seed (after pesticide treatment) with double the normal dose. Slurry-inoculated with Rhizobium culture should never be mixed with super-phosphate as this is acidic and will kill the bacteria. In case of PSB, in acidic or alkaline soils, treated seed may be sown after coating with calcium carbonate or gypsum. The culture use is specifically recommended in areas where the crop is being grown for the first time. In such areas, culture application should be repeated for two to three years continuously. In rice fallow areas, the use of culture is very beneficial; because of anaerobic conditions in rice fields, there is mortality of native rhizobia. PSB can be mixed with Rhizobium. The bio-fertiliser should be mixed in equal quantity and applied as mentioned above. The response to PSB is very good in acidic soils. Krishanasamy *et al.* (1985) reported that $ZnSO_4$ resulted in the highest uptake of NPK in cowpea. Application of 5 kg Zn + 1.5 kg B + 0.5 kg Mo + 40 kg S/ha to blackgram has significantly enhanced the available nutrient status of the post harvest soil (Poongothai and Chitdeshwari, 2003). Influence of Rhizobium and organic amendments to mungbean and urdbean registered significant enhancement in plant biomass, root nodulation and grain yield over control (Nagarajan and Balachander, 2001). Field studies on the effect of biofertilizers and inorganic nutrition on blackgram grown in rice fallow showed that all the treatments except mere inoculation of biofertilizers exerted a conspicuous effect on yield and other growth parameters. The increase in seed yield due to spraying of urea (2 %) at pre-flowering and podfilling stages and DAP (4 %) spray at pod filling stage were 40.2, 34.2 and 31.1 %, respectively, over unfertilized treatment (Table 10). Addition of 50 kg P_2O_5 /ha alone or along with VAM inoculation enhanced the seed yield by 25.2 % and 30.8 %, respectively. Starter N @ 20 kg/ha also showed significant beneficial effect on yield and other physiological parameters.

Table 10. Yield of blackgram as affected by various fertilization treatments

Treatment	Seed Yield (kg/ha)	Marginal B:C ratio (Value:1)
No fertilization	719	-
Rhizobium inoculation (0.5 kg/ha)	802	20.7
VAM (5 kg/ha)	779	12.0
Rhizobium + VAM	820	11.2
50 kg P_2O_5 ha	962	2.6
VAM + 50 kg P_2O_5 ha	1039	3.2
Starter N (20 kg N/ha)	955	10.6
Urea spray (2%) at pre flowering stage	1204	71.3
Urea spray (2%) at pod filling stage	1043	40.5
DAP (4 %) spray at pre-flowering stage	1043	40.5
CD at 5 %	235	-

Source : Raju and Nagaswara Reddy (1999)

Proper pest and nutrient management are very vital for enhancing the production of any crop. Appropriate nutrients, such as sulphur, zinc and phosphorus, help in improving plant biomass for pulses and result in better yields. Currently, only 15 per cent of the total area under pulses is irrigated, compared to an average of 46 per cent for food grains in general. Providing scheduled and controlled irrigation increase yield. Irrigation requirements for pulses are much lower than those for other crops and could be provided through sprinklers, etc.

Phosphorus and Potassium Requirement: A case study from Uttar Pradesh

Uttar Pradesh is one of India's major pulse-producing states, with 12% of the total area of 21.6 million hectares (M ha) and 18% of total production. Although pulse productivity in Uttar Pradesh is 45% higher than the country's average (609 kg/ha), on-station agronomic and nutrient management experiments have revealed a great opportunity for improvement. A site-specific nutrient management (SSNM) strategy was evaluated (Series A), as well as the effectiveness of applying potassium (K), sulfur (S), and magnesium (Mg) through SKMg double salt (Series B) and the results obtained are given below.

Series A: Grain yield of pulses increased tremendously with adequate and balanced use of nutrients. The highest chickpea yield of 4.07 t/ha was obtained with nitrogen (N), phosphorus (P), and K (30-60-90 kg N-P_2O_5-K_2O/ha) plus 40 kg S/ha, 20 kg zinc sulfate ($ZnSO_4$)/ha, 10 kg manganese sulfate ($MnSO_4$)/ha, and 5 kg borax (B)/ha application. The state recommendation (SR) of 20-60-30 kg/ha produced only 2.4 t/ha, or 41% less than the SSNM treatment. Biological N-fixation parameters in chickpea, such as nodules per plant, fresh weight of nodules, dry weight of nodules, N content of nodules, and leghaemoglobin contents, were 39%, 11%, 12%, 34%, and 11% higher under SSNM compared to the SR. Similar trends were recorded for N-fixation parameters of pea and lentil. The maximum economic yield of 3.52 t/ha was recorded for pea with the SSNM treatment above, which was 73% higher than SR of 20-60-40 kg/ha. Lentil yield under this SSNM treatment was 2.23 t/ha, which was 36% higher than SR.

Series B: Yields of chickpea, pea, and lentil obtained with 15-60-40 kg/ha plus 40 kg S/ha and 20 kg Mg/ha with K, S, and Mg supplied via SKMg double salt were 57%, 60%, and 34% higher than with N application alone.

A large potential exists for improving pulse crop productivity with adequate and balanced fertilization. Current fertilizer recommendations need revision and should consider macronutrients, secondary, and micronutrients. SKMg double salt is an effective source for Uttar Pradesh pulse crops. Efforts are being made to have these results accepted as part of the effort to upgrade fertilizer recommendations.

Practices of INM: The black gram requires 15-20 kg N and about 50 to 60 kg P_2O_5 per hectare which should be basal placed but when it is to be grown as mixed crop with other crops, the fertilizer requirement of the main crop should be considered. Application of 70-80 kg Diammonium phosphate (DAP)/ha is more convenient as it meets adequately the total requirement of N and P_2O_5. Molybdenum also helps in increasing crop production. However, response to P_2O_5 application is higher in red soils followed by lateritic soils. Potassic fertilizers should be applied as per soil test value. In case, soil test facilities are not available, one can apply 30-40 kg K_2O / ha. The fertilizers should be drilled at the time of sowing in such a way that they are placed about 5-7 cm below the seed. Application of S @ 20 kg/ha improved the grain yield by 226 kg/ha compared to control. The nutrient use efficiency was 11.3 kg grain kg S^{-1} applied and gypsum is the most efficient source, followed by single super phosphate.

Greengram is generally grown without application of fertilisers. However, experiments conducted at various places have revealed that yield of mungbean can be increased to a great extent with the application of fertilisers. Responses of mungbean to N are generally absent or negative. Rhizobium culture responses to application are quite variable. Responses are comparatively higher in summer season in comparison to *kharif*. In general, a starter dose of 10-20 kg N/ha is recommended. Responses of mungbean to phosphorus are quite conspicuous. Significant increase in the yield of mungbean was showing due to application of 80 kg P_2O_5/ha. The response of mungbean on red soils followed by that on laterite soils. Responses of mungbean to potassium are negligible or absent owing to the high potassium levels in soil. Overall, application of 100 kg Diammonium phosphate (18-46-0) per hectare is generally recommended for mungbean. This fertilizer is comparatively cheaper on nutrient basis and can supply required quantity of N and P to the crop. Fertilizer placement should (total quantity) be done at the time of sowing.

Chickpea is generally cultivated with application of low doses of organic manure and chemical fertilizers resulting in low yields. In some states like Madhya Pradesh and Maharashtra, gram is not manured, particularly when the previous crop is manured. A basal dressing of farmyard manure (FYM) is provided in the black soils of Andhra Pradesh and in the loamy soils of West Bengal and Gujarat at a rate of 10, 4 and 5 cartloads, respectively. Total uptake of nitrogen by a chickpea crop has been estimated to vary 60-145 kg / ha, depending upon growing conditions of the crop. Positive response to starter nitrogen dressing of about 15-25 kg N/ha was observed in soils of low fertility status and poor in organic matter. Symbiotic N fixation apparently seems to be effective enough in most of these areas to meet the major nitrogen need of the crop. In areas where nodulation has been either very poor or has completely failed, significant response to increasing rates of N application have been obtained. Soil and foliar application of more N reduced the symbiotic N-fixation. Phosphorus uptake has been found to range from 5 to 10 kg ha depending upon crop growth conditions. No response to phosphorus might be due to high phosphate-fixing capacity of soil. Lack of response to P indicate that chickpea may be more efficient in uptake and utilisation of soil phosphorus. Application of phosphate is rare except in loamy to heavy loam soils of Punjab. The sources of P_2O_5 viz; dicalcium phosphate and superphosphate, were found to be equally efficient. In general, application of phosphate fertilizers increased the yield of chickpea in Andhra Pradesh, Bihar, Mysore, Punjab, Uttar Pradesh, West Bengal and Gujarat. It was found that application of P_2O_5 stimulated vegetative growth, seed setting and seed size, but not the time of flowering and maturity. Chickpea when grown on high pH soils rich in calcium carbonate, showed typical symptoms of iron defieciency. The deficiency can be corrected by a foliar sprays of 0.5% W/V ferrous sulphate solution at the beginning of reproductive growth and a fortnight later. Zinc deficiency can be corrected by a foliar spray of 0.5% W/V zinc sulphate solution or by applying 25 kg Zinc Sulphate/ha. Application of Molybdenum increases the grain yield. Copper, manganese and zinc also have a favourable effect on the crop when applied in combination.

Use of culture and pelleting materials in pigeonpea have revealed that inoculation increased the grain yield of pigeonpea. Inoculation with effective Rhizobium strains improved plant vigour. Roots of pigeonpea under uninoculated condition also produced nodules by natural Rhizobium present in the soil. Many of the natural Rhizobium present in the soil may belong to the strain, which induces nodule development but they fix little or no nitrogen. Adequate supply of phosphorus is significant for higher yields of pigeonpea. In India, responses to phosphorus have been generally positive and in some areas, highly significant. Pigeonpea crop planted in June responded up to 40 kg P_2O_5 ha whereas crop planted in mid-April responded up to 60 kg P_2O_5/ha. However, further increase in the level of phosphorus did not increase the yield. Pigeonpea normally does not respond to potassium application unless grown on soils low in available potassium. The recommended nutrient for pigeon pea is 20 kg K_2O ha over 20 kg N + 40 kg P_2O_5 ha. Most of the pigeonpea cultivars show susceptibility to zinc deficiency. Soil application of 2-4 ppm zinc or foliar spray of 0.5 per cent zinc-sulphate with 0.05 per cent lime has proved effective in controlling zinc deficiency.

Region specific INM practices: Some INM options for various pulses and cropping systems involving pulses in semi arid and sub humid regions are presented in Table 11. In general, it is observed that, FYM, biodigested slurry, neem cake, sugarcane pressmud and compost are used as organic source for pulse production in various regions and the quantity differs with fertility status of soil. Application of Rhizobium and VAM also forms part of INM package in selected locations.

Table 11. Recommended INM practices in pulses and cropping system involving pulses in semi arid and sub humid regions of India

Location	Crop/cropping system	Soil	IPNS practice	Reference
I. Semi Arid				
New Delhi (Delhi)	Pigeonpea	Sandy loam	40 kg P_2O_5/Ra + PSB+VAM	Pramod Kumar *et al.* (2012)
	Pigeonpea+ mungbean	Sandy loam		
Modipuram (UP)	Sorghum-chickpea	Sandy clay	6 t of FYM + 50 % of RDF/ha	Gangwar and Yogendra Singh (1992)
Arjia (Rajasthan)	Maize + pigeonpea	Clay loam	Substitution of 50 % N with FYM	Anonymous (1994)
Jaipur (Rajasthan)	Greengram	Loamy sand	30 kg S/ha	Prajapat *et al.* (2012)
Junagadh (Gujarat)	Chickpea	Medium black	20 kg N + 40 kg P_2O_5 ha + Rhizobium inoculation	Vadavia *et al.*, 1991
Indore (MP)	Soybean-wheat	Vertisols	Substitution of 50 % N with FYM	Rao and Singh (1993)
Akola (Maharashtra)	Blackgram, greengram and groundnut	Vertisols	Rhizobium inoculation+ Molybdenum @ 4 g kg as seed treatment	Daterao *et al.* (1992)
	Soybean	Vertisols	150 % recommended dose of N and P through urea and SSP	
Secundrabad (AP)	Rice-chickpea-groundnut	Sandy loam	100 % NPK + Rhizobium and Phosphobacteria @ 2 kg/ha each	Chandrakar *et al.*, 1983
Marateru (AP)	Blackgram	Vertisols	50 kg P_2O_5 + VAM inoculation	Raju and Nageswara Reddy (1999)
Coimbatore (TN)	Blackgram	Black soil	75 % P_2O_5 as	Thiyageswari and

			Tunisia rock phosphate + vermicompost @2 kg/ha + Phosphobacteria @ 2 kg/ha	Perumal (2000)
Pudukottai (TN)	Blackgram	Acidic soil	Biodigested slurry @ 5 t/ha + Rhizobium	Nagarajan and Balachander (2001)
	Greengram	Acidic soil	Compost @ 5 t/ha + Rhizobium	
Raipur (Chattisgarh)	Soybean	Vertisols	30 kg P_2O_5 through rock phosphate +PSB+ Rhizobium inoculation + VAM	Sarawgi *et al.* (2012)
II. Sub humid				
Varanasi (UP)	Rice-lentil	Sandy c lay loam	50 % N through FYM + 50 % prilled urea to rice	Singh *et al* (2001)
Sehore (MP)	Pigeonpea and blackgram	Vertisols	FYM or Sugar pressmud @ 5 t/ha with Rhizobium inoculation	Namdeo and Gupta (1992)
	Soybean	Vertosols	FYM @ 6 t/ha or sugar pressmud @ 5 t/ha with Rhizobium inoculation	
Bhopal (MP)	Soybean-wheat	Vertisols	FYM @ 8 t/ha + 12, 15 and 3 kg N, P and Zn/ha to soybean, 90, 22 kg N and P to wheat	Anonymous (1995)
Udham Singh Nagar (Uttarakhand)	Soybean	Clay loam	25 kg N + 5 t FYM or 25 kg N + 1 t neem cake/ha	Saxena *et al.* (2001)
Bapatla (AP)	Soybean	Sandy loam	Biogas slurry @ 15 t + 50 kg N as urea/ha	Varuna and Reddy (1999)

The results of AICRP-DA in INM project indicated that use of 50 % of organic in combination with 50 % of inorganic not only enhanced the productivity of pulses in different cropping systems but also improved the soil fertility of the rainfed soils. With integrated use of organic and inorganic nutrients under different intercropping systems not only sustainable productivity was noticed (Table 12), but also reduction in cost of production was ensured. Seed treatment along with Rhizobium inoculation gave added advantages as it enhanced biological nitrogen fixation, which met 80-90 % N requirement of pulse crop and also left a residula amount of 30-40 kg N/ha for the succeeding crop (Singh and Dwivedi, 2006). Studies conducted at Kanpur indicated that Rhizobium bio inoculants saved 50 % of recommended doses of fertilizers like N, P and S under rainfed and irrigated conditions of clay loam soils (TAR-IVLP, 2005).

Table 12. Effect of integrated nutrient management on yield and sustainability of pulses in intercropping systems

Centre	Crops/Cropping systems	Nutrient Management	Yield (kg/ha)		SYI	
			Main crop	Intercrop	Main crop	Intercrop
Faizabad	Maize + Pigeonpea (5)	100 % recommended N	1224	1068	0.38	0.31
		50 % recommended N	948	718	0.25	0.16
Phulbani	Pigeonpea+rice (5)	45 kg N/ha (urea)	606	1576	0.17	0.52
		20 kg N as FYM + 25 kg N/ha (urea)	785	1585	0.26	0.52
Ranchi	Rice + blackgram (3)	40 kg N/ha (urea)	1944	820	0.74	0.73
		15 kg N as compost + 20 kg N/ha (urea)	1691	916	0.62	0.82
	Rice + lentil (5)	100 % RDF as organic	2547	618	0.30	0.14
		50 % N (urea) + 50 % N (FYM)	2701	903	0.33	0.31
		100 % RDF as inorganic	2285	692	0.24	0.22
		15 kg N (Greenleaf) + 20 kg N/ha (urea)	2274	728	0.21	0.26
Bijapur	*Rabi* sorghum + chickpea (7)	100 % RDF (inorganic)	1235	785	0.57	0.62
		15 kg N (compost) +20 kg N/ha (urea)	1354	872	0.65	0.69
Bellary	*Rabi* sorghum + chickpea (7)	15 kg N (Greenleaf) + 20 kg N/ha (urea)	773	370	0.03	0.23
		15 kg N (green leaf) + 10 kg N/ha (urea)	719	353	0.01	0.21

Values under parenthesis indicates number of years of experimentation
Source : Ramakrishna *et al.,* (2007)

Secondary and micro nutrients management: Application of secondary and micro nutrients to pulse crops is an economically viable proposition to farmers. The monetary benefit a farmer gets from application of sulphur and micro nutrients to pulse crops is proportional to the amount of seeds produced per kg of applied nutrients and this varies from crop to crop and region to region. Combined application of S @

30 kg/ha and Zn @ 5 kg/ha proved to be most remunerative and economical for greengram in silt loam soils of Faizabad (Table 13). In sandy loam soils of Allahabad, application of 30 kg $ZnSo_4$ + 2 kg B/ha recorded higher grain yield (1047 kg/ha) of blackgram and was more remunerative with maximum B:C ratio (1.88). Application of S through SSP @ 30 kg S/ha in summer greengram gave maximum net returns (Table 14).

Table 13. Effect of S and Zn levels on yield and B:C ratio of summer greengram

Treatment	Yield (q/ha)	B:C ratio
No sulphur, No zinc	5.77	0.35
Zinc @ 5 kg ha-1	7.30	0.44
Sulphur @ 15 kg ha-1	6.51	0.49
S 15 +Zn 15	8.80	0.69
S 30 + Zn 0	7.13	0.60
S 30 + Zn 5	10.19	0.92
S 45 + Zn 0	7.25	0.59
S 45 + Zn 15	10.27	0.90
CD (P=0.05)	0.88	-

Source : Masood Ali and Venkatesh (2009)

Table 14. Yield and economics due to sulphur application in summer greengram

Treatment	Yield (t/ha)	Net profit (Rs/ha)
No fertilizer	0.85	-
NPK	1.41	5886
NPK + S through SSP	1.67	9106
NPK + S through K2SO4	1.54	7367
NPK + S through elemental sulphur	1.55	7101

Source : Masood Ali and Venkatesh, (2009)

Obviously, integrated nutrient management involving organic, inorganic and bio-fertilizers can play a significant role in timely supplying of adequate quantity of secondary and micronutrients to pulses.

Conclusions

Nutrient imbalance is one of the major abiotic constraints limiting productivity of pulses. The inbuilt mechanism of biological N_2 fixation enables pulse crops to meet 80-90 per cent of their nitrogen requirements, hence a small dose of 15-25 kg N/ha is sufficient to meet out the requirement of most of the pulse crops. However, in emerging cropping systems like Rice - Chickpea, a higher dose of N (30-40 kg/ha) had shown a beneficial effect. Phosphorus deficiency in soils is wide spread and most of the pulse crops have shown good response to 20-60 kg P_2O_5/ha depending upon nutrient status of soil, cropping system and moisture availability. Response to potassium application is location specific. In the recent years, use of sulphur (20-30 kg/ha) and some of the micronutrients such as Zn, B, Mo and Fe have improved productivity of pulse crops considerably in many pockets. B and placement of phosphatic fertilizers and use of bio-fertilizers enhanced the efficiency of applied as well as native P. Foliar nutrition of some micronutrients proved quite effective. The amount and mode of application is determined by indigenous nutrient supply, moisture availability and genotypes. Balanced nutrition is indispensable for achieving higher productivity. At the same time, in view of the increasing nutrients demand, there is immense need to exploit the alternate source of nutrients viz., organic materials and bio-fertilizers, to sustain the productivity with more environment friendly nutrient management systems. Considering the measures listed above, India has the potential to produce over 37 million tonnes of pulses. If India desires to achieve its true potential in pulse production, there is an urgent need to focus on integrated nutrient management.

Future thrust

Few researchable issues which needs immediate attention are;

- Identification of cultivars with high nutrient use efficiency under moisture stress conditions as most of the pulses are grown in rainfed regions with minimal care.
- Devising an infield mechanism through intensive complimentary cropping systems for supplying the required quantity of primary, secondary and micro nutrients, are needed to fix the demand-supply gap of nutrients in pulse based cropping systems.
- Quantification of direct, residual and cumulative effect of nutrients in cropping/ farming system mode is essential.
- Matching of nutrient release from organic sources and peak demand by pulse crops is essential in order to develop source (organic)-soil-crop stage specific integrated nutrient management packages.
- Evaluation of customized fertilizers in pulse production in order to ensure their balanced use.

References

Ahlawat, I.P.S and Ali, M. 1993. In: Fertilizer management in food crops (Eds. H.L.S. Tandon). pp. 114-138

Amarender Reddy, A.2009.Pulses production technology: Status and Way Forward, Economic and Political Weekly 44 (52) : 73-80.

Anonymous. 1995. Status of centre-state coordination in Agricultural research, education and extension in Region VII Agenda notes XVI meeting, 21-22 November 1997, Panaji, Goa, ICAR, New Delhi, pp. 142

Anonymous.1994. Research Highlights. Directorate of Research, Rajasthan Agricultural University, Bikaner. pp. 73-74

Chandrakar, B.S., Singh, K and Ram, G. 1983. Influence of blue green algae on rice and its residual effect on yield of blackgram. Oryza 20 : 69-71.

Daterao, S.H., Vinay Babu, B., Hanwante, P.R and Solanke, V.M.1992. The effect of biofertilizer (Rhizobium) in combination with molybedenum on yield and nitrogen economy in kharif legumes. PKV Research Journal 16(1) : 137-139.

Ganeshamurthy, A.N., A. Subbarao and T.R. Rupa. 2003. Integrated Nutrient Management in pulse based cropping systems In: Pulses in new perspective (Eds: Masood Ali, BB Singh, Shiv Kumar and Vishwa Dhar), Indian Society of Pulses Research and Development, Kanpur pp: 287-300.

Gangwar, K.S. and Yogendra Singh.1992. Integrated nutrient management in fodder sorghum (Sorghum bicolour) -gram (Cicer arietinum) cropping sequence under dryland conditions. Indian Journal of Agronomy 37(1) : 107-109.

Masood, Ali and Venkatash, M.S. 2009. Improving yield and quality of pulses through secondary and micronutrients, Indian Journal of Fertilizers 5(4) : 117-123.

Nagarajan, P and Balachander, D. 2001. Influence of Rhizobium and organic amendments on nodulation and grain of blackgram and greengram in acid soil. Madras Agricultural Journal 88 (10-12) : 703-705.

Namdeo, S.L. and Gupta, S.C. 1992. In: National Seminar on Organic Farming (Eds: M.M. Rai and L.N. Verma), Ravi Printing Press, Jabalpur. pp. 151-161,

Prajapat, K, Shivran, A.C. Choudhury, G.L. and H.R. Choudhary. pp 2012. Influence of planting pattern and sulphur on mungbean (Vigna radiate) and sesame (Sesame indicum) intercropping under semi-arid region of Rajasthan. Indian Journal of Agronomy 57(1) : 89-91.

Kumar, Pramod, Rana K.S. and Rana D.S. 2012. Effect of planting systems and phosphorus with bio fertilizers on the performance of sole and intercropped pigeonpea (Cajanus cajan) under rainfed conditions. Indian Journal of Agronomy 57(2) : 127-132.

Raju, R.A and Nageswara Reddy, M. 1999. Effect of biofertilizers and inorganic nutrition on blackgram (Phaseolus mungo) grown in rice fallows, Indian Journal of Agronomy 44(4) : 782-786.

Rama Krishna Reddy, Y.S, Subba, G., Ravindra Chary, G and Prasad, Y.G. 2009. Improving productivity of grain legumes under moisture stress conditions In: Legumes for ecological sustainability (Eds. Masood Ali, Sanjeev Gupta, P.S. Babu and Naimuddin), Indian Society of Pulses Research and Development, IIPR, Kanpur, India, pp 92-117.

Rao, C.R. and Hansa Modi. 2009. Potential for enhancing the productivity of pulses. Indian Journal of Fertilizers 5(2) : 41-44, 47-50 & 67.

Rao, U.M.B and Singh, S.P.1993. Crop response to organic sources of nutrients in dryland conditions. pp 287-304. In: Recent Advances in Dryland Agriculture (Ed., L.L. Somani), Scientific Publishers, Jodhpur, India.

Reddy, A A 2004: "Consumption Pattern, Trade and Production Potential of Pulses", Economic & Political Weekly, 30 October, pp 4854-60.

Sarawgi,S.K., Shrikant Chitale, Alok Tiwari and Sandeep Bhoi.2012. Effect of phosphorus application along with PSB, Rhizobium and VAM on P fractionation and productivity of soybean (Glycine max). Indian Journal of Agronomy 57(1) : 55-60.

Saxena, S.C., Manral, H.S. and Chandel, A.S. 2001. Effect of organic sources of nutrients on soybean (Glycine max). Indian Journal of Agronomy 46(1) : 135-140.

Singh, G.R., Parihar, S.S and Chaure, N.K. 1999. Direct and residual effect of organic manures on rice (Oryza sativa) -gram (Cicer arietinum) cropping sequence, International Journal of Tropical Agriculture 17 (1-4) : 195-198.

Singh, M.V.2008. Micronutrient deficiencies in Indian soils and field usable practices for their correction. (www.iiss.nic.in/recommendations1.pdf), IISS, Bhopal p. 18.

Singh, S.K., Verma, S.C and Singh, R.P. 2001. Effect of integrated nutrient management on yield, nutrient uptake and changes in soil fertility under rice (Oryza sativa) -lentil (Lens culinaris) cropping system. Indian Journal of Agronomy 46(2) : 191-197.

Singh, V.K. and Dwivedi, B.S. 2006. Yield and N use efficiency in wheat and soil fertility status as influenced by substitution of rice with pigeonpea in a rice-wheat cropping system. Australian Journal of Experimental Agriculture 46: 1185-1194.

Tandon, H.L.S. 1995. Sulphur fertilizer for Indian Agriculture-A Guidebook (2nd Edition), Fertilizer development and Consultation Organization, New Delhi. pp.101

TAR-IVLP. 2005. Technology assessment and refinement through Institute Village linkage programme 9Completion report, 1999-2004. Agro ecosystems Directorate (Rainfed), NATP, CRIDA, Hyderabad. pp 74-78.

Thiyageshwari and Perumal, R.2000. Changes in available phosphorus and grain yield of blackgram (Vigna mungo) under integrated nutrient management in Inceptisol. Agropedology, 10 : 40-43.

Vadivia, A.T., Kalaria, K.K., Patel, J.C. and Baldha, N.M.1991. Influence of organic, inorganic and biofertilizers on growth, yield and nodulation of chickpea. Indian Journal of Agronomy 36:263-264.

Varuna, V and Reddy, S.N.1999. Influence of integrated supply of nitrogen through organic and inorganic sources on growth, nutrient uptake and yield of soybean. Journal of oil seeds research 16(2) : 337-339.

Vision, 2030. 2011. Indian Institute of Pulses Research (ICAR), Kanpur pp.42

❑❑❑

Chapter - 9

Integrated Water Management

Ashwani Kumar and Rajbir Singh
Directorate of Water Management, Bhubaneswar - 751 023, Odisha, India

Pulses in India are considered a residual crop and grown under rainfed conditions in marginal/ less fertile lands. About 90 percent of total pulse production comes from 97 million ha of rainfed areas in our country. Presently, Madhya Pradesh, Maharashtra, Uttar Pradesh and Rajasthan contribute 85% of the total area and production of pulses in the country. Pulse crops usually face soil water deficits sometime during their growth cycles. Thus, drought stress is a common phenomenon and ranks high among all abiotic stresses limiting yields of pulses (Ali and Mishra, 2000). Hence, there is an urgent need to increase the pulse productivity to increase the income of farmers and reduce malnutrition. Productivity of pulses in rainfed agriculture has remained low and unstable owing to climate and soil related constraints (Swindale, 1982). Soil moisture is the foremost factor that determines the productivity of rainfed pulses where rainfall is the only source of water. To maintain required soil moisture in rainfed lands, every action must be focused to conserve as much quantity of rainfall in the soil as possible (Rao, 1997). Unfortunately, farmers following rainfed agriculture in our country are not practicing any sort of moisture management techniques though we have a handful of techniques suitable for rainfed conditions.

The variability in yields of pulses is mainly due to susceptibility of the cultivars of different pulses to abiotic stresses. The major abiotic stresses affecting these legumes is moisture stress (drought and water logging) limiting the productivity of food legumes, especially under rainfed conditions, however, the extent of damage is greatly influenced by the stage of the crop growth, location and season. In the moisture stresses, drought situation is much more common both during *kharif* and *rabi* in contrast to water logging. Drought adversely affects the nodulation and biological nitrogen fixation (BNF), even when the native population of relevant efficient rhizobia is high or rhizobial inoculation has been applied. The pulse crops face two types of drought: intermittent drought being caused by breaks in rainfall, and terminal drought occurring at the end of the growing season due to depletion of stored soil moisture. Intermittent drought is more common in *kharif* crops while terminal drought is mostly associated with the *rabi* pulses and late-sown *kharif* crops. Another water stress is flooding and water logging which results in extensive destruction or impairment of crops. Temporary or continuous flooding water is very common in pulses which reduces the shoot and root growth, dry matter accumulation and finally crop yield. Trung *et al.* (1985) reported that mung bean plants waterlogged for 3 days at different stages recorded reduced seed yield by 82 per cent at flowering stage, 60 per cent at pod formation stage, 40 per cent at pod development stage and 17 per cent at grain maturity stage. Water logging resulted in disappearance of leghaemoglobin in root nodules and cessation of nitrogen fixation activities, reaching almost zero during the 3 days of waterlogging.

Water management options for pulses

Although the pulse crops are predominantly grown under rainfed conditions, they respond to irrigation very well. On the basis of front line demonstrations conducted on the farmers' fields during 1993-98, chickpea, field pea, pigeon pea and common bean, on an average, recorded 28.4, 22.5, 9.6 and 83.5 per cent increase in the grain yield, respectively, due to irrigation (TMOP 2000). Currently only 15% of the total area under pulses is irrigated as compared to an average of 46% for all food grains. There seems to be good scope of increasing the irrigated area under pulses by their introduction in the non-traditional areas and cropping systems where irrigation facilities are available. Summer pigeon pea + green gram – wheat sequence in irrigated tracts of Uttar Pradesh, Punjab, Haryana, and Rajasthan; *rabi* pigeon pea in Orissa, Bihar, Andhra Pradesh, Chhattisgarh, Tamil Nadu, Karnataka and West Bengal; green gram in the spring and summer fallow in the northern India and *rabi* sowing in southern India; extension of common bean as a *rabi* crop in Plains of north-eastern India; possibility of cowpea as summer crop; introduction of rice bean in certain states; *utera* cropping of urd bean in rice fallow of Orissa, Assam, West Bengal, Maharashtra and Andhra Pradesh, are some of the niches for increasing the irrigated area under pulse crops.

Water requirement of a pulse crop usually varies from location to location due to variation in agro-climatic and soil conditions. Even at the same location, a wide variation in water requirement of a pulse crop may exist due to variation in varieties, method of irrigation, irrigation efficiency, scheduling of irrigation, nutrient and weed management and other cultural practices. It also varies with the season, when the same variety of a pulse crop is grown under two different seasons. The factors affecting the uptake water also account for variation in water requirements of pulse crops. Providing scheduled and controlled irrigation can lead to increase in yields. Under the All India Coordinated Research Project on Water Management (AICRPWM), researches have been carried out to estimate the water requirement of different pulse crops. The findings revealed that water requirement of pulses is location- and season-specific and also varies with the method and scheduling of irrigation. Several researchers have made rigorous efforts to determine optimum irrigation schedules for different pulse crops, covering various agro-climatic regions, seasons, and situations of water availability in the country. In addition to the time of application, another important aspect of water management in pulses is the depth of water application. It should not be applied very deep, as it will create anaerobic conditions for Rhizobium and other microorganisms. Waterlogging causes compaction of soil and does not allow air to penetrate into soil thus adversely affects the root nodule bacteria and the entire system of symbiosis. The water use vis-à-vis irrigation requirement of important pulse crops based on IW/CPE (irrigation water/ cumulative pan evaporation) ratio and critical growth stages has been given in Tables 1 and 2, respectively.

Table 1. Optimum schedule of irrigation (based on IW/CPE ratio) at different locations in India

Crop/ Location/ AER*	Soil type	No. of irrigations (depth, cm)	Optimum IW/CPE ratio	Grain yield (t/ha)	Irrigation water requirement (cm)
***Kharif* pigeon pea**					
Morena, MP (4)	Sandy loam	3 (7.5)	0.90	2.00	22.5
Kota, Rajasthan (5)	Clay	4 (6)	0.80	1.45	24.0
Navsari, Gujarat (19)	Clay	2 (8)	0.60	2.06	16.3
Rahuri, Maharastra (6)	Clay	6 (7)	0.80	4.35	42.0
Summer mung bean					
Pantnagar, Uttaranchal (9)	Silt clay loam	5 (6)	0.60	1.42	30.0
Bhavanisagar, Tamilnadu (8)	Silt loam	5 (4)	0.60	1.32	22.0
***Kharif* mung bean**					
Dharwad, Karnataka (6)	Clay	1 (6)	0.60	1.32	6.0
Summer urd bean					
Bhavanisagar, Tamilnadu (8)	Sandy loam	4 (9)	0.60	1.14	36.0
Chalakudy, Kerala (19)	Silty loam	5 (3)	0.75	0.65	15.0
Common bean (Rajmash)					
Pusa, Bihar (13)	Sandy loam	2 (6)	0.80	1.19	12.0
Faizabad, UP (9)	Silt loam	3 (5)	0.70	1.56	15.0
Lentil					
Pusa, Bihar (13)	Sandy loam	2 (6)	0.60	1.60	12.0
Jabalpur, MP (11)	Clay	3 (7.5)	0.60	1.62	22.5

*AER: Agro-ecological zone of India; NBSS&LUP, Nagpur, (1992)

Source: Annual report of AICRP on WM; 1990-2009 and Yadav *et al.* (2000)

Table 2. Grain yield, irrigation water requirement and water use in major pulse crops at optimum irrigation schedule (based on growth stages) at different locations

Crop/Location/ AER	Soil type	No. of irrigations (depth, cm)	Growth stages	Grain yield (t/ha)	Irrigation water requirement (cm)
Kharif pigeonpea					
Morena, MP (4)	Sandy loam	2 (7)	Veg., grain dev.	1.11	14.0
Faizabad, UP (9)	Silt loam	1 (6)	Branch.	1.78	6.0
Parbhani, Maharastra (6)	Black clay	3 (6)	Veg., flow.pod form.	1.64	18.0
Sriganganagar, Rajasthan (2)	Loam	3 (6)	Veg., pod form, grain dev.	1.24	18.0
Chickpea					
Sriganganagar, Rajasthan(2)	Loam	2 (6)	Veg., pod form.	2.06	12.0
Parbhani, Mahrastra (6)	Medium black	2 (6)	Flow., pod form	0.98	12.0
Delhi (2)	Sandy loam	1 (6)	Pod form.	2.20	6.0
Morena, MP (4)	Sandy loam	1 (7)	Grain dev.	2.68	7.0
Kota, Rajasthan (5)	Clay	1 (6)	Flow	2.41	6.0
Rahuri, Maharastra (6)	Black clay	2 (8)	Branch., flow	2.82	16.0
Hisar, Haryana (2)	Black clay	1 (7)	Pod form.	2.43	7.0
Navsari, Gujarat (19)	Clay loam	3 (7)	Branch., flow., pod form.	2.01	21.0
Powarkheda, MP(10)	Clay loam	3 (6)	Branch., flow., grain dev.	1.90	18.0
Lentil					
Delhi (2)	Sandy loam	2 (7)	Seedl., flow.	3.40	14.0
Pantnagar, Uttaranchal(9)	Sandy loam	2 (6)	Veg., pod form	2.20	12.0
Ludhiana. Punjab (2)	Sandy loam	2 (7)	Veg., flow.	1.99	14.0

Veg.=Vegetative; Flow.=Flowering; Form.= Formation; Dev.= Development; Branch.=Branching; Seedl.=Seedling

*AER: Agro-eclogical zone of India; NBSS&LUP, Nagpur, (1992)

Source: Annual report of AICRP on WM; 1990-2009 and Yadav *et al.,* (2000)

Chickpea

Chickpea crop meets its high water requirement from conserved moisture in deep soils. Chickpea is capable of drawing water from depths greater than 150 cm (Singh and Bhushan 1979) but a major proportion of total extractable water comes from the top 60 cm of the soil profile which has the main concentration of active roots (Gupta and Agarwal 1977). Extractable moisture in the soil profile can be increased by management practices that encourage root growth or by using genotypes with deeper and more prolific root systems. On black cotton soils, chickpea responded to 2 irrigations in the presence of 50 or 100 kg/ha diammonium phosphate (Bihari *et al.* 1993). The importance of phosphorus application on low phosphorus soils in this respect has been highlighted by Singh and Bhushan (1979).

Chickpea is also grown under irrigated conditions in the Indian subcontinent particularly in areas where the water holding capacity of the soil and the seasonal precipitation are low and evaporative demand of the atmosphere is high. Saxena and Yadav (1976) and Saxena (1980) have reviewed the crop responses to supplemental irrigation in different production areas in India, where the evaporative demand is high, for example in Hyderabad, frequent irrigation doubles yields. On the other hand at Hisar, where evaporative demand is less, irrigation tends to cause excessive growth leading to lodging and a decrease in yield. Supplemental irrigation to release the crop from soil moisture stress at critical stages of crop growth has been found to result in substantial yield increases in chickpea (Saxena 1984). Irrigation early during vegetative growth on light soil with a low water holding capacity or later during vegetative growth on heavier deep soils, gave increased yields (Saxena 1980). The late flowering and pod setting stages appear to be the most sensitive to soil moisture stress (Palled et al, 1986). A frequent low volume irrigation regime, in areas where temperatures rise beyond the optimum during growth and development, has the advantage of lowering the soil and crop canopy temperatures with favourable effects on nodulation and nitrogen fixation (ICRISAT 1981) and crop yield, as is often the case in peninsular India. Irrigation need of the crop may vary with sowing dates due to a change in day length and temperature at different growth stages of the crop. In sandy loam soils at Faizabad (Uttar Pradesh), irrigation scheduled at 0.2 and 0.4 IW/CPE ratio produced similar yields in early sowing (30 October) whereas in late sowing conditions (30 November), 0.4 IW/CPE ratio was superior to 0.2 IW/CPE ratio (Singh and Dixit 1992).

Lentil

Lentil, an important crop of winter season, is mainly grown on l[illegible]g, inundated and marginal lands in the states of MP, UP, Bihar, West Bengal, Orissa, Haryana, Rajasthan, Assam and Punjab. Lentil crop is grown in a rain-free season on the conserved soil moisture, hence, the response to irrigation is dependent upon the amount of soil moisture conserved. This, in turn, depends on the rainfall in the previous season, soil depth and soil texture. On sandy loam soils, having low water holding capacity, positive responses to irrigation, ranging from 1 to 3, are very common. Flowering appears to be the critical stage for moisture. However, lentils are highly sensitive to over-watering. On fine-textured soils of low infiltration rate, and also in uneven fields, temporary waterlogging can cause serious damage to the crop growth and yield. Improved field layout with an even surface, gentle slope and adequate drainage are essential prerequisites for successful lentil cultivation in irrigated agriculture.

High temperatures early in the crop season, coupled with the problem of salinity, necessitate frequent irrigation. To maintain adequate aeration and facilitate application of irrigation, sowing on the ridges is practiced. Since the salinity hazard is highest on the top of the ridge, sowing on the side of the ridge proves advantageous. Also sowing on the inner sides of two contiguous ridges, using the furrow in between for irrigation and leaving every alternate furrow without irrigation, should enable the cropped area to become relatively salt free. In lentils, the highest grain yield of 1547 kg/ha was obtained when irrigation was scheduled at 0.6 IW/CPE ratio coupled with 70 kg P_2O_5/ha, which was 70 per higher than control at the same level of irrigation (Venkateswarlu and Ahlawat 1993).

Field pea

The pea crop subjected to moisture stress at various phenological growth stages (vegetative, flowering and pod-filling) revealed that moisture stress at flowering appeared to be the most critical in influencing its yield. The highest grain yield, consumptive use (261 mm) and WUE were recorded with no moisture stress, followed by moisture stress at vegetative stage (Kasturikrishna and Ahlawat 1999). Application of one irrigation either at flowering or at pod formation was superior to applying it at branching stage in black clay loam soils at Kota (Bhatia 2000). This crop is susceptible to soil moisture stress at flowering stage and the grain yield depends on the availability of soil moisture at the reproductive stage. Application of one irrigation at pre-flowering stage to pea resulted in 62 per cent increase in yield over control (Singh *et al.* 1978). Application of 20 kg N + 40 kg P_2O_5/ha in conjunction with 2 irrigations at branching and flowering stages produced higher grain yield than 10 kg N + 20kg P_2O_5/ha with the same level of irrigation in field peas (Ali 1990).

Pigeon pea

Pigeon pea is mainly grown in the monsoon season under rainfed conditions. It is cultivated in areas receiving mean annual rainfall of 600 to 1400 mm. Being a deep rooted crop, it is capable of extracting moisture from deeper layers in the soil. The consumptive use of pigeon pea varies with the varieties and agro-climatic conditions. The long-duration varieties require more water as compared to short-duration varieties. The consumptive use of *kharif* pigeon pea has been estimated at 406 mm at Hisar, 416 mm at Sriganganagar and 491 mm at Pantnagar (Agarwal *et al.* 1997). It responds to irrigation even under rainy season, particularly under long period of moisture stress and during post-rainy season in long-duration varieties.

Based on climatological approach (IW/CPE), it has been reported that three irrigations each of 7.5 cm depth, at IW/CPE ratio of 0.9 produced the highest grain yield (2.33 t/ha) of pigeon pea in sandy loam soil at Morena in Madhya Pradesh and the response to irrigation declined gradually with delayed sowing. Contrary to these results, ten irrigations each of 6 cm depth, at IW/CPE ratio of 0.75 produced the highest grain yield (4.34 t/ha) and the water requirement of the crop was found to be 600 mm (AICRPWM, 1997) at Rahuri in Maharashtra. In clay soils of Navsari, Gujarat and Kota in Rajasthan, the optimum IW/CPE ratio was 0.6, requiring 2 and 4 irrigations, respectively. In general, irrigation scheduling based on critical growth stages required less water but produced grain yield at par with IW/CPE ratio approach.

At Sriganganagar (Rajasthan), two irrigations applied at branching and flowering stage were comparable to irrigation at 125 mm CPE (Agarwal *et al.* 1997). Furthermore, in sandy loam soils at New Delhi, two irrigations, at pre-flowering and flowering stages, increased the yield by 25.9 per cent over control (Agarwal *et al.* 1997). Rana and Malhotra (1989) noted that an early sown pigeon pea (May 5) produced higher grain yield in comparison to late sown (May 25), when first irrigation was given either 2 or 3 weeks after sowing. Rao *et al.*, (1983) reported that irrigation at different stages caused three times increase in grain yield of pigeon pea and the response to irrigation was significantly higher in medium duration cultivar (C 11) than in long duration cultivar (NP 5 (WR) 15), however, the delay in seeding significantly reduced the yields both under control and irrigated conditions.

Pigeon pea is generally grown on Vertisols in post-rainy season and these usually have water content close to field capacity in their deeper layers. The beneficial effects from irrigation or late rainfall, therefore, depend upon the wetting of the surface region of the soil, which alleviates the water stress around the shallow roots. It is also likely that irrigation prevents, reduces or delays the appearance of soil cracks as the soil dries out. Even though both soil moisture stress and soil cracking aggravate with time, yet the response to supplemental irrigation decreased when sowing was delayed from September to November. This suggests that some other factors, such as temperature, photoperiod or ruptured roots due to soil cracking maybe of equal or greater importance than moisture in determining the yield of the post-rainy season pigeon pea crop.

In winter pigeon pea grown at Navsari (Gujarat), responsed to irrigation up to 0.3 IW/CPE ratio with respect to seed yield and 0.6 IW/CPE ratio with respect to stalk yield. This indicates that plants under no irrigation suffered from moisture stress while at 0.6 IW/CPE ratio the tendency was more towards vegetative growth. The consumptive use of water increased from 89.0 mm under control to 323.5 mm under 0.6 IW/CPE. The consumptive use was more under shallow tillage (212.5 mm) than deep tillage (201 .0 mm). On the contrary, at Khandha (Gujarat), winter pigeon pea recorded significant increase in seed yield with the increase in irrigation level from 0.5 IW/CPE ratio (1.87 t/ha) to 0.9 IW/CPE ratio (2.44 t/ha) and the water requirement also increased from 320 mm to 480 mm (Patel and Patel 2000).

Green gram

Green gram takes 2.5 to 3 months to mature and can be grown as summer and *Kharif* season crop in northern India and also as winter season crop in southern India. Rainy season green gram hardly requires irrigation, except under prolonged drought but *rabi* and summer green gram require adequate irrigation. Soil moisture content of 50 per cent was considered optimum for the maximum production of dry weight, protein content and number of nodules (Varma and Rao 1975). Summer green gram responded favorably to 2 irrigations at Delhi and Kanpur (Uttar Pradesh) and 3 irrigations at Hisar (Haryana) and Sriganganagar (Agarwal *et al.* 1997).

When irrigations were scheduled on IW/CPE ratio concept, differential irrigation regimes did not influence the grain yield of *kharif* crops at Dharwad (Karnataka), because of 14.9 cm rainfall received during the crop season 1997(AICRP WM, 1997). One irrigation of 6 cm depth, besides a common pre-sowing irrigation of 8 cm depth, scheduled at IW/CPE ratio of 0.4 was sufficient to produce optimum grain of 1.44 t/ha, being at par with other IW/CPE ratios. Similarly, at Hisar (Haryana) and Delhi, summer green gram could be irrigated either at IW/CPE of 0.4 or at 300 mm CPE for higher yields. The application of irrigation at 300 mm CPE resulted in 75 mm saving of irrigation water and 41 per cent higher yield than frequent watering of crop at 200 mm CPE which can mainly be attributed to better physiological functions, flowering, and pod setting. This study suggested that mild moisture stress is essential to stimulate flowering, better pod formation and higher harvest index of green gram (Pannu and Singh 1993). At Sriganganagar (Rajasthan), irrigation schedules based on critical growth stages were worked out, indicating there in the requirement of 3 irrigations for summer crop and 2 irrigations for *kharif* crop (Yadav *et al.* 2000). Mohanty and Sharma (1985) recommended one irrigation at the pre-flowering stage (30 days) or 2 irrigations at pre-flowering and grain-filling stages when water is a limiting factor. At Navsari, (Gujarat), the highest grain yield of summer green gram cv. 'K 851' was recorded with irrigation at 0.75 IW/CPE ratio, however, it was at par with 0.60 IW/CPE ratio. The consumptive use of 304 mm was recorded with 6 irrigations scheduled at 0.75 IW/CPE ratio while the maximum WUE of 3.69 kg/ha mm^{-1} was recorded with 0.45 IW/CPE ratio (Dabhi *et al.* 1998). At New Delhi, the consumptive use of 238 mm under black polythene mulch and 312 mm in unmulched green gram was noted under summer crop. Of this water depletion, 40-50 per cent was from 0-30 cm top soil layer and 60-70 per cent of the total depletion was during seedling to flowering stage (Gupta *et al.* 1998).

Bhatia and Chowdhury (1977) reported that summer green gram (cv. 'Pusa Baisakhi' 'and 'Type 1') required 2 irrigations (140 mm of irrigation water), applied at 50 per cent soil moisture depletion from its maximum root zone depth, in addition to pre-sowing irrigation, for obtaining high yields on sandy loam soil at Delhi. At Jammu, 5 irrigations scheduled at an interval of 15 days to summer mung bean outyielded 4 and 3 irrigations. This was due to the favourable effect of 5 irrigations on growth and yield attributes. Consumptive use of water increased with increase in number of irrigations, being the maximum (35.06 cm) in 5 irrigations. The Water use efficiency (WUE) was, however, maximum with 3 irrigations (3.06 kg/ha mm^{-1}) (Soni and Gupta 1999). Pal and Jana (1991) observed the highest grain yield of green gram (2122 kg/ha) with 2 irrigations in combination with 60 kg P_2O_5 ha, which was at par with 2 irrigations integrated with inoculation (2168 kg/ha).

Black gram

Black gram, mainly grown as rainfed crop, responds to irrigation in *kharif* and summer seasons in the absence of sufficient rains. At Bhawanisagar, in sandy loam soils, the optimum grain yield of 1.08 t/ha of *kharif* crop was obtained under 8 irrigations, each of 4 cm depth scheduled at IW/CPE of 0.6. But in similar soil type at Chalakudy, the crop required 5 irrigations each of 3 cm depth, scheduled at IW/CPE of 0.75 for optimum yield. A common pre-sowing irrigation was, however, required for proper germination at both the locations.

Kidney bean

Kidney bean is grown in *kharif*, but development of new cultivars has made its cultivation possible during *rabi* season also, when there is less incidence of pests and diseases and virtually no adverse effect of high rainfall. Kidney bean is a shallow-rooted crop and sensitive to over-supply of water. Good crop can be produced with very little water applied during the season. Light irrigations are advisable, since flooding not only results in wasteful use of water but may also cause excessive wine growth and delayed maturity. At Pusa, in sandy loam soil, two irrigations, each of 6 cm depth, at IW/CPE of 0.8, besides a pre-sowing irrigation of 6 cm depth, produced optimum grain yield of 1.19 t/ha. Similar trend was observed at Faizabad in silt loam soils where the crop required three irrigations, each of 5 cm depth, at IW/CPE of 0.7 to produce optimum grain yield of 1.52 t/ha^{-1}.

Common bean (Rajmash)

Dhar (1998) recorded the highest grain yield with 0.6 IW/CPE ratio (2 irrigations) at Pantnagar. However, the highest WUE (74.2 kg/ha/cm^{-1}) was recorded with 0.4 IW/CPE ratio. Four irrigations at 25, 50, 75 and 100 days after sowing produced higher straw yield (3858 kg/ha) over other treatments but grain yield was significantly lower compared to 2 irrigations. The grain yield of winter season crop of common bean grown at Pusa (Bihar) increased with increment in irrigation levels up to 2 irrigations. Three irrigations each of 6 cm depth, either scheduled at IW/CPE of 0.8 besides a pre-sowing irrigation of 6 cm or the same number of irrigations scheduled at 25, 50 and 75 days after sowing (DAS) produced the grain yield of 1.29 t ha^{-1} at Pusa in Bihar (Bhatia 2000). Similar trend was observed at Faizabad in silty loam soil, where the crop required 3 irrigations each of 5 cm depth, scheduled at IW/CPE ratio of 0.7 to produce grain yield of 1.52 t/ha. At Rahuri (Maharashtra), common bean produced the maximum grain yield with 2 irrigations, at branching and at flowering during rainy season. However, it was on par with irrigation scheduled at flowering and 75 mm CPE. The consumptive use of water increased with frequency of irrigation, whereas WUE was maximum with one irrigation at flowering. Combination of 120 kg N/ha and scheduling of irrigation at 75 mm CPE resulted in the highest grain yield (Wani *et al.* 1998). At Powarkhera, on the basis of three years experimentation on Rajmash, irrigating the crop with four irrigation starting at a interval of 20 days along with 120 kg N/ha proved the most effective treatment for highest grain yield (8.75 q/ha) and net returns (Rs. 34129. ha^{-1}) in Tawa command. It is recommended that rajmash should be irrigated with four irrigations at 20, 40, 60 and 80 days after sowing along with 120 kg N/ha. At Jammu, 5 cm depth of irrigation at every 15 days interval exhibited highest yield of 9.45 t/ha during 2009(Annual report, 2010, AICRPWM). Inter-row mulching of paddy straw @ 3 t/ha at 20 days after sowing should be practiced to maximize the water use. Five irrigations *viz.,* 3 cm within one week, 5 cm each after 3rd and 5th week and 6 cm each after 7th and 10th week after sowing were adequate to meet the water requirement of the crop (Annual report, 2010, AICRPWM).

Sprinkler irrigation for successful production of pulses

Sprinkler system of irrigation has tremendous scope for irrigating the pulses in different agro-ecological regions. Gravity flow system of irrigation is the common method of irrigation in all the command areas of irrigation project in India. Where supplies are planned for dry crops like millets and pulses, it would be economically rewarding if sprinkler irrigation can be introduced in these areas. The area available under this category is estimated to be about 10 m ha based on the assumption that at least 50 % of the 20 m ha under CAD schemes (Anonymous, 1998) can be brought under sprinkler irrigation through irrigated low duty crops like pulses. This would also substantially contribute to higher levels of pulse production on sustainable basis.

The research work on the use of sprinkler irrigation system has been mainly confined to the area of design of layout, operational uniformity, savings in water for different crops. Most of the studies are restricted to cereals and other crops and unfortunately pulses are neglected lot as for as area under sprinkler is concerned and if some studies have been done, they are restricted to only one or two crops like gram and green gram. Studies conducted at Hisar in gram showed that growing gram under sprinkler irrigation system resulted into 51 percent higher yield and 56 percent saving in water compared to surface irrigation (Mane and Ayare, 2007). Similarly, gram irrigated with sprinkler system at 0.75 IW/CPE resulted into 31 percent increase in yield and about 11 percent saving in water compared to surface irrigation at S.K.Nagar, whereas corresponding increase in yield was 57 percent and 69 percent saving in water at Vyara in Gujarat (Anonymous, 1998). Further, irrigating summer cowpea under sprinkler irrigation at 1.0 IW/CPE resulted into 3 percent increase in yield but there was a saving of 19 percent water compared to surface irrigation at Navsari (Anonymous, 1998). Growing of green gram under sprinkler system resulted into 69 percent increase in yield at Akola (Anonymous, 1998). Researches conducted on sprinkler irrigation system in pulses at different centers of AICRP on Water Management (AICRP WM) have clearly indicated an increase in yield (15-37%) and savings in water (10-51%). At Pantnagar, sprinkler irrigation system was found better in pea and resulted into 4 percent increase in yield and 55 percent higher water use efficiency. Based on the finding, it was found that under *tarai* region, where soils are associated with shallow to deep water table conditions; it was advisable to follow rice- vegetable pea crop rotation for higher profit. Soil remained wet after the harvest of rice crop and pea can be sown from mid of October to mid November and two light irrigations should be recommended by following the sprinkler method of irrigation (Annual report, AICRPWM, 2010). At Sriganganagar, sprinkler irrigation in gram at IW/CPE 0.6 gave significantly higher seed yield (2.11 t/ha) with water use efficiency of 8.62 kg/ha/mm while the lowest WUE of 6.40 kg/ha mm was observed with flood irrigation. Based on the finding, it was recommended that two irrigations, each of 60 mm depth, through sprinkler at 60 & 110 days after sowing should be recommended to obtain higher yield of gram and result into saving of 26 percent irrigation water (Annual report, AICRPWM, 2010). There is ample scope to introduce sprinkler irrigation system in important pulse crops like gram, pigeon pea, lentil, beans etc. Efforts have been made under National Pulse Development Programme (NPDP) to distribute sprinkler sets along with other critical inputs so that the farmers can be made aware of the technology.

Strategic approach for efficient water management in pulses

Pulses are predominately grown under rainfed conditions, however, these crops respond to irrigation well. Judicious use of water can make a significant impact in improving the productivity of pulses to a great extent and strategic planning is required for boosting the productivity of pulses for enhancing the per capita availability of pulses among the masses. Following points should be take into consideration while framing long term strategy for boosting pulse productivity in the country.

i) **Introduction of pulses in irrigated tracts :** Concerted efforts should be made to increase the area under pulses in irrigated commands by introducing them in the cropping systems either as sole crop or as intercrop. For instance, introduction of summer pigeon pea + green gram-wheat sequence in irrigated tracts of Uttar Pradesh, Punjab, Haryana, and Rajasthan; green gram in the spring and summer fallow in the northern India and *rabi* sowing in southern India. Intercropping and growing short duration varieties between *kharif* and *rabi* season, by relay cropping and intercropping, will ensure further utilization of existing agricultural land. Efforts should be made to diversify some of the land with pigeon pea based cropping system in Indo-Gangetic Plains.

ii) **Introduction of pulses in pressurized irrigation systems** : There is a general perception that pressurized irrigation system like sprinkler is most suitable for wheat, bajra, sorghum, vegetables and micro-irrigation system is suitable for wide spaced crops like sugarcane, cotton, vegetables and fruits. Recent experiments have clearly shown the beneficial effects of sprinkler irrigation on pulses like gram, pea etc and micro irrigation in inter-cropping of pulses like green gram, black gram and rajmash in wide spaced crops like cotton, sugarcane and orchards. There is good scope of introducing pulses in cropping system where pressurized irrigation system is more prevalent. Research efforts are being undertaken to standardize the scheduling of pigeon pea under drip irrigation system so that it can be introduced in the cropping system which are being irrigated with drip system. Therefore, efforts should be initiated for introducing these crops in pressurized irrigation system.

iii) Intercropping of pulses in wide by spaced crops : Concrete efforts should be initiated for introducing inter-crop of pulses in wide spaced crops like cotton, sugarcane and other horticultural crops. For instance, autumn sugarcane intercropped with *rajmash* has proved profitable as well as sustainable due to additive nutrient synergies in the system and positive associative effects of *rajmash* on sugarcane yield attributes and yield. Moreover, intercropped *rajmash* in autumn sugarcane improved soil fertility through entire leaf fall at pod maturity besides root mass addition. Similarly, intercropping of green gram in cotton under drip irrigation in Parbhani has proved quite effective in improving profitability and sustainability. Introduction of *kharif* pulses like green gram, black gram, pigeon pea, moth, and rabi pulses like gram, pea, lentil in orchards can not only improve soil fertility but also help in getting higher yield. With the increase in area in horticultural crops as envisaged with the start of National Horticulture Mission, there is ample scope for introduction of pulses in this system. Thus, these intercropping systems hold great promise in increasing the land utilization efficiency, reducing the production cost, economizing the use of market purchased costly inputs and making production system sustainable which would in turn ensure household economic and nutritional security. Sincere initiation should be undertaken for increasing area under pulses as intercrops in wide by spaced crops.

iv) Introduction of pulses in Rice-fallow : Ardent efforts should be made in utilizing the residual moisture in paddy fallow for raising pulses. India has the potential to increase acreage by encouraging production of pulses in rice fallows. A substantial part of rice fallow land can be targeted for cultivation of pulses during the *rabi* season. Replacement of upland paddy with pulses is another viable option which has the potential to give better net returns to farmers. Growing pulses like broad bean, green gram, black gram, especially where soils are light texture, should be encouraged. Similarly, cultivation of pulses as low duty drop in tank and well achkat areas/commands should be encourgaed for enhancing the productivity of pulses. Similarly, introduction of rajmash in Hirakund command during post-rice season should be encouraged for increasing the area and productivity. Further, sincere efforts should be made to introduce *rabi* pigeon pea in Andhra Pradesh, Chhattisgarh, Orissa, Tamil Nadu, Karnataka and West Bengal; *utera* cropping of urd bean and rice bean in rice fallows of Orissa, Assam, West Bengal, Maharashtra and Andhra Pradesh. Similarly, introduction of pulses particularly pigeon pea on bunds and dykes should be popularized in traditional rice growing belts.

v) **Providing supplemental irrigation in high rainfall areas :** Efforts should be made for harvesting rain water in different agro-ecological regions receiving high rainfall (more than 1000 mm) which can be efficiently utilized for supplemental irrigation during post rainy season for enhanced productivity of *rabi* pulses like gram, field pea, rajmash etc. Similarly, run-off harvesting should be encouraged in different regions for providing supplemental irrigation for low duty crops of pulses.

vi) **Run-off harvesting in semi-arid regions**: Runoff harvesting should be done through *nadi/tanka/khadin* in Rajasthan, *bhandaras* in Maharashtra, *bandhis* in Madhya Pradesh and Uttar Pradesh and *ahars* in Bihar. Similarly, anicut, checkdam and percolation tanks constructed across the ephemeral stream to intercept seasonal rainfall are most suitable for Deccan Plateau areas in south-eastern Rajasthan, Gujarat, Maharashtra, Madhya Pradesh and Gujarat. During *rabi* season, pulse based intercropping systems like wheat + chickpea, wheat + lentil, mustard + chickpea, mustard + lentil are found most successful in these regions.

vii) **Raised-sunken bed technology for high rainfall areas**: The technology of raised and sunken beds should be popularized in high rainfall vertisols in Central plateau for irrigating pulses in MP, Maharashtra, Chhattisgarh, and Orissa. The land modification in terms of raised sunken system should be initiated in soybean, pigeonpea, blackgram-rice during *kharif* season and chickpea-wheat during *rabi* season for higher yield and net return.

viii) ***in-situ* moisture conservation in arid and semi-arid regions:** Sincere attempts should be made to demonstrate in-situ harvesting of rain water to the farmers by simple technologies like summer ploughing, leveling of fields and making of bunds in arid and semi-arid regions. Similarly, efforts should be made to demonstrate how the conserved moisture can efficiently be utilized for cultivation of *rabi* pulses like gram, lentil etc.

ix) **Demonstration of efficient technologies for in-situ harvesting of water**: Awareness should be made by demonstrating the concept of conservation furrows in medium rainfall on light to medium soils in Andhra Pradesh and Karnataka region, Haveli system in Malwa plateau in Madhya Pradesh and Maharashtra for harvesting in-situ harvesting of precious water and utilizing for cultivation of different pluses (castor + pigeonpea; gram, soybean etc) in these regions.

x) **Demonstration of Broad bed furrow (BBF) system:** Broad bed furrow system of planting should be demonstrated for cultivation of maize + pigeon pea, sorghum + pigeon pea and soybean + pigeon pea during *kharif* season and maize + chickpea, sorghum + chickpea during *rabi* in vertisols of semi-arid regions.

xi) **Capacity building and human resource development for higher water use efficiency**: Efforts should be made to train the extension agencies and field functionaries at various levels to improve water use efficiency with higher productivity of pulses per drop of water. They should be given sufficient training and orientation to perform better consistent with the prescribed objectives.

Future concerns

For improving the WUE in pulse crops, biological, environmental and management factors and their interactive effects may be manipulated/ considered simultaneously rather than in isolation. The problems of water management in pulse crops are highly location and season-specific. Therefore, there is a need to develop and demonstrate location- and season-specific practices of improved on-farm water management on farmers' fields of different agro-climatic zones of the country. If India desires to achieve its true potential of producing over 37 million tons of pulses by 2020, a focused and integrated approach is needed to address these issues. Some important action points for higher productivity of pulses are :

- There is immense need to exploit genetic engineering as well as biotechnology to develop plant types which exhibit good response to higher doses of inputs so that these biotypes can be adjusted in irrigated tracts for higher productivity and compete well with the cereals.
- There is a urgent need to introduce pulses in irrigated tract for higher productivity. Efforts should be taken to diversify some of the land the pigeonpea based cropping systems in the Indo-Gangetic Plains.
- There is a vast scope for introduction of sprinkler irrigation for higher productivity of pulses. Irrigation requirements for pulses are much lower than other crops and could be provided through sprinklers efficiently.
- Appropriate strategies and planning should be framed for increasing area under micro irrigation system for enhanced water use efficiency on the one hand and higher productivity on the other hand. Incentives should be given to promote pulses in newly planted orchards under National Horticulture Mission.
- All round efforts should be made for utilizing the residual moisture in paddy fallow for raising pulses. A substantial part of rice fallow land in eastern India (about 12 million hectares) can be targeted for cultivation of pulses during the *rabi* season.
- Replacement of upland paddy with pulses is another viable option which has the potential to give better net returns to farmers. Similarly, cultivation of

pulses as low duty crops in tank and well commando should be encouraged for enhancing the production of pulses.

- Watershed management approach has shown great promise of increasing crop yield along with resource conservation. These approaches need upscaling to cover more and more areas to boost the productivity of pulses.
- The water management technologies consisting of optimum irrigation schedules, appropriate method of irrigation and judicious use of other production inputs have to be demonstrated through 'On-farm demonstrations' on farmers fields.

References

Anonymous, 2009. The Hindu Survey of Indian Agriculture, Chennai, India.

Anonymous, 1998. Sprinkler irrigation in India. Indian National Committee on Irrigation and Drainage (INCID), New Delhi.

Agarwal M.C., Dhindwal, A.S., Jaiswal, C.S., Prabhakar, A. and Aujla M.S 1997. Status of research on agricultural water management of northern region. Annual Report of Directorate of Water Management Research (ICAR), Patna. pp. 23-25.

AICRPWM (All India Coordinated Research Project on Water Management). Annual Reports for the year 1991-92 to 2009-2010.

Ali M. and Mishra J.P. 2000. Nutrient management in pulses and pulse based cropping systems. Fertilizer News, 45(4) : 61-69.

Ali, M. 1990. Consolidated report on Kharif/Rabi/Summer Pulses Agronomy. All India Coordinated Pulse Improvement Project, Directorate of Pulses Research. DPR, Kanpur, pp. 43, 179 and 183.

Bhatia P.C and Chowdhury, S.L. 1977. Annual Report of All India Coordinated Pulse Improvement Project, IARI, New Delhi, pp. 10-11.

Bhatia, R.K. 2000. Irrigation schedules for pulses in India. In Extended Summaries of International Conference on Managing Natural Resources, IARI. February 14-18, 2000, New Delhi, pp. 508-509.

Bihari, B., Kushwaha, H. S. and Vaidya, M.S. 1993 Response of chickpea (Cicer arietinum) to irrigation and fertilization. Indian Journal of Agronomy 37 : 110 111.

Dabhi, B.M., Patel, J.C., Patel H.C. and Raj V.C. 1998. Scheduling of irrigation in summer greengram (Phaseofus radiatus L.) under Vertisols of south Gujarat. In Extended Summaries of 1st International Agronomy Congress on Agronomy, Environment and Food Security, November 23-27, 1998, Indian Society of Agronomy, New Delhi, pp. 142.

Dhar, S. 1998. Effect of irrigation schedules on moisture extraction pattern, water use efficiency and productivity of rajmash (Phaseolus vulgaris L.). In Extended Summaries of 1st International Agronomy Congress on Agronomy, Environment and Food Security, Indian Society of Agronomy, New Delhi. November 23-27, pp. 270-271.

Gupta, R.K. and Agarwal, G.G. 1977. Consumptive use of water by gram and linseed. Indian Journal of Agricultural Sciences 47 : 22-26.

Gupta, V.K., Garg, R.N. and Singh G. 1998. Soil moisture distribution in mung bean and mustard crop as affected by mulches and ridge sowing. In Extended Summaries of 1st International Agronomy Congress on Agronomy, Environment and Food Security, Indian Society of Agronomy, New Delhi. November 23-27, 1998, p. 224. :

ICRISAT (International Crops Research institute for the Semi-Arid Tropics) 1981. Annual Report 1979-80 of ICRISAT, Patancheru: ICRISAT, pp. 81-82.

ICRISAT (International Crops Research Institute for the Semi-Arid Tropics) 1983. Annual Reports for 1978-83 of ICRISAT, ICRISAT, Patancheru. pp. 34-36.

Kasturikrishna, S, and Ahiawat, I.P.S. 1999. Growth and yield response of pea (Pisum sativum) to moisture stress, phosphorus, sulphur and zinc fertilizers. Indian Journal of Agronomy 44 : 588-596.

Mane, M.S and Ayare, B.L. 2007. Principles of sprinkler irrigation. Jain Brothers, New Delhi.

Mohanty, R.N. and Sharma, J.P. 1985. Effect of irrigation on growth and yield of summer greengram. Indian Journal of Agronomy 30 : 386-388.

Pal, A.K. and Jana, P.K. (1991). Effect of phosphorus, Rhizobium inoculation and irrigation on growth, yield and WUE of summer greenaram (Phaseolus radiatus). Indian Journal of Agronomy 36 : 536-540.

Pannu, R.K and Singh, D.P. 1993. Effect of irrigation on water use, water use efficiency, growth and yield of mung bean. Field Crops Research 31 : 87-100.

Patel, D.B and Patel C.L. 2000. Effect of irrigation and planting geometry on growth and yield of Winter pigeonpea (Cajanus cajan). In Extended Summaries of National Symposium on Agronomy : Challenges and Strategies for the New Millennium, Junagadh:GAU. November 15-18, 2000, P. 145.

Rana, K.S. and Malhotra O.P. 1989. Response of summer pigeonpea to irrigation and date of sowing, Indian Journal of Agronomy 34 : 249-250.

Rao., I., Madhusudana V.N., Venkataratnam, N., Faris, D.G. and Sheldrake, A.R. 1983. Response to irrigation in post-rainy season pigeonpea. International Pigeonpea Newsletter 2 : 35-36.

Rao S. N. 1997. Watershed development needs new look. The Hindu Business line dated 20.03.1997.

Saxena, MC. 1980. Recent advances in chickpea agronomy. In Proceedings of the International Workshop on Chickpea Improvement, Patancheru : ICRISAT. February 28 — March 2, 1979, pp. 96-98

Saxena, M.C. 1984. Agronomic studies on winter chickpeas. In Proceedings of the Workshop on Ascochyta Blight and Winter Sowing of Chickpeas (Eds M.C. Saxena and K.B. Singh), Aleppo: ICARDA. pp.123-137.

Saxena, M.C. and Yadav, D.S. 1976. Some agronomic considerations of pigeonpea and chickpeas. In Proceedings of International Workshop on Grain Legumes, Patancheru: ICRISAT. pp. 31.62.

Singh, G. and Bhushan, L.S. 1979. Water use, water use efficiency and yield of dry land chickpea as influenced by P-fertilization and stored soil moisture and crop season rainfall. Agricultural Water Management 2, 299-305.

Singh, K.P., Sharma, H.C. and Singh, T. 1978. Response of pea T-163 to irrigation. Indian Journal of Agronomy 23 : 98-101.

Singh, V.K. and Dixit, R.S. 1992. Effect of moisture regime and sowing date on chickpea. Indian Journal of Agronomy 37 : 739-743.

Swindale L.D. 1982. Distribution and use of arable soils in the semiarid tropics. In: Managing Soil Resources.Transactions of the 21 st International Congress of Soil Science. 8-16 Feb, 1982. IARI, New Delhi, India. pp 67-100

TMOP (Technology Mission on Oilseeds and Pulses) 2000. Transfer of technology. In Expert Committee Report on Pulses, Department of Agriculture and Cooperation, Ministry of Agriculture. New Delhi: pp. 47-50.

Trung, B.C., Yoshida, S. and Kobayashi, Y. 1985. Influence of excess soil moisture on the nitrogen nutrition and grain productivity of mung bean. Japanese Journal of Crop Science 54 : 79-83.

Varma, A.K. and Rao, N.S.S. 1975. Effect of different levels of soil moisture on growth, yield and some physiological aspects of nodulation in greengram. Indian Journal of Agricultural Sciences 45,11-16.

Venkateswarlu, U. and Ahlawat, I.P.S. 1993. Effect of soil moisture regime, seed rate and phosphorus fertilizers on growth, yield attributes and yield of late sown lentil (Lens culinaris). Indian Journal of Agronomy 38 : 236-243.

Wani, A.G., Tumbare, A.D. and Dhonde, M.B. 1998. Response of rainy season frenchbean (Phaseolus vulgaris) to irrigation regimes and nitrogen. Indian Journal of Agronomy 43 : 694-699.

Yadav, R.L., Singh, S.R., Prasad. K., Dwivedi, B.S., Batta, R.K., Singh, A.K., Patil, N.G. and Chaudhary, S.K. 2000. Management of irrigated agro-ecosystem. In Natural Resource Management for Agricultural Production in India (Eds J.S.P. Yadav and G.B. Singh), Indian Society of Soil Science. New Delhi : pp. 775-870.

□□□

Chapter - 10

Weed Management

Anil Dixit
Directorate of Weed Science Research (ICAR), Mahrajpur, Jabalpur, Madhya Pradesh.

Symbolic to its nomenclature, 'PULSE' (P = people, U = umbrella, L = livestock, S = soil and E = energy) is a superb group of agricultural crops extending an energy umbrella to people (as a major source of dietary protein), livestock (as a green nutritious fodder and feed) and soil (as a builder and restorer of soil fertility through biological N_2 fixation). On account of these qualities, pulses have occupied a unique position in the Indian farming system. Crop diversification with greater inclusion of legumes is essential to alleviate declining factor productivity, input use efficiency and increasing cost of cultivation for realizing sustainable use of natural resources. Pulses are increasingly becoming a component of the intensive cropping systems as intercrop, substitute crop or catch crop. In upland situations, pulses play an important role as intercrops and ensure more income and efficient use of resources. While analyzing the reasons of low productivity of pulses in various parts of the country, it is observed that non availability of high yielding disease resistant varieties, cultivation on marginal and sub-marginal lands under deficit moisture and fertility and poor pest management are the major constraints in realizing optimum yield potential.

An analysis of relative contribution of various inputs in pulse production was made under All India Coordinated Pulse Improvement Project. It was observed that in *kharif* pulses, weed management was the most important input followed by fertilizer and insect pests and disease control (Ali and Lal, 1989). Lack of adequate weed management practices caused 30-40% reductions in the yield of *kharif* pulses and 25% in case of *rabi* pulses. The challenge before Indian agriculture is to feed the second most populous nation in the world accounting for about 17% of world's population supported by only on 2.3% land area. The demand for food grains in the year 2020 is projected at around 300 mt comprising of 112 mt of rice, 82 mt of wheat, 39 mt of coarse cereals and 28 mt of pulses. The gap in demand at production levels of 2009-10 would be 15 mt in rice, 2 mt in wheat, 2 mt in coarse cereals, 12 mt in pulses and 4 mt in oilseeds (Table 1).

Table 1 : Gap in demand of food grains (Million tonnes)

Category	Demand by 2020	Production during 2008-09	Gap in demand
Rice	112	97	15
Wheat	95	93	2
Pulses	28	16	12
Oilseeds	35	31	4
Coarse cereals	39	37	2

Major pulses in India

Pulse crops viz., pigeon pea (*Cajanus cajan*), green gram (*Vigna radiata*), urdbean (*Vigna mungo*) and cowpea (*Vigna unguiculata*) are usually grown during rainy season. Chickpea (*Cicer arietinum*), field pea (*Pisum sativa*), lentil (*Lens culinaris*), lathyrus (*Lathyrus sativus*) and rajmash (*Phaseolus vulgaris*) are major winter season crops grown under rainfed and irrigated conditions. However, under assured irrigation, green gram, black gram and cowpea are also taken after harvest of rapeseed-mustard and potato during summer in northern India. Some of the pulses like lentil, lathyrus, green gram and black gram are also grown as *utera* in rice-fallows in parts of Madhya Pradesh, Chattisgarh and Andhra Pradesh.

Weed flora

Since pulses are grown under different agro-ecological situations, the nature and intensity of weed flora varies to a great extent. During rainy season, severe infestations of carpet weed (*Trianthema monogyna*) occur in semi arid tract of the country. *Commelina benghalensis* and *Digera arvensis* are of secondary importance. Among grassy weeds, *Echinochloa colona, Dactyloctenium aegyptium, Digitaria sanguinalis, Panicum maximum* invade the crops heavily during *kharif* season. The profuse and rank growth of *Cyperus rotundus* is most common in irrigated summer and rainy season pulses and offers *rhizospheric* competition through its chain of underground tubers. *Saccharum spontaneum* and *Sorghum halepense* interfere with the crop production through their bulk of under ground rhizomes. *Celosia argentea* is a major problem in rainfed condition in light textured sandy loam soils of the country. In winter pulses, the occurrence of *Chenopodium album, Sinbaria pinnatifoda, Fumaria parviflora* are found in plenty in irrigated conditions. *Asphodelus tenuifolius* emerges in different flushes and pose problems mostly in rainfed chickpea and lentil throughout northern India. This competes with the crop mainly for nutrients and moisture. *Carthamus oxycantha* has been reported as most troublesome weed in field pea as harvesting and threshing becomes problem due to its spiny nature in Jalaun area of Uttar Pradesh. *Convolvulus arvensis* poses problems by trailing with pea and lentil plants. *Phalaris minor* and *Avena ludoviciana* are important grassy weeds posing problems in winter pulses growing under irrigated conditions. *Cuscuta spp.*, is an important parasitic weed causing a lot of damage in mungbean and urdbean in coastal Andhra Pradesh, Tamil Nadu and parts of Madhya Pradesh in rice-fallows. The reports of its infestation are also available in other pulses like chickpea and lentil.

Most of the area under pulses is rainfed. Weeds compete with crop plants for nutrients, moisture, sunlight and space, and thus, reduce crop yields which are variable (Table 2) with crop and also associated with the nutrients and moisture provided to crop plants. In general, the loss in crop yield is more during rainy season.

Emergence of weeds and crops generally occures simultaneous by in the field, but weeds because of their better adaptation and survival mechanism under adverse conditions and fast growing habit, supersede the crop plants. They not only compete with crop plants for nutrients, moisture, sun light but space but also secrete *toxic allelochemicals* which affect the growth and development of crops adversely. Weeds indirectly reduce the yield potential by serving as alternate host to a number of crop pests. As a result of the severe crop-weed competition, the quantity and quality of the produce are reduced to a great extent. The degree of loss depends upon the nature and magnitude of weed infestation.

Emergence of weeds in pulses begins simultaneously with the crops leading to severe competition between weeds and the crop right from the very early stage. Keeping the field free of weeds throughout the cropping season no doubt gives higher crop yield but it may not be economical. Removal of weeds at any time during growing season may not be beneficial. Stage of weed removal is as important as removal per se. Therefore, it is necessary to identify the critical period of crop-weed competition to render the weed control practices more effective and economical. This provides information on the active duration when the presence of weeds makes their effect deleterious on crops. Critical period varies in different pulse crops (Table 2). Weeds that emerge beyond competitive period do not affect the productivity but these weeds should be removed before settings of seeds to avoid further weed multiplication.

Table 2 : Critical period of crop-weed competition and losses due to weeds in different pulse crops

Crops	Critical period (Days after sowing)	Reduction in seed yield (%)
Pigeon pea	15-60	20-40
Mungbean	15-30	25-50
Urdbean	15-30	30-50
Cowpea	15-45	15-30
Chickpea	30-60	15-25
Peas	30-45	20-30
Lentil	30-60	20-30
French bean	30-60	15-30
Fababean	30-60	40-75

Weed Management Practices

Weed management in pulses requires an integrated approach that utilizes effective cultural, mechanical, biological, ecological and chemical methods in a mutually supportive manner into the crop production system with due consideration of economic, environmental and sociological consequences. Various methods of weed management such as preventive, cultural, mechanical, chemical etc. have been used with different degree of successes in different agro-ecological zones and productions systems, which are as follows.

Prevention

It is one of the potential methods to reduce weed infestation in fields. Use of clean seeds coupled with avoiding use of soil and sand from weed infested area in clean areas. Use of well-rotten and decomposed organic manure, restricting cattle movement from weed infested area to clean area, cleaning irrigation and drainage channels, fences, roads and bunds are some of the preventive measures. Soil solarization is another method which destroys the weeds during fallow summer season.

Cultural

Despite the great progress made in agriculture, manual and mechanical methods continue to be important weed management practices in many regions of the world. Cultural methods are used to complement manual and mechanical methods. Cultural practices (Table.3) have pronounced effect on crop-weed interference. They are manipulated in such a way that they become more favorable for crop growth and less to weeds. They are not only eco-friendly but also reduce the use of costly herbicides.

Mechanical

Mechanical weed control involves removal of weeds with various tools and implements including manual removal by hand weeding. Inter-cultural operations are performed primarily to destroy the weeds present in the field and create favorable soil conditions for crop growth. Hand hoeing and manual weeding are the most common practices performed for weed control in almost all pulses. One or two hand weedings at critical crop-weed stage provide satisfactory control of weeds in most of these crops. However, hand weeding twice did not control the perennial weeds *viz.*, *Cyperus rotundus* and *Cynodon dactylon* due to their re-emergence. Climate and soil type play an important role in the possibilities for mechanical weed control. Monitoring the early development of weeds is necessary for timing of weed harrowing at the optimum stage. The success of mechanical weeding depends upon the stage of weeds, crop geometry and climatic conditions. Hand weeding may also be used after mechanical inter row weeding to deal with weeds left in crop rows. Repeated intercultural operations under those conditions will not only discourage germination of weed seeds (which normally occurs in the soil layer) but also aid in conserving precious soil moisture. Any practice that provides vigorous uniform crop establishment will usually assist in reducing weed prevalence.

Table 3 : Classification of cultural practices potentially applicable in an integrated weed management system based on their prevailing effect (modified from Barberi, 2001)

Cultural practice	Impact	Example
Crop rotation	Reduction in weed emergence, Improving soil fertility	Alternation between winter and spring-summer crops
Primary tillage	Reduction in weed emergence, Farm hygiene	Deep ploughing, alternation between ploughing and reduced tillage
Seedbed preparation	Reduction in weed emergence, Establishing good crop stand	False/stale seed-bed technique
Cultivation	Reduction in weed emergence	Post-emergence harrowing or hoeing, ridging
Cover crops	Reduction in weed emergence	Cover crops grown in-between two cash crops and used as green manure or dead mulch
Intercropping	Reduction in weed emergence, Improvement in crop competitive ability, Cover risk, Improve soil fertility	Cover crop used as living mulch, intercropped cash crops
Thermal treatment	Reduction in weed emergence, Pest control	Pre-emergence or localized post-emergence flam weeding
Mulching/soil solarization	Reduction in weed emergence	Use of black or transparent films (in glass house or field)
Crop genotype	Improvement in crop competitive ability	Use of cultivars characterized by quick emergence, high growth and soil cover rates in early stage
Sowing/planting	Improvement in crop competitive ability	Use of transplants, higher seeding rates, lower inter-row distance, anticipation of, or delay in sowing,/transplant date
Fertilization	Reduction in weed emergence, improvement in crop competitive ability	Use of slow nutrient releasing organic fertilizers and amendments, fertilizer placement
Irrigation	Reduction in weed emergence, improvement in crop competitive ability	Irrigation placement (micro/trickle-irrigation)

Use of intercrops as tool for weed management

Weeds constitute one of the major constraints to increase agricultural production in many areas practicing intercropping. Finding an intercrop which will suppress the growth of the weeds but not the crop has been difficult. Weed suppressing ability of intercrop is dependent upon such factors as the component crops and cultivars selected, crop density, relative proportion of the component crops, their spatial arrangement and the fertility and moisture status of the soil (Moody and Shetty, 1981). Suppression of weeds in intercropping may have a beneficial effect on the subsequent crop as well.

Moody (1977) demonstrated that the presence of cowpea resulted in fewer weed problems in subsequent crops.

Very few herbicides are recommended and used to control the weeds in intercropping system since it has been difficult to find compounds that will control a broad spectrum of weeds without causing damage to the component crops. Weed management includes weed suppression through intercrops, use of herbicides and weeding tools in pulse based intercropping systems. Considering the diversity of weeds problem, no single method of weed control could provide desired level of efficiency under all situations. Herbicides should be used as supplement at as low rates as possible. The integration of chemicals with mechanical and cultural practices has given a substantial base in improving the weed control efficiency. The combinations of cultural and chemical methods have been found effective in reducing the weed infestation.

In chickpea, infestation of *Chenopodium album*, the major weed, can be minimized by delaying the planting (Table 4). Among the herbicides, application of *pendimethalin* at 1.5 kg a.i. /ha was found most effective. Under late planting conditions in chickpea and lentil, delayed first weeding (45-60 days after sowing) was more effective in comparison to weeding done at 30 days. Increasing the seed rate by 25% is also helpful in suppressing the weeds in peas (Singh *et al.*, 2001).

Intercropping systems have also been found to suppress weeds through formation of good canopies due to competitive planting pattern and have thus provided an opportunity to utilize inter crops as a tool of weed management (Rao and Shetty, 1976). The suppression of weed growth in intercropping system is mainly due to increased leaf area index and light interception. Importance of short duration and quick growing intercrops with wide by spaced long duration and tall growing crops towards weed suppression resulting in saving of weeding requirement in base crops, have been well recognized and documented.

Table 4 : Total weed dry weight and grain yield of chickpea as affected by planting dates and weed control measures

Weed control treatments	Dose (kg ai/ha)	Weed dry weight (g/m^2)			Grain yield (kg/ha)			
		Nov. 5	Nov. 20	Dec. 5	Nov. 5	Nov. 20	Dec. 5	Mean
Fluchloralin	1.0	54.1	41.3	35.9	1192	1308	1055	1185
Fluchloralin	1.5	36.7	31.7	21.6	1387	1507	1189	1361
Pendimethalin	1.0	42.0	34.8	30.5	1247	1293	1031	1190
Pendimethalin	1.5	26.8	23.6	15.3	1438	1577	1327	1447
Metribuzin	0.25	50.9	41.3	35.6	1048	1132	980	1053
Metribuzin	0.50	32.1	28.4	23.2	1256	1310	1076	1211
Weed free	-	0		0	1590	1729	1354	1558
Weedy	-	140.7	116.0	89.3	481	631	427	513
Mean	-				1205	1311	1055	

C.D. (5%) for weed control : 44 C.D. (5%) for planting dates: 81

Pigeon pea is slow growing, wide by spaced and usually intercropped with short duration crops like sorghum, green gram, black gram, cowpea, sesame etc. under rainfed situations. Sorghum + pigeon pea intercropping system is extensively practiced in Karnataka, Maharashtra and Andhra Pradesh and is known to reduce weed intensity apart from efficient utilization of land space. Intercropping of soybean, black gram (Ahlawat *et al*., 1982,) and cowpea (Shetty and Rao 1981; Venkateshwarlu and Ahlawat 1986) with pigeon pea have been found to reduce density and dry matter accumulations of weeds to the extent of 45-50 per cent. Paired row planting of pigeon pea with three rows of soybean (2:3) and pigeon pea + soybean (1:2) recorded significantly higher harvest index and less weed biomass/m^2 than the sole pigeon pea and soybean planting. Similarly, intercropping of green gram, pearl millet and sesame with rainfed pigeon pea led to weed suppression to the extent of 36.9, 50.8 and 36.6 per cent, respectively (Table-5).

Table 5 : Weed smothering efficiency of different intercrops.

Intercropping systems	Weed dry weight (kg/ha)	Weed smothering efficiency (%)
Pigeon pea sole	2672	-
Pigeon pea + green gram(1:2)	1684	36.9
Pigeon pea + pearl millet (1:2)	1314	50.8
Pigeon pea +sesame (1:2)	1694	36.6

Source: Tiwari and Rathi (1997)

Use of herbicides

Weed control through manual/mechanical though very effective, has certain limitations such as non availability of labour during peak period, high labour cost, and unfavorable environment particularly in rainy season etc. Under such conditions, use of herbicides is advantageous and economical. In the use of herbicides for weed control particularly in pulses, some of the important considerations that have to be kept in mind are, they should not be harmful to the soil micro flora that build up soil fertility and the nodule-inducing organisms which fix atmospheric nitrogen, the harvested produce should not contain any residue that may prove toxic to man or animal and they should not leave any residues in soil which would be toxic to crops grown in rotation.

Based on the research work done at different AICRP-WC Centres and elsewhere in the country, some herbicides have been found selective, effective and economical in different pulse based cropping systems (Table 6).

Table 6 : Promising herbicides for different pulse crops.

Herbicide	Dose(kg ai/ha)	Crops	Weeds controlled
Pre-plant incorporation			
Fluchloralin	0.75-1.0	All pulses	BLW and Grasses
Trifluralin	0.75-1.0	All pulses	BLW and Grasses
Pre-emergence			
Alachlor	1.5-2.0	Rainy season pulses	Grasses
Linuron	1.0-1.5	Chickpea, peas, lentil	BLW and Grasses
Metolachlor	1.0	Pigeonpea, green gram, blackgram	Grasses
Metribuzin	0.50-0.75	Peas	BLW and Grasses
Oxyfluorfen	0.10-0.20	Blackgram, green gram	BLW and Grasses
Pendimethalin	1.0	All pulses	BLW and Grasses
Post-emergence			
Clodinafop- propargyl	0.060	Chickpea, peas, lentil, mustard	Grasses
Quizalofop ethylz	0.050	Chickpea, peas, lentil, mustard	Grasses
Imazethapyr	0.10	Blackgram, greengram and pigeon pea	BLW and Grasses
Metribuzin	0.3-0.5	Peas	BLW and Grasses

BLW-Broad leaved weeds

Integrated weed management

Integrated weed management is gaining importance due to its high sustainability index. This is an integration of effective, dependable and workable weed management practices, such as cultural, mechanical and biological that can be used economically by the producer as a part of sound farm management systems. This approach takes into account the need to increase agricultural production in totality, reduce economical losses, risk to human health and potential damage to flora and fauna, besides improving the safety and quality of environment. Integrated chemical + mechanical/manual weeding has been found very much effective in almost all pulses and pulse based cropping systems in controlling the weeds. Pendimethalin at 0.75-1.0 kg/ha + one hand weeding at 30-45 days after sowing is very effective in controlling weeds in chickpea, pigeon pea, green gram, black gram, pea and lentil crop. Inclusion of intercrops involving food and fodder crops, viz. green gram, black gram and cowpea (green fodder), in-between rows of pigeon pea reduces weed problem along with bonus yield of intercrop (Kumar and Singh 2010).

Control of parasitic weeds

Among parasitic weeds, *Cuscuta spp.* is a major problem in pulses (green gram, black gram, lentil, and chickpea) and oilseeds (niger and linseed) causing a lot of damage in rice-mungbean/ urdbean cropping system in coastal Andhra Pradesh, Tamil Nadu and parts of Madhya Pradesh. Its infestation has also been reported in other pulses *viz.* lentil, chickpea and pigeon pea. The parasitic weeds can be effectively managed through crop rotation by rotating host crop with trap crops, as they induce germination of parasitic weed seeds but they themselves are not parasitized. Due to lack of suitable host, *Cuscuta* seedlings will emerge and die. Application of *pendimethalin* at 1.0 kg/ha as pre-emergence herbicide has been found effective in reducing Cuscuta infestation in pulses.

Future outlook

The weed problem has, however, aggravated following the introduction and large scale cultivation of high yielding varieties and hybrids which are responsive to fertilizers and irrigation. Over time the techniques that have been followed for the management of weeds have changed substantially. Herbicides are increasingly being considered, as an efficient and economical alternative to manual and mechanical methods of weed management. They have a special place, when it comes to the management of weeds, which are morphologically similar to crop plants as well as those that are parasitic and perennial in nature. They are cost effective, labour saving and also address issues like human drudgery and gender equality favourable. Currently, in India the herbicide use is concentrated only in few crops (rice, wheat and tea) and in few regions, which is expected to intensify manifold in the near future. This is based on the belief, that labour would become far more expensive and would be in short supply in the coming years. Further; the increased extension activities would create greater awareness among farming community, of the advantages of herbicide use. In such a scenario it is pertinent to educate the farmers, extension personnel as well as the general public alike, regarding the safe use of pesticides in general, and herbicides in particular.

Herbicide use as per lebel is unlikerly to cause any negative implication on user on environment. However, in view of the potential negative impact of herbicides, it is reasonable to advocate an integrated strategy, involving all other management methods. Weed management through bio-control agents (insects pests, disease causing pathogens) and use of crops resistant to weeds (through allepathy), appear very novel and desirable. There is need of greater impetus on chemistry, biotechnological manipulations, and field management strategies. The understanding of allelopathy may facililtate understanding of biological control mechanisms and on the contrary it my facilitate formation of strategies for biological control to facilitate weed management. But, despite decades of intensive research, no substantial headway has been made in this area.

In all probability, therefore, herbicides are expected to continue to play a vital role in the overall scheme of things. More research is needed towards better integration of herbicides with other methods in pulses. Despite the best efforts, many of the developed countries have not been very successful in cutting down the consumption of herbicides for various reasons.

Herbicide tolerant technology should be another important element in the development of integrated ween management in pulses. Imparting herbicide resitance to normally herbicides susceptibable crops to produce herbicide resistant crops (HRCs) has been the most extensively exploited area of plant biotechnology. to date, herbicides have been tailored to be used with particular crops, rather than the crps being bred to tolerate to be used with particular crops, rather than the crops being bred to tolerate the herbicide.

As it has been seen time and again that technologies per se are not harmful but it certainly depends on how they are used. As Jonathan Gressel, an Israeli weed scientist puts it "the plough can be used constructively or destructively depending upon the direction it is used on the slope".

References

Ali, M. and Lal, S., 1989.Priority inputs in pulse production. Fertilizer News. 34 (11) : 17-21.

Ahlawat I.P.S., Singh, A and Mohta, N.K., 1982. Studies on weed control in solid and mixed stands of pigeon pea and soybean: Indian Journal of Agronomy 27 : 191-194

Barberi, P. 2001.Weed management in organic agriculture: Are we addressing the right issue? Weed Res. 42 : 177-193

Kumar N. and Singh K.K. 2010. Weed management in pulses: why, when and how. Indian Farming, July 2010. pp 9-12.

Moody, K., 1977. Weed control in multiple cropping. In proceedings of the symposium on cropping system research and development for the Asian Rice Farmers, IRRI, Philippines. Sept. 1996. pp281-294.

Moody and Shetty, S.V.R., 1981. Weed management in intercropping systems. In proceedings of International Workshop on Intercropping. ICRISAT,.Hyderabad. pp.229-237.

Rao, M.R. and Shetty, S.V.R., 1976. Some biological aspects of intercropping systems on crops weed balance: Indian Journal of Weed Science.8:32-43.

Singh, N.P., Sahu, J.P.Sharma B.B. and Singh, V.K. 2001. Integrated Weed Management in peas. Annual Report of Research on *Rabi* G.B.P.U.A. & T., Pantnagar. Pulses 2000-2001,

Tiwari, A.N. and Rathi, K.S., 1997. Integrated weed management for sustainable production of pigeon pea based cropping system In : Proc. Brighton crop protection Conference, Brighton, England, pp.167-172.

□□□

Chapter - 11

Integrated Pest Management

O.P. Sharma, O M Bombawale and Om Pal Singh
National Centre for Integrated Pest Management, PUSA campus, New Delhi - 110 012

In the past, focus on implementing plant protection measures was given to cereals, largely because of their central role in food security, however, recent shortages in legumes has warranted an all out attention on edible pulses. For more than four decades, the concept of integrated pest management (IPM) - which combines a range of controls, including conserving natural predators, using pest resistant crop varieties, intercropping and rotation, with judicious use of pesticides - has struggled to gain ground against conventional agricultural ideology and practices. But it is now increasingly challenging the methods of high-input agriculture and policies arising with implementation of General Agreement for Trade and Tariffs (GATT) & World Trade Organization (WTO) that cannot be ignored. In fact, a major factor responsible for sustaining soil productivity in this country has been the highly diversified nature of the cropping patterns which either include a pulse crop or a legume crop as one of the components. India is the largest producer of pulses in the world with 25% share in the global production. While chickpea is the topper among pulses occupying 39% of pulse area, pigeon pea is next with 21% area. The major pulse producing states are Maharashtra (20%), Madhya Pradesh (17%), Rajasthan (11%), Uttar Pradesh (11%) and Andhra Pradesh (11%) together accounting for 70% of the total production of 14.76 m.t from an area of 23.63 m ha (www.agricoop.nic.in, 2007-08). In general, the pulse production is not keeping pace with the domestic requirements, hence, the country has to import 1.5-2.8 m.t per year, which is a matter of concern. Keeping this in view, the IPM programme covers the two important pulse crops *viz.*, chickpea and pigeon pea, across different cropping systems which will take care of changing scenario emerging out of climatic changes as well as cropping patterns. The present strategy, which is based on a holistic approach and changing ecosystem, will help in increasing production through reduction in pest infestation that would eventually reduce the need for the import of these two vital commodities.

Pulse crops have certain unique features which together make them indispensable and difficult to replace. Firstly, pulse crops play an important role in the agricultural economy of India by virtue of their ability to fix atmospheric nitrogen in symbiotic association with *Rhizobium.* Secondly, pulse crops have deep penetrating root systems which enable them to utilize the limited available moisture more efficiently than many other crops. Consequently, the farmers have chosen to grow pulse crops under highly diversified conditions. In general, they are more drought-resistant than any cereal crop, though the degree of tolerance varies from crop to crop. Thirdly, the pulses have played a very important role in human diet in our country, as a source of protein. Because of their high protein content which varies from 20 to 30% and is almost three times that of cereals, it also invites a number of pre and post harvest pest and disease problems.

Factors limiting pulse production

The major cause for low production of pulses in India are: (i) ecological factors, (ii) lack of appropriate pulse production and protection technologies, (iii) poor post-harvest technology, (iv) less thrust on basic research, (v) inadequate supply of quality seed to farmers, and (vi) socio-economic constraints. Insect pest problems and viral diseases transmitted by them are major bottle necks in realising higher yields (Dubey and Sharma, 2002).

Pulses in general are susceptible to a large number of diseases and insect pests which cause heavy losses. Through there are several diseases which attack pulse crops, the major diseases are root/collar rots, vascular wilt, aerial blight and grey-mould in chickpea; vascular wilt, stem blight and sterility-mosaic in pigeonpea; yellow-mosaic, powdery mildew and *Cercospora* leaf-spot diseases in green gram, black gram and cowpea; and pea powdery-mildew in peas. For yield stability and wider adaptability of genotypes, it is important that varieties with multiple resistance to these major diseases are identified and incorporated in the breeding programmes, adopted, and popularized. Such multiple resistant varieties are needed more urgently among the pulse crops.

The pulse crops are attacked by a wide range of insect pests both in the fields at various stages of crop growth and storage. Among them those pests which attack the reproductive stages are of major economic importance, *viz.* the pod-borers. Most of the pests attack the crop at reproductive stage causing direct losses and generally the plants capacity to withstand the attack at this stage is limited. The development of varieties resistant to insect attack has not met with significant success and there is a long way to go before minimizing crop losses through the use of resistant varieties in the farmers' fields becomes a reality. Although a number of plant protection schedules, based on the use of insecticides and microbials have been recommended for minimizing the losses, they have not proved to be adequate. The management practices employed to contain these at this stage rely mostly on insecticides and/or a combination of different groups of insecticides. This has lead to the problems of residues, resurgence and resistance in insects. There is, therefore, an urgent need that an efficient integrated pest management module based on the judicious use of insecticides, biological pest control and appropriate cropping systems is developed for pulse crops for wider adoption in the country.

Extent of losses caused due to key pests

In pulse crops, it is particularly difficult to estimate losses caused by pests as these crops have good ability to compensate for pest damage even if it occurs in the podding stage. On an average 30-80% crop losses occur in pulses due to ravages of insect pests valued at Rs. 4000 - 5000 crores (Asthana *et al.*, 1997). Chickpea and pigeonpea can withstand defoliation up to 60 and 50% respectively without significant effects on grain yields. In case of pulses, pod borer, *Helicoverpa armigera* (Hubner) has been reported to cause yield losses varying from 4.2 to 39.7 per cent in 12 major chickpea growing states. The average loss due to insect pest damage in chickpea has been estimated at 29.2 per cent at the national level. In case of pigeonpea, losses often exceed 50 per cent, while in chickpea, avoidable losses were estimated at 34.7 and 28.7 per cent in Bihar and Andhra Pradesh, respectively (Krishnamurthy Rao and Murthy, 1983; Rao *et. al.* 1990). Estimated losses (%) due to specific pests in chickpea crop are as follows:

- Gram pod borer 10-90%
- Gram cutworm 5-30%
- Termites 5-15%
- Semilooper 0-10%
- Wilt/rootrot 20-25%
- Ascochyta blight 5-10%
- Botrytis 5-10%

Key pests and diseases

The keypests and diseases of pulse crops in different zones are highlighted in (Table 1).

Srivastava (1964) in a review on the insect pests of pulse crops reported 150 species of insects, with 25 of these being of major significance. A number of pests such as plume moth (*Exelastis atomosa)* and podfly (*Melanogromyza obtusa)* on pigeonpea; *Gujhia* weevil (*Tanymecus indicus)* and *Holotrichia insularis* on chickpea; horn caterpillar (*Agrius convolvuli)* and *Anarsia ephippias* on green gram and balckgram; *Spodoptera litura* and *Ophiomyia phaseoli* on peas have the status of key pests. Of late Lal *et al.* (1985) reported nearly 200 species on pigeonpea, 60 on mungbean, urdbean

Table -1 Key pests of chickpea, pigeonpea, mungbean and urdbean in India.

Zones	Insect Pests	Diseases
Chickpea		
North East plain zone	*Agrotis ipsilon, Autographa* spp and *Helicoverpa armigera*	*Fusarium oxysporum* f.sp. *ciceris, Rhizoctonia solani, R. bataticola, Ascochyta rabei* and *Botrytis cinerea*
North West plain zone	*Odontotermes, Helicoverpa armigera* and *A. nigrisigma*	*Ascochyta rabei* , *Botrytis cinerea* and *F. oxysporum* f.sp. *ciceris*
Central zone	*Odontotermes, Aphis craccivora* and *Helicoverpa armigera*	*F. oxysporum* f.sp. *ciceris, R. solani, R. bataticola*
South zone	*Helicoverpa armigera*	*F. oxysporum* f.sp. *ciceris, R. solani, R. bataticola*
Pigeon Pea		
North East plain zone	*Helicoverpa armigera, Maruca testulalis, M obtusa, Apion spp. Grapholita critica, Lampedes boeticus, Mylabris pustulata, Melanogromyza obtusa, Clavigralla gibbosa* and *Exelastis atomosa*	*Fusarium udum, Sterlity Mosaic, Phytopthora drechsleri var. cajani, Alternaria alternata, A tenuissima*
North West plain zone	*H. armigera, M. testulalis, G. critica, M. pustulata, L. boeticus, E. atomosa and C. gibbosa*	*F. udum, Alternaria alternata, and A tenuissima*
Central zone	*H. armigera, C. gibbosa, L. boeticus and G. critica*	*F. udum and Sterlity Mosaic*
South zone	*H. armigera, M. testulalis, M. pustulata, and Tanaostignoids sp.*	*Fusarium udum, and Sterlity Mosaic*
Mungbean and Urdbean		
North East plain zone	Mungbean : *Empoasca kerri, Bemesia tabaci, Ophiyamyia phaseolin, Madurasia testulalis* Urdbean : *Diacrisia obliqua, Amrasca moorie, B. tabaci*	Mungbean Yellow Mosaic Virus*Cercospora canacensMacrophomina phaseolina*
North West plain zone	Mungbean & Urdbean: *Caliothrips indicus, E. kerri, B. tabaci, Madurasia obscurella, M. testulalis*	Mungbean Yellow Mosaic Virus
Central zone	Mungbean & Urdbean: *Caliothrips indicus, E. kerri, B. tabaci, Madurasia obscurella, M. testulalis*	Mungbean Yellow Mosaic Virus*Cercospora* spp*Macrophomina phaseolina*
South zone	Mungbean & Urdbean *:* *E. kerri, B. tabaci, Madurasia obtusa, M. testulalis*	Mungbean Yellow Mosaic Virus*Erysiphe polygoniMacrophomina phaseolina*

and pea, and 24 on chickpea and lentil. Among these, 34 insects were serious pests on one or more of these crops. The insects which have become serious pests, include podbug, *Calvigralla gibbosa* Spinola and various species of leaf webbers and binders especially *Cydia critica* on pigeonpea; glaerucid beetle, *Madurasia obscurella* and jassid, *Empoasca kerri* on mungbean; various species of hairy caterpillars and whitefly, *Bemisia tabaci* on black gram; leafminer, *Acrocercopsphaeospora*; blue butterflies, *Euchrysops cnejus* and *Lampides boeticus*, and *Epilachana* beetles, *Epilachana dodecastigma* and *E. vigintioctopunctata* on cowpea; leaf miner, *Chromatomyia horticola* and semilooper, *Thysanoplusia orichalcea* on peas. Polyphagous pests like cutworms, *Agrotis ipsilon* and *Ochropleura flammatra*; hairy caterpillars, *Amsacta moorei*, *Amsacta albistriga* and *Spilosoma obliqua* have also become serious on one or more of these crops (Arora and Dhaliwal, 1996). *H. armigera* has been a menace in the pulse growing regions and of late is causing considerable damage to mungbean also during the *kharif* season. *Maruca vitrata* is another pod borer causing serious damage in India and Sri Lanka. Pod sucking bug, *Clavigralla scutellarius*, is also gaining importance in southern India causing serious losses in terms of shriveled grains. Podfly, *Melangromyza obtusa* (Malloch) which is serious all over Asia has emerged as the key pest in Vietnam (Shanower *et al.*, 1999) as well as in central and eastern India.

Current scenario of pests and management options and need of hour

In the recent past (2002-07), *Helicoverpa* population, which used to be major production constraint has gone down due to disruption in generation by transgenic cotton. However 2008 and 2009 crop seasons have witnessed intense comeback of *Helicoverpa* on pigeonpea and chickpea in India. Host plant resistance (HPR) has played a crucial role in combating various plant diseases and presently represents an important component of integrated pest management. In absence of high levels of host-plant resistance to *Helicoverpa* and its resistance towards a number of pesticides, there is are urgent need of coupling their use with other IPM options for effective and economical crop management. In major pulse growing areas, population of *Helicoverpa* has declined resulting in less use of pesticides. This tendency has resulted in upsurge of secondary insect pests such as podfly, which are difficult to locate. Similar is the case with podbugs, aphids and thrips. In severe cases, Thrips are causing drop of flowers. Despite of available wilt and Sterility mosaic disease (SMD) resistant varieties, their incidence is on increase. *Trichoderma* has been accepted as an important component as seed protectant resulting in availability of a good number of commercial preparations (164). Use of quality *Trichoderma* formulation has not only increased plant stands but has also helped in vigorous plant growth. There has been a problem with SMD wherein only HPR is working. It has been observed that SMD affected plants also harbor powdery mildew disease. In central India, climate change has resulted in increase in incidence of *Cercospora* leaf spot causing drop in flowers to the tune of 70%.

HaNPV has been found to be an effective tool for management of *Helicoverpa* in initial stages (up to 3rd instar). Due to limitations in availability of neem seed for use as extract, neem oil is readily available in various formulations and serves as anti-feedant as well as repellant. Despite availabilities of technologies, the coverage under IPM at present is estimated to be less than 4% of the total cultivable area in the country. There is need to integrate and interweave compatible plant protection strategies to achieve success of IPM in chickpea.

The success story of practical implementation of Integrated Pest Management technology in cotton based cropping system in Southern Maharashtra requiring only one spot application of pesticidal spray (Chlorpyriphos, Spark, Match, Endosulfan, Cartap hydrochloride, Acephate, Ethofenprox, Karate, Curacron and Ethion) has paved the way for adopting similar approaches in other parts of country. The IPM technology has not only brought down use of pesticides from 4 to 1 but has resulted in higher benefit ratio of 4.79 in comparison to 2.37 in conventional practices. Review of recommendations hosted on different State Govt. sites indicates gross deficiencies in terms of holistic approach, which need attention. Indonesia is well known example of success in the history of plant protection, wherein Govt withdrew the subsidies on pesticides and promoted ecological approach to manage rice pests. This has prevented the pesticides, which not only killed pests, but also beneficial organisms (such as parasitoids, predators, bees, earthworms and birds). It is worth mentioning that predators and parasitoids (*Campoletis chlorideae, Apenteles rufricus, Bracon hebetor* and *Carcelia illota*) are capable of reducing 60% yield losses. The implementation of IPM programme and its success has served as a model for other countries.

NCIPM's efforts based on cotton based cropping system in rainfed areas has resulted in conservation of naturally occurring predators and egg parasitoids as well as utilization of local bio-diversity in a community approach. Modifications in habitat by introducing growing of coriander or linseed have enhanced natural defenders population along with a niche to predators and modifying microclimate resulting in disease suppressive soils. Experience has not only enhanced natural defenders population but also helped on utilisation of locally available bio-diversity (neem as biopesticides) apart from extra remunerations and creating employment to rural youth. Based on past experience NCIPM has taken large scale demonstrations of IPM in pigeonpea as well as chickpea in 20000/ha and 16000/ha, respectively, in five pulse growing states with the help of DAC under A3P programme.

Integrated Pest management

Cultural practices

Cultural control methods are really crop production practices that make the crop environment less suitable for pest development. Pulses are generally grown in intercropping system, especially in rainfed areas. Apart from increasing profitability and resource use efficiency, they also act as a safeguard under adverse weather conditions. Decades of research has resulted in the development of remunerative cropping systems involving major pulses. Some of the important crop rotations are: pigeonpea-wheat, rice-chickpea/lentil, pigeonpea/rice-wheat-mungbean etc. Intercopping of chickpea with mustard in 4:1 row ratio, mungbean with sunflower in 6:2 ratio, chickpea with linseed in 4:2 ratio, potato + rajmash in 3:2 ratio and pigeonpea with urdbean in 2:1 ratio have been found efficient and highly productive. Mungbean and urdbean have been found most compatible for intercropping with coarse cereals and spring planted sunflower and sugarcane. Allelochemicals from sorghum roots have been found effective against soil borne pathogens such as Fusarium. Summer legumes significantly improve yield and nitrogen (N) economy of the succeeding rice crop as compared to fallow. Increasing N availability is considered as one of the important beneficial effects of pulses on the succeeding non-legume crops (Kumar and Singh, 2009).

Cultural controls are selected based on knowledge of pest biology and development. Being mainly rainfed, cultural practices do not play a significant role in pest management. The adjustment of sowing dates seems to offer little scope due to soil moisture utilization for having good plant population. However, sowing in the month of October was found to be optimum and resulted in less pod damage in chickpea. Most often timing of sowing will not result in least pod damage in these crops and does not coincide with that required for optimum grain production or pod set.

Seed treatment (as per local need) with fungicide is recommended with preference to microbial over chemicals or insecticides. Seed treatments are relatively low cost and initially protect the seed and seedlings from seed-borne, soil-borne pathogens and cutworms. In the nematode endemic areas only, apply neem cake @ 50 kg/ha or else apply nematicides (as needed or recommended). Since, control of nematodes is often linked with disease prevention or reduction hence it may be given due attention. In areas where termite and black cutworms (*Agrotis ipsilon*) are a problem, seed treatment with chlorpyriphos 20 EC @ 8 ml/kg of seeds or presowing mixing of soil with chlorpyriphos dust, HCH 10% dust, Sevidol 4% dust @ 25 to 30 g is recommended.

Wider spacing also reduced insect pest and foliar diseases incidence. Intercropping of mustard, wheat, barley, linseed, safflower and sorghum offered less pod damage in chickpea while traditionally intercropped cereals sorghum and millet did not affect the extent of damage. Similarly, intercropping of coriander or linseed with chickpea reduced pod damage due to pod borer. Field sanitation and soil tillage help to minimise pod infesting pests as well as root and stem diseases.

Inter-cultivation of nector rich plants like coriander helped support and sustain the natural enemy population such as *Ichneumonid parasite* and *Campoletis chlorideae*. Conserving other natural enemies also such as chrysopa, stink bugs, predatory wasps and spider by using need based eco-friendly insecticides also controls damage to pulse crops. Intercropping with linseed, mustard, wheat barley, coriander etc., minimizes the damage caused by *H. armigera* and nematodes.

Host plant resistance

It is the most promising means of pest management, but has been over used. Several promising selections have been made for resistance to the pod borer, which is the key pest in both the crops. Gene erosion led to a limiting factor that most pest resistant selections were not high yielding, further, majority of the pod borer resistant lines were highly susceptible to *Fusarium* wilt which is a serious problem in endemic areas. Varietal phenology and growth habit offers pest avoidance in Northern India as the cold winter suppresses their activity. In pigeonpea, determinate types are known to suffer less damage by podfly while indeterminate types are less prone to pod borer attack. Not much progress has been made due to the development of resistance in insects to various groups of insecticides and their interaction with the host plants, changing pest scenario and upsurge of new pests. Some of the resistant varieties are listed in table-2 .

Physical control

Physical controls by means of hand collection and shaking of plants can help farmers develop *Helicoverpa* nuclear polyhedrosis virus (HaNPV) in his own courtyard. Using mulches to smother weeds and providing row covers to protect plants from insects are other examples. Promoting picking of insect pests by birds is perhaps the most simple pest control method—is another physical control practiced during era of traditional agriculture.

Pheromone traps (@ 5-10/ha) should be used for monitoring of adult male populations of *H. armigera* and timing of initiation of curative plant protection practices. A threshold level of 5 - 6 moths/trap is indicative of its peak activity period and indicator of initiation of management practices.

Installation of bird perches of 3-5 feet @ 20 per ha has encouraged predation by insectivorous birds such as *Bubulcus ibis, Acridotherus tristis, Passion domesticas, Psittacula krameri* etc.

Biological Control

Biological controls include natural enemies, such as insect predators, parasitic wasps, or nematodes. In IPM programs, native natural enemy populations are conserved, and non-native agents may be introduced. Several natural enemies have been reported effective on both the crops (Table-3). These are capable of restricting the population of most potential pests to non-damaging levels. However, their potential has not been exploited to the core. Research on natural enemies of pests in these crops has been mostly focused on *Helicoverpa*. The egg parasitoids, *Trichogramma* spp., avoid these crops. Augmentative biocontrol with *Campoletis chlorideae*, *Eucelatonia bryani* and *Carcelia illota* holds promise. Many disease organisms that attack pests are exploited as biological control agents and are commercially developed for application to crops, but more needs to be done to bring down the cost without which it will not find users.

There is an urgent needs for making quality bioagents locally available in time. Genetically transformed crops with pest resistance holds great promises in the biological control category.

Table 2. Resistant / tolerant varieties of pulses.

Variety	Resistant	Area to which recommended
Chickpea		
Avrodhi	Wilt	North west plain zone
Bharati	Wilt & Root rot	Whole zone
C-235	Ascochyta blight	North west plain zone
Pusa 209	*Helicoverpa*	North and Central zone
ICCV 88 - 510	*Helicoverpa*	North and Central zone
JG 89 - 2	*Helicoverpa*	North and Central zone
PDG 90 - 2E	*Helicoverpa*	Northern hills
ICC 506	*Helicoverpa*	Bihar, Madhya Pradesh, Maharashtra
ICCV 7	*Helicoverpa*	Maharashtra
Dulia	*Helicoverpa*	Maharashtra and Northern Telangana of A.P.
H 355	Wilt & Root rot	Madhya Pradesh, Maharashtra & Rajasthan
PDE 5	*Helicoverpa*	Maharashtra and Northern Telangana of A.P.
PBG 1	Ascochyta blight	North and Central zone
GNG 469	Ascochyta blight	Whole zone
GNG 663	Maruca testulalis	Maharashtra, Northern Telangana of A.P. and Madhya Pradesh
Gaurav	Ascochyta blight	North and Central zone
AKG 33	*Helicoverpa*	Maharashtra, Northern Telangana of A.P. and Madhya Pradesh
JC 315	*Helicoverpa*	Central India and North India
JG 74	*Helicoverpa*, Wilt	Central India
Pigeonpea		
ICPL 1	*Helicoverpa*	Gujarat, Maharashtra and A.P.
ICPL 269	*Helicoverpa*	Maharashtra and Northern Telangana of A.P.
ICPL 372	Pod fly	U.P, West Bengal, A.P, Tamil Nadu and Gujarat.
ICPL 329	Pod fly	U.P, West Bengal, A.P, Tamil Nadu and Gujarat
Maruthi ICPL 8863	Wilt	Central and South India
ICPL 85066	Pod fly	Gujarat, Maharashtra and A.P.
Hy 2C	*Helicoverpa*	Madhya Pradesh, Karnataka and A.P
BDN 1 - 2	*Helicoverpa*, Wilt	Madhya Pradesh, Karnataka and A.P

Contd.

Bahar	Podborers, Sterlity Mosaic	Central India
Birsa Arhar 1	Wilt	Central and eastern zone
PDA 89 - 2E & 92 - 3E	Pod fly	U.P, Maharashtra and M.P
PDA 93 - 1E	*Helicoverpa*	U.P, Maharashtra and M.P
PDA 88 - 2E	Pod fly	Orissa, Maharashtra and M.P.
MA 91 - 2	Pod fly	Orissa, Maharashtra and M.P
Mukta	Pod fly	U.P, West Bengal, A.P. , T.N, and Gujarat.
UPAS 120	*Helicoverpa*	U.P, Maharashtra and M.P.
JA 3 & JA 4	*Helicoverpa*	Central India
C11	Wilt	Whole zone
Vishakha TT-6	Wilt	Central India
Mungbean		
PDM-11	Yellow Mosaic Virus	Eastern U.P., Bihar, West Bengal , Orissa, South West U.P., M.P., Maharashtra and Gujarat
ML-337	Yellow Mosaic Virus	Eastern U.P., Bihar, West Bengal , Orissa, Punjab, Haryana, Rajasthan, Delhi, Western U.P., South West U.P., M.P., Maharashtra and Gujarat
PDM-54 & PDM- 84-139	Yellow Mosaic Virus	Eastern U.P., Bihar, West Bengal and Orissa
Pant Mung-3	Yellow Mosaic Virus	Punjab, Haryana, Rajasthan, Delhi & Western U.P.
Mum-2	Yellow Mosaic Virus	Punjab, Haryana, Rajasthan, Delhi & Western U.P.
MH-88-111	Yellow Mosaic Virus	Punjab, Haryana, Rajasthan, Delhi & Western U.P.
TARM-1	Yellow Mosaic VirusPowdery mildew	South West U.P., M.P., Maharashtra , Gujarat, Andhra Pradesh, Karnataka and Tamilnadu
PDM-54	Yellow Mosaic Virus	Andhra Pradesh, Karnataka and Tamilnadu
Pusa-9072	Yellow Mosaic Virus	Andhra Pradesh, Karnataka and Tamilnadu
Urdbean		
PUD-1 and PDU-88-3	Yellow Mosaic Virus	Eastern U.P., Bihar, West Bengal , Orissa, Punjab, Haryana, Rajasthan, Delhi , Western U.P., South West U.P., M.P., Maharashtra and Gujarat,
WBU-108	Yellow Mosaic Virus	Punjab, Haryana, Rajasthan, Delhi & Western U.P.
LBG-17	Powdery mildew	Andhra Pradesh, Karnataka and Tamilnadu

Table 3. List of naturally occurring predators and parasitoids.

Crop	Pest	Natural enemy	Stage
Pigeonpea	Helicoverpa armigera	Aspergillus niger, Aspergillus falvus, Copidosoma floridanum	Larval stage Larval stage Larval stage
	Melanagromyza obtuse	Euderus agromyzae, Erytoma spp., Euderus spp., Ormyrus orientalis	Larval/pupal stage Larval/pupal stage Larval/pupal stage Larval/pupal stage
	Eucosma Clavigralla gibbosa Nezara viridula	Apanteles taragamae Gryon spp. Gryon spp.	Larval/pupal stage Egg Egg
Chickpea	Helicoverpa armigera	Campoletis chlorideae, Apanteles ruficrus, Bracon (Habrobracon) hebetor, Carcelia illota, Nuclear polyhedrosis virus (HaNPV)	Larval stage Larval stage Larval stage Larval stage Larval stage
	Autographa nigrisigna	Unidentified tachinid Parasite	Larval stage Larval stage
Mungbean and Urdbean	Helicoverpa armigera	Spiders and Lacewings Trichogramma sp. parasite wasps Microplitis and Netelia (wasps) Tachinid flies	Egg Egg Caterpillar Caterpillar
	Mirids	Beauvaria	Adults
	Loopers	braconid wasps	Larval stage
	Spodopatera lutura	As for Helicoverpa	
	Nezara viridula	Trissolcus basalis.	Egg
		Trichopoda giacomellii	Larvae and adult

Microbial Control

Field application of NPV, Bt., were found effective against lepidopterous pests. Application of nuclear polyhedrosis virus (HaNPV) @ 250 LE (for chickpea) and 500 LE /ha alone or along with tinopol @ 0.1% reduced the effect of UV radiation in pigeonpea. Spraying should be carried out in the evening hours. By spraying B.t. formulations @ 1.0-1.5 kg/ha.

Chemical Control

Chemical controls are used to keep populations below economically damaging levels when pests cannot be controlled by other means. Pesticides with the least negative impacts on non-target organisms and the environment are most useful in IPM programs. Several groups of insecticides have been found effective against major pests (Table 4). Due to the development of resistance, resurgence and secondary outbreaks, a combination of recommended insecticides has also been found effective, particularly the combination with neem based insecticides, bio pesticides like Bt., and NPV had a synergistic effect and suppressed the pest population to a greater extent than when used alone. However, their injudicious uses by farmers have altered the benefits. The effective application of insecticides and economic utility based on the need is now possible with control droplet application and ULV equipments. New generation pesticides with low environmental impacts are being developed, but unfortunately the progress is moving at a snail's pace. Addition of salt (0.5% NaCl) as an adjuvant can improve chemical control of mirids at lower chemical rates. Pesticides are most effective if applied before larvae enter the pods.

Constraints and policies in adoption of IPM Technology

Field failures and development of resistance by Helicoverpa has brought an end to the "pyrethroid era". A decision concerning control that was easy to make during two decades earlier has now become difficult. To maintain pest populations below an economic injury level needs switching method of control to an alternative insecticide. Experiences show that pulses (Bundelkhand of UP and *Tal* areas of Bihar) production remains dependent on the vagaries of rain God. Most of the chemical pesticides cannot be used or are ineffective due to shortage of water. IPM adoption studies revealed that farmers, in general, are aware of biopesticides and IPM tools such as pheromones and microbial pesticides but due to their non-availability, they are not able to use them. The other factors are:

i. IPM technologies are available only for few pests and they have not been field tested in combination with other pests

ii. The IPM techniques, which have been successfully employed, are mainly the use of resistant varieties and biopesticides (HaNPV). However, even these have not been tried and received with confidence in all zones as compared to use of chemical pesticides which provides immediate solutions.

iii. The inputs such as Rhizobium cultures, pheromone traps, and HaNPV are not readily available and need proper planning and storage at local level. The user confidence in these approaches is still rather low and convincing them is difficult because of quality problems. Government infrastructure is lagging behind in quality testing facilities to prevent duplicate or misbranded or substandard pesticides.

Table 4. List of commonly used pesticides in pulses crop.

Name of the pest	Recommended Pesticides	Dosage
Insect pests		
1. Helicoverpa pod borer	Profenphos 50 EC	2.0 ml/l
	Methomyl 40 SP	0.6 ml/l
	Thiodicarb 75 WP	0.6 ml/l
	Indoxacarb 14.5 SC	0.3 ml/l
	Spinosad 45 SC	0.1 ml/l
	Emamectin benzoate 5 SG	0.2 g/l
	Flubendiamide 480 SC	0.1 ml/l
	Rynaxypyr 20 SC	0.15 ml/l
	Novaluron 10 EC	0.75 ml/l
	Chlorpyriphos 20 EC	2.5 ml/l
	Quinolphos 25 EC	2.0 ml/l
	Acephate 75 SP	1.0 g/l
	HaNPV @ 250 to 500 LE/ha	1.0 ml/l
	NSKE 5%	50 g/l
	Neem oil 3000 ppm	20 ml/l
2. Spotted pod borer	Profenphos 50EC + DDVP 76 EC	2.0 ml + 0.5 ml /l
	Methomyl 40 SP + DDVP 76 EC	0.6 ml + 0.5 ml /l
	Thiodicarb 75 WP + DDVP 76 EC	0.6 ml + 0.5 ml /l
	NSKE 5% + DDVP 76 EC	50 g + 0.5 ml /l
3. Pod fly	Acephate 75 SP + Jaggery (1%)	1.0 g + 10 g/l
	Imidacloprid 17.5 SL+ Jaggery (1%)	0.2 ml + 10 g/l
	Thiomethoxam 25 WG + Jaggery (1%)	0.3 ml + 10 g/l
	Thiodicarb 75 SP+ Jaggery (1%)	0.6 ml + 10 g/l
4. Spodopatera lutura	As per Helicoverpa	
5. Pod sucking Bugs	Acephate 75 SP	1.0 g/l
	Methomyl 40 SP	0.6 g/l
	Quinolphos 25 EC	2.0 ml/l
	Chlorpyriphos 20 EC	2.5 ml/l
6. Plume moth	-----do----	-----do----
7. Defoliators	-----do----	-----do----
8. Leaf Webbers	-----do----	-----do----
9. Eriyophid mite	Dicofol 18.5 EC	2.5 ml/l
	Oxydemeton methyl 25 EC	2.0 ml/l
	Dimethoate 30 EC	1.7 ml/l
10. Two spotted Spider mite	-------do-------	-------do-------
11. Aphids	Dimethoate 30EC	1.7 ml/l
	Acephate 75 SP	1.0 g/l
	Quinolphos 25 EC	2.0 ml/l

Contd.

12 Cow bugs	--do---	-----do----
13 Scale insects	---do---	-----do----
14 Thrips	---do---	-----do----
15 Blister beetle	Acephate 75 SP Methomyl 40 SP Quinolphos 25 EC Chlorpyriphos 20 EC	1.0 g/l 0.6 g/i 2.0 ml/l 2.5 ml/l
16 Blue butterfly	----do---	-----do----
17 Termites	Chlorpyriphos 20 EC as a soil drench	5 ml/l
18 Ash weevil	Chlorpyriphos 20 EC	2.5 ml/l
Plant diseases		
1. Fusarium wilt	Trichoderma harzanium / Trichoderma viride Carbendazim 50 WP + Thiram 80 WP Spot soil drenching with Carbendazim 50 WP	Seed treatment @ 10 g/kg of seeds Seed treatment @ 1+3 gm/kg of seeds 1.0 g/l of water
2. Phytopthora blight	Metalaxyl 35 WS of seeds	Seed treatment @3g/kg
3. Bacterial Leaf Spot and Stem canker	Streptocycline	0. 25g/l
4. Sterility mosaic virus	Dicofol 18.5 EC Oxydemeton methyl 25 EC Dimethoate 30 EC	2.5 ml/l 2.0 ml/l 1.7 ml/l
5. Cercospora leaf spot	Mancozeb 75 WP Carbendazim 50 WP	2.0 g/l 1.0 g/l
6. Alternaria blight / leaf spot	---- do ----	---- do ----
7. Powdery mildew	Sulphur 80 WP Triadimefon 25 WP Carbendazim 50 WP Bacillus subtillis	4.0 g/l 1.0 g/l 1.0 g/l
8. Foliar rust	Mancozeb 75 WP	2.0 g/l
9. Soft rot	Spot soil drenching with Carbendazim 50 WP	1.0 g/l
10. Yellow mosaic virus	Dimethoate 30EC Acephate 75 SP Quinolphos 25 EC	1.7 ml/l 1.0 g/l 2.0 ml/l
Storage pests		
1. Bruchids	Triazophos 40 EC (In field at harvest) Mustard or coconut oil (smearing)	2.0 ml/l 10 l/kg of seeds

iv. There is urgent need to promote potential bioagents (fungal and bacterial). The success of the use of biocontrol agents especially Trichoderma and Gliocladium preparations as well as biopesticides has so far been demonstrated on a limited scale and not immediately available to farmers. There is need for scientifically exploring and promoting possibilities of using botanicals such as Pongamia, Mahua and Palmrosa oil as insecticides.

v. There is no regulation in pricing structure of bio-pesticides and bio-fertilizers.

vi. Although new green molecules have been introduced in the country, there is no rationale in price as a result farmers are forced to rely on old molecules with higher environment impact quotient (EIQ) and half life.

There is a misunderstanding that biotechnology can take care of all pest problems. In fact, biotechnology and IPM, especially bio based products, can co-exists and be of greater help to each other. Much of the research, development, technology transfer, education and training undertaken by the crop protection industry supports implementation of IPM. However, more needs to be done on research and development of IPM technologies and promote their use at field level. Education and dissemination of technology over larger areas should be taken up with the support of NGOs & private pesticide industries and their network. IPM being knowledge based, more attention may be given through Self Help Groups (SHGs), extension workers and village elders

References

Asthana, A.N., Lal, S.S. and Viswa Dhar, 1997. Current problems in pulse crops and future needs. In : National seminar on plant protection towards sustainability, Dec. 22 - 24. Plant protection association of India, Hyderabad, pp. 1-3.

Dhaliwal, G.S. and Arora R. 1996. Changing Status of insect pests and their management strategies. In: Changing scenario of Indian Agriculture. K.S. Gill, G.S. Dhaliwal and B.S. Hansra Eds., Commonwealth Publishers, New Delhi.

Dubey, O. P. and Sharma, O. P. 2002. Policy and perspective of plant protection practices of major pulses. Proceedings of National organised by Ind. Soc. of Pulses Research, Kanpur (eds. Ali, Masood; Chaturvedi, S.K. and Gurha, S.N.) IIPR, Kanpur. 2002. pp. 87-91

Krishnamurthy Rao, B.H. and Murthy, K.S.R.K. (eds) 1983. Proc. National. Seminar Crop Losses due to Insect Pests. Indian J. Ent. (Special issue, Vol I &II), Hyderabad.

Kumar, Shiv and Singh Mohan, 2009. 25 years of Pulses Research at IIPR (1984-2009). Indian Institute of Pulses Research. pp 211

Lal,S.S., Yadava C.P.and Dias, C.A.R. 1985.Assesment of crop losses caused by Helicoverpa armigera, FAO Plant Protection Bulletin 33 : 27-35.

Ramachandra Rao, G. G., Raghavaish, G. and Nagalingam, B., 1990. Effect of botanicals on certain behavioural responses and on growth inhibition of tobacco caterpillar Spodopatera litura. In : Proceedings of national symposium on problem and prospects of botanical pesticides in integrated pest management, CTRI, Rajamumdry, A.P., India. pp. 12

Shanower T.G. and Romeis, J. 1999. Insect pests of pigeonpea and their management. Annual Review of Entomology 44 : 77-96.

Srivastava B. K. 1964. Pests of pulse crops. In Entomology in India (ed.) N.C. Pant Silver Jubilee number of Indian J. Ent. vol. 83-91.

❑❑❑

Chapter - 12

Post Harvest Losses of Pulses, Value Addition and Management

R.T. Patil* and Dattatreaya M. Kadam
*Principal Scientist, CIAE, Bhopal, M.P.
Central Institute of Post Harvest Engineering & Technology,
P.O. PAU Campus, Ludhiana - 141 004, Punjab

Pulses are the major source of lysine rich protein, which is supplementary to cereal based proteins (metheonine). Leguminous species producing edible seeds occupy prominent place in Indian dietary habits. They are, even today, a primary source of protein for the majority of population in the country. A variety of pulse crops such as peas, lentils, urad, tur, moong, yellow peas, chicken pea, pigeon pea and beans are grown under different agro-climatic conditions, and India is recognized globally as a major player contributing to about 25 per cent of the world's production. The present demand for pulses in the country is estimated to be about 15 million tonnes, while the production hovers around 12 to 14 million tones. The post harvest losses in the legumes take place at farm as well as in storage as reported by DMI (2002) are detailed in Table 1, 2 and 3.

Table 1. Post harvest losses in pulses (%)

Pulse Crop	Threshing	Winnowing	Transport field to threshing yard	Transport from Threshing yard to store	Storage	Total
Pigeonpea	0.61	0.43	0.58	0.23	0.35	2.20
Chickpea	0.77	0.78	0.81	0.82	0.56	3.74
Greengram	0.63	0.61	0.67	0.19	0.29	2.38
Blackgram	0.65	0.62	0.70	0.30	0.30	2.46
Lentils	2.21	1.01	2.20	0.64	0.64	7.14

Table 2. Post harvest losses in pulses (%) during farm operations as per ICAR study (2007)

Pulse Crops	Harvesting	Collection	Threshing	Winnowing	Drying	Packaging	Transport	Total
Pigeonpea	0.5	0.4	0.7	1.0	0.5	0.2	0.1	3.4
Chickpea	0.7	0.6	0.8	0.4	0.6	0.2	0.2	3.4
Blackgram	1.1	1.0	1.6	0.3	0.5	0.2	0.1	5.0
Greengram	0.9	0.6	1.6	0.3	0.3	0.1	0.2	4.1

Table 3. Total post harvest losses in pulses (%) as per ICAR study (2007)

Pulse crops	Loss during Farm Operations	Storage Loss	Total Loss
Pigeon pea	3.4	2.0	5.4
Chick pea	3.4	0.9	4.3
Black gram	5.0	1.1	6.1
Green gram	4.1	1.4	5.5

Pulses protein

Pulses contain 17 to 25% of protein by weight. The protein content of pulses is about double that of the cereals and more than meat and eggs. The pulses contain lipids and fats in low quantities. Pulses are good source of thiamin, niacin, calcium and iron. In dry pulses very low amounts of ascorbic acid is present, but when these are soaked in water, during their germination, ascorbic acids are made available.

Man and other mono-gastric animals get protein from pulses. The proteins of pulses are mainly glocolin but some species of pulses may also contain albumin. The standard of protein depends on the composition of amino acids. The protein efficiency ratios of pulses range from 0.7 to 2.1, which is less than egg and milk protein. While cooking the nature of pulses proteins gets improved. Pulses contain more lysine while sulfur containing amino acids are less. Since the cereals have low levels of lysine, when these are mixed with pulses, the PER value of diet improves. Man gets proper quality of protein through cereal and pulses diet, which is easily digestible.

It is a general belief that pulses are comparatively hard to digest by mono-gastric animals. Various studies have shown that well cooked pulses' digestibility is upto 90 per cent. Pulses cause flatulence. They contain about 60% carbohydrate mainly starch, which in general is well-absorbed and utilized. The fat content of common '*dal*' is between 1 to 2 per cent.

Processes used for pulse milling

Pulse milling is an important agro-based industry in India next only to rice and wheat milling. Unlike wheat or rice milling, processing of pulses to produce *dal* is unique and indigenous to India. This is because a substantial quantity of pulses (more than 10 million tonnes) is consumed in this country as *dal* (de-husked split pulse). Pulses are de-hulled particularly to improve their appearance, texture, cooking quality, palatability and digestibility. The split pulses are not affected by insect pests during storage but whole grains are much more susceptible to their attack. There are about 10,000 *dal* mills in the country and only 8 to 10 per cent of these are located in the rural areas, where pulses are largely produced (Table 4). Estimates indicate that about 30% of the production of pulses is retained by the farmers and is processed in rural sector using traditional techniques. There are several small-scale millers who process one to five quintals of pulses per day either by custom milling or for trade purposes as a family occupation. However, with the advent of organised large scale milling systems, most of these processors have been thrown out of their profession since their product is inferior in quality and yield. For this, reason the *dal* available in the market comes entirely from the large-scale mills.

The grain characteristics of pulses are such that husk is 10 -16 % and hence the potential *dal* recovery of 84 - 90% is possible. However, the actual *dal* recovery is 70 - 80%. It is essential that *dal* recovery be higher as it will improve the quality because of higher protein concentration in peripheral region. About 70% of pulses produced are milled in commercial *dal* mills. The conventional pulse milling methods are time consuming, labour intensive and the rate of recovery is low because the commercial mills are scaled up versions of cottage milling. The available commercial milling gives low recovery (65-72%) and high milling losses. The milling losses alone contribute to 15-20%, which comes to around 1.5 million tonnes. The high milling losses are added to the price of finished product. There is wide difference between the prices of pulses and finished *dal* due to the middlemen, and, hence, both the growers and consumers are at loss while the middlemen ensure a hefty profit. There exists a strong need to make farmers also a processor of produce, to improve their earning through value addition. The modern but small scale processing therefore can help curb the losses and increase farmers income.

Process outline

Milling of pulses means removal of the outer husk and splitting the grain into two equal halves. The kernel of some pulses holds the husk tightly. Therefore, de-husking of these pulses is relatively difficult and they are called hard to mill pulses. To loosen the husk, pre-treatment with oil and conditioning of grains by alternate drying/wetting is done. It facilitates de-husking and splitting of pulses. A large amount of abrasive force is applied to the conditioned pulses for de-husking of the grains. It results in high losses in the form of brokens and powder. Consequently, the yield of split pulses in traditional mills is only 65 to 70 per cent in comparison to 82 to 85 per cent potential yield.

Table 4: Status of dal milling industry in the country

State	Type of Mills	Capacity q/day	Type of machinery used by mills across the country	Pulses processed
Andhra Pradesh	Large	100 to 200	Graders, rollers, vertical chakkies	Urad, Moong, *Tur*
	Small	15 to 30	Vertical chakkies (plate mill), roller aspirator	
Bihar	Large	150	Graders Rollers, Vertical chakkies	Lentil, Khesari
Gujarat	Large	200 to 300	Smooth and rough emery rollers, size graders, impact splitters	*Tur*
	Small	15 to 20	Emery rollers and stone chakkies	
Haryana	Large	300 to 500	Size grader, rollers (pitting), under runner disc sheller and mixer-cum-polisher	*Tur*, Gram, Peas
Madhya Pradesh	Medium	10 to 80	Agra type *dal* mill roller, size grader, mixer, vertical plate mill and polisher	*Tur*, Gram Moong, Urad
Maharashtra	Large	10 to 20 q/h	Mostly modernized (rollers and chakkies), mechanical driers	All pulses
	Small	1.0 to1.5 q/h	Combined emery rollers and horizontal chakkies, CFTRI designed pulse dehusker	
Odhisha	Small mills	10 to 15	Emery roller grader and polisher	Moong, Urad Horse gram
Rajasthan	Large	100	Emery rollers, Disc shellers reel sieves	Gram Moong Moth bean
	Small	15 to 20		
Tamil Nadu	Large	50 to 100	Emery rollers, reel sieves water mixer and polishers	Gram, Moong *Tur*, Horse gram,
	Small	5 to 20	Plate mill / roller, grader and aspirator	
Uttar Pradesh	Large	20 to 50	Agra type machinery under, runner disc sheller, vertical plate mill tapered rolls	Gram, Moong, Peas, *Tur*, Lentil, Urad (splits)
	Small	1 to 2 q/h	Units from Kanpur vertical plate mills / burr mill	
West Bengal	Large an Small	2 to 50	Emery coated roller either tapered or cylindrical cleaning and grading units and polishers	Urad, Lentil, Peas, Gram, *Khesari*

Important unit operations of pulse (Pigeon pea) milling

Cleaning/grading of raw grains: In this process the raw grains received in a milling plant are first cleaned and graded. In cleaning, various inorganic and organic impurities present in the grain mass are removed with the help of air draft and screens. Destoners are also used to separate mud and stones. The cleaned grains are then graded as per their size mostly by a reel grader (Fig. 1). The presence of impurities lower the quality of *dal* and the impurities may also damage the milling machines.

Conditioning: A layer of gum between cotyledons or kernel and outer husk is present in most of the pulses. This gum layer may be thin or thick which in turn governs the degree of adherence of the seed coat to the cotyledons. The nature of gums influences the adherence of husk to kernel while the amount of gum effects the duration and severity of the conditioning process. The main objective of the conditioning is to loosen the husk to facilitate its separation from the kernel, thus reducing the milling losses. The conditioning process is done by various methods. There is no standard method. Conditioning of grains can be achieved by water treatment, hydrothermal treatment, use of salts and chemicals and use of heat alone. Insufficient conditioning results in incomplete loosening of husk, therefore, greater abrasive/scouring forces are necessary for complete removal of the husk. This causes higher milling losses in the form of brokens and powder with lower recovery of *dal*.

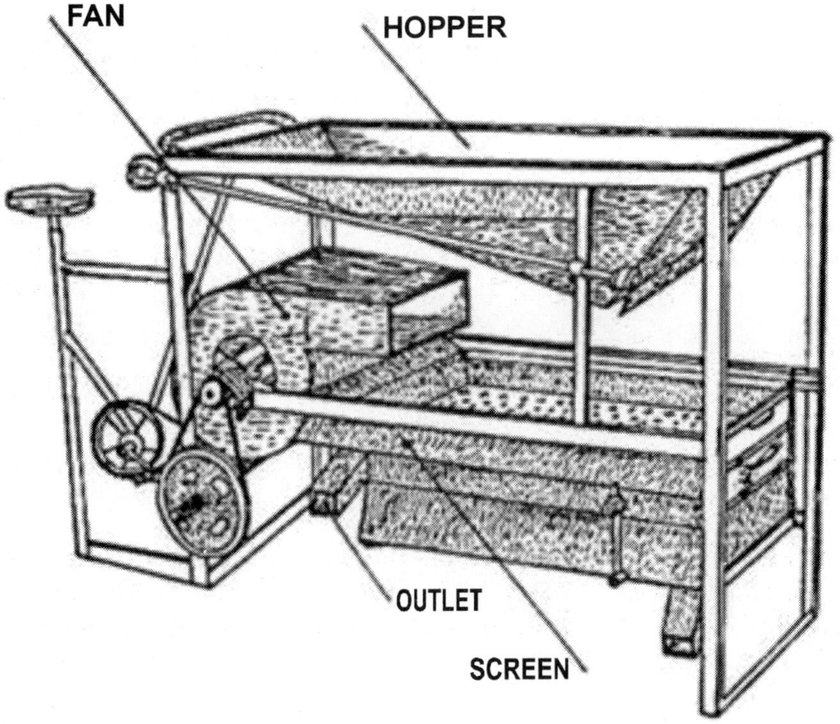

Fig. 1: Low cost raw pulse cleaner cum grader

Dehusking and splitting: The conditioned grains are subjected to abrasive/ scouring forces for removal of husk and for splitting of cotyledons into two equal halves. Dehusking and splitting are two separate but the most important and major unit operations of any pulse milling process. Dehusking of pulses is generally carried in an abrasive roller mill. In some pulse milling plants, vertical stone *chakki* is also used to dehusk and split the grains. For splitting of the dehusked and moistened grains, vertical disk burr mill is used or the grains are allowed to fall on a hard or cemented surface from sufficient height or by centrifugal force. Due to impact with hard surface, the dehusked grains are split.

Polishing of *dal*: This is done to impart desirable shine and lusture to *dal*. During this process a pre-desired quantity of edible oil and water is mixed with *dal* by passing them through a screw conveyor. The presence of oil and water imparts desirable colour and shine to milled pulses.

Grading of *dal*: Separation of *dal* is carried mainly as per their size and soundness. The separation is usually achieved by rotating reel graders, in which separate compartments have different opening sizes of screens. Various sizes of *dal fall* through the openings of these screens. Graded *dal* are bagged separately as per their grades.

Domestic level processes

In eastern parts of India, alternate soaking and sun drying method is in use. Pulses are soaked in water for 6-8 hours, water is then drained off and the grains are spread in thin layer and sun dried for two days. The dried pulse is then milled by hand operated 'stone *chakki*'. One woman can produce about 5 kg of *dal* per hour. Two passes of pulses are normally required for husk removal up to 80-90%.

In the central and northern India, where the dry method of milling is in use, pulses are mixed with edible oil and dried under sun for a day or two. Then moisture is added in small amounts. It is then heaped for few hours and then milled in hand operated 'stone *chakki*'. One labour can produce about 30-40 kg of *dal* in eight hours.

There is a different wet milling method being practiced in southern India. The grains are steeped in water for few hours, the water is then drained off and red earth is added to it. The grains are then dried under sun in thin layers for 2 days. Such conditioned grains are milled by "manual *chakki*". Dehusking and splitting takes place simultaneously. The husk in all the above methods is removed by winnowing.

Dal produced by the dry method is said to cook better while *dal* produced by wet method tastes better but takes longer time to cook. Processing by home and cottage scale methods is not efficient. These methods of processing cause cup formation, more breakage and powdering of cotyledons. Further, these methods are time consuming and the *dal* prepared is not properly dehusked and edges of cotyledons break during splitting.

Commercial scale methods

Wet milling method : Comparatively smaller units are adopting this method of processing. This method is mostly practiced in southern India.

The milling quality of above method was found to be good and on an average 72 per cent *dal* recovery is reported. Cooking time mainly depends upon steeping time. More the steeping time better is the dehusking results but the resultant *dal* will take longer time to cook. Treatment with red earth imparts yellow colour. It also provides mild abrasiveness and thus helps in the removal of husk from cotyledons. Wet method of *dal* milling is labour consuming and longer processing time of 4-5 days is the characteristic of this method. A batch of 6-7 tonnes is possible to process by this method.

Raw Pigeon pea

↓

Cleaning by sieves chaff, dirt etc.

↓

Soaking in water for 4-12 hours in cement tanks, then water is drained off

↓

Mixing of red earth to form a paste @ 2—3% level. Grains are then heaped for 16 hours.

↓

Drying under sun in thin layer for 2-4 days, heaped in night to retain heat

↓

Dehusking and splitting by emery coated vertical *chakki* about 95-98% of the grains are dehusked and splitted.

↓

Gota ← Grading by sieves ← husk, powder, *chuni* & small brokens

↓

Pigeon Pea Dal

Dry milling method

This method of milling of pulses is mostly adopted in the central and northern India.

Raw pigeon pea

↓

Cleaning and grading by sieves → chaff, dirt etc.

↓

Pitting or scratching by emery roller → dust, *chuni*

↓

Grains with cracks in the husk are mixed with edible oil like linseed @ 1.5 to 2.5 kg/t of pulses in an *anger* mixer, then held for 12 hours

↓

Oiled grains are spread in 5-10 cm thick layers for sun-drying, heaped in nights, After 2-4 days of drying on last day the grains are sprayed with 3-5% water and heaped overnight

↓

Milling by 1st roller. About 40-50% of grains are dehusked in the process

↓

Dhal separating sieves → husk, *chuni*, and brokens

↓

Whole dehusked grains

↓

Sun drying and addition of water

↓

Dehusking and splitting in 2nd roller

↓

Sieving

↓

Grade I Dal

Dry methods of milling vary widely from region to region. It has been observed that edible oil requirement in pulse processing varies from 150 to 500 grams per 100 kg of grain processed. The expenditure on oil alone constitute about 38% of the total processing cost. The energy requirement of dry milling method is also more. Repeated dehusking in roller mill together with repeated material handling for dehusking, grading aspirating and polishing results in consumption of more energy per unit of *dal* processed.

Milling of pulses in Central Region (Jabalpur region)

Pigeon pea

↓

Cleaning with sieves → chaff, dirt etc.

↓

Carborundum roller scratching → → *Chuni*, small brokens

↓

Linseed oil mixing @ 150 ml/l00 kg of grain, held for overnight

↓

Sun drying for 10 hours

↓

Carborundum roller, 2-4 passes

↓

Cleaning → ret, *chuni*, small brokens

↓

Grain and *gota*

↓

Linseed oil mixing by auger

↓

@ 100 ml/100 kg t grain and water @ 250 ml/100 kg of grain, held overnight

↓

Sun drying in thin layer of 1" thickness for 10 hours

↓

Sheller (burr mill)

↓

Aspirator cleaner broken

↓ →

Pigeon Pea *Dal*

CFTRI *Dhal* Milling Technology

This method eliminates mixing of oil and water for loosening of husk. It is claimed that it can give an average yield of 80% *dal* in lesser time and at lesser processing cost. Clean and graded grains are conditioned by dry heat treatment in two passes in LSU dryer with 120°C hot air. After each pass through dryer the grains are tempered for 6 hours in tempering bin. The flow chart of CFTRI method is given below.

Pigeon pea

↓

cleaning by rotary reel cleaners

↓

Pre conditioning: Pulses are conditioned in 2 passes through LSU dryer with 120°C hot air. After each pass the hot pulses are tempered in bins for 6 hours.

↓

Dehusking : By roller machine. Then moisture is added to whole dehusked pulses.

↓

Lump breaking: Some of the moistened gota form lumps. These are fed to lump breaker

↓

Conditioning and Splitting: Gota is conveyed to LSD dryer and exposed to hot air for few hours. Emery roller splits the hot conditioned and dried gota

↓

Grading → brokens

↓

Pigeon Pea *Dal*

The CFTRI method requires high electrical loads for operation of dryers.

Pantnagar *Dal* Milling Technology

In this process the pigeon pea grains are cleaned and graded, then these are passed through an emery roller mill for scratching of grains. The pitted or scratched grains are mixed with 6% sodium bicarbonate solution. The treated grains are tempered for 4 hours in shade. Then the grains are dried under sun to 9.5% moisture content. The conditioned grains are then milled by a concentric cylinder roller mill. The milled product is cleaned and graded with the help of blower, cyclone separator and reel graders. It is reported that this method gives-80% recovery of *dal*. Steps involved in *dal* milling process as follows:

Whole pigeon pea seed

↓

Cleaning

↓

Grading

↓

Soaking for 1 h in 6% sodium bicarbonate solution

↓

Tempering

↓

Drying

↓

Tempering

↓

Milling on a concentric double cylinder machine → Powder

↓→ Husk + fine powder to cyclone separator

Cylindrical screen separator

↓

Non husked whole grain | Husked whole seed | Split *dal* | Broken seeds

Husked whole seed ↓

Splitting

↙ ↘

Dal **Broken seed**

The Pantnagar technology was reported to be on par with the CFTRI technology in *dal* recovery and dehusking. It is not dependent on weather and has a high de-husking efficiency (over 97%) with a single pass. It requires less energy 45 kWh/t compared to 85 kWh/t for the CFTRI process, and 60 kWh/t as compared to the traditional process.

CIAE method

In order to eliminate the use of edible oil in pre-treatment process the following process was adopted. First the pigeon pea, green gram or black gram is cleaned and graded. The grains are then fed to the roller mill developed in CIAE for scratching or pitting. During this operation scratches are developed on husk of grains. The scratched grains are cleaned. The grains are then soaked in tap water at ambient temperature for 30 minutes in case of pigeon pea and about an hour in case of black gram and green gram. Afterwards water is drained off, and the soaked grains are dried to 9-10% moisture content. Such conditioned grains are fed again to the roller mill to produce dehusked, split cotyledons. For dehusking and splitting of pulses, peripheral speed of 13.5 to 14.0 m/s of the abrasive roller were found to give highest milling efficiency.

Pretreatment factors affecting *Dal* recovery

Water: Soaking of pigeon pea in water helps loosen the binding action of the gum between seed coat and cotyledons. Increasing the moisture helps soften the seed coat. Soaking time varies from 2 to 14 h in villages. In large *dal* mills, the seed moisture level is increased by sprinkling water or sometimes alternate wetting and drying.

Oil: This treatment allows oil to penetrate the pigeon pea seed through the husk into the cotyledon layer and the cotyledons are released under the mild heat of the sun from the seed coat. Treatment of seed for several days can totally loosen the husk.

Chemicals: The uses of sodium bicarbonate, calcium hydroxide, sodium carbonate, and sodium chloride as a pre-treatment have been reported. Pre-treatment with 6% sodium bicarbonate solution was most effective, resulting in 78% *dal* recovery.

Heating: It is an efficient way to loosen the seed coat. In traditional methods, roasting on sand at 100-125°C for 5-10 min is done. Conditioning of the pigeon pea seed is done in a chamber with hot air (120-180°C), where the seed is exposed to a temperature of 70-95°C. The heat is transferred through a counter-current flow technique. In the conditioning chamber, hot air enters through the seed mass and moves upwards, while the seeds move down by gravity.

Quality standards of milled pulses

The quality standards of milled pulses are given in Table 5 and 6 below.

Table 5. Quality standards of milled pulses

Quality character	Upper limit for each split pulse				
	Arhar	Moong	Urad	Chana	Masur
Moisture (%)	14	14	14	16	14
Foreign matter (%)	2	2	2	2	2
Other edible grains (%)	0.5	4	4	4	2
Damaged grains (%)	5	5	5	5	5
Weevilled grain (%)	3	3	3	3	3
Uric acid (mg/kg)	100	100	100	100	100
Mycotoxin including aflatoxin (mg/kg)	30	30	30	30	30
Rodent hair and excreta (pcs/kg)	5	5	5	5	5
Total of foreign matter, other edible grains and damaged grains (%)	6	8	8	7	7

Post harvest management of pulses

The current pulses processing is a batch process, involving excessive material handling, which can be reduced by introducing continuous type processing system. The sun drying is generally practiced in *dal* industry which is not reliable and during rainy days the mill has to be closed. Therefore, some economical and reliable mechanical drying system is needed for that batch cum bin type *dal* dryers are commercially available. The *dal* is produced in the growing season but the mills have to run for the whole year (300 days) hence, improved marketing system where the millers could purchase cleaned and graded raw material, will also reduce the cost of processing of *dal*. In '*dal*' milling, a lot of dust is generated inside the mill. Hence, development and use of appropriate dust collecting and diffusing devices is essential as dust inside closed mill causes pollution and is hazardous to workers. Overall, there is a need for standardization of milling processes and equipment and standards for milling techniques and machinery.

Table 6. Dietary nutrients of pigeon pea

S. No	Constituents	Green Seed	Mature Seed	Dal
1	Protein (%)	21.0	18.8	24.6
2	Protein digestibility (%)	66.8	58.5	60.5
3	Trypsin inhibitor (mg-1)	2.8	9.9	13.5
4	Starch (%)	48.4	53.0	57.6
5	Starch digestibility (%)	53.0	36.2	-
6	Amylase inhibitor (mg-1)	17.3	26.9	-
7	Soluble sugars (%)	5.1	3.1	5.2
8	Flatulence factors (g /100 g soluble sugar)	10.3	53.5	-
9	Crude fiber (%)	8.2	6.6	1.2
10	Fat (%)	2.3	1.9	1.6
Minerals and trace elements (mg/100 g dry matter)				
1	Calcium	94.6	120.8	16.3
2	Magnesium	113.7	122.0	78.9
3	Copper	1.4	1.3	1.3
4	Iron	4.6	3.9	2.9
5	Zinc	2.5	2.3	3.0
Vitamins (mg/100 g fresh weight of edible portion)				
1	Carotene (Vit A per 100 g)	469.0		
2	Thiamin (Vit B1)	0.3		
3	Riboflavin (Vit B2)	0.3		
4	Niacin	3.0		
5	Ascorbic acid (Vit C)	25.0		

Factors affecting pulse milling outturn

Theoretically the endosperm of the pulse accounts for maximum 87-89% of the whole grain legumes. But in practice it is not achievable. Some of the important factors responsible for lower recovery of *dal* are given below.

A. Grain parameters are (i) Type of grain, (ii) Species or strain of grain legume, (iii) Chemical constituents, (iv) Moisture content of the grain endosperm, (v) Shape of the grain, (vi) Size, bold, small, (vii) Seed coat surface texture, and, (viii) Thickness of the gum layer.

B. Machine parameters are (i) Size of roller, (ii) Peripheral speed of the roller, (iii) Texture of the abrasive material, (iv) Hardness of the abrasive material, (v) Clearance between the roller and cage, (vi) Inclination of the roller, and, (vii) Outlet position

Precautions

The following are the general precautions to be taken in post harvest management and processing of legumes.

- Harvest the legume pads at proper stage of maturity i.e. Harvest the crop when 80% of the pods are mature (turned yellow).
- Harvest during conducive weather conditions.
- Threshing and winnowing on cemented (*pucca*) floor.
- Market the grain after grading to get remunerative prices in the market.
- Use proper and scientific method of storage.
- Package properly to protect the quality and quantity of produce during transit

Nutrient losses on dehulling

During dehulling outer layers of cotyledons are scarified resulting in 10% yield loss as powder fraction. The outer portion of cotyledons is a rich source of protein, sugar, fiber, and ash but poor in starch. Further, considerable amounts of calcium (about 20%) and iron (about 30%) are removed, but it does not affect protein quality and amino acids. In the post harvest value chain, out of 100 units produced, 9 units are used for seed purposes, 14 units are consumed whole, about 9.5 units are lost during threshing, transport, primary processing and storage. Hence, 67.5 units out of 100 are available for milling. and after the milling loss, 62.5 units are available in the food chain and 37.5 units go out of the value chain.

Value added products from pulses

Roasting of pulses or puffed legumes : The puffed chickpea and peas are widely used in India as snacks. The chickpea or peas are soaked in water for about 15-20 minutes. The water is drained and wet grains are kept in a closed vessel for the moisture to equilibrate in the grain. The wet grains are puffed in a hot iron vessel containing sand at 190-200^0 C for 60-80 sec. The grains puff and the husk is split off. For efficient puffing, the temperature just before explosion must be sufficient to create a high enough water vapour pressure without burning the pericarp and the temperature increase must be fast enough to build up the required pressure before the water evaporates. Moisture content of the kernel has a pronounced effect on popping behaviour. The kernels, which are too dry do not generally pop properly.

Parching of pulses : Legumes such as Bengal gram and peas are parched to give highly acceptable products. Bengal gram is tied in a moist cloth and kept overnight before it is parched. Peas are soaked in water for 5 minutes, dried partially in the sun for 15 minutes and then parched. Salt and turmeric powder are sometimes added to the steeping water or the grains smeared with the paste of the same before they are parched. Parchased is done in a hot iron vessel containing sand at 190-200°C for 60-80 seconds. Parching Bengal gram is a nutritious snack in the rural areas and helps reduce protein calorie malnutrition in children.

Preparation of papad : *Papad* also known as *appalam* or *papadam* in South India, is essentially a thin wafer like product, usually circular in shape rolled from a dough made out of pulse, cereals, edible vegetable oil, alkaline and mucilaginous additives. It is normally consumed in toasted or fried form. The steps involved in the processing of *papad* are given below:

All the ingredients except water are mixed well. The dough is prepared after adding water. The prepared dough is kneaded well and divided into balls of 15 gm each and pressed to 0.8 cm thickness with 8-9 cm diameter *papad.* The *papad* is dried under the sun for 10 minutes or in shed to reduce the moisture content to less than 14.0 per cent. It is packed in polyethylene pouches.

Preparation of instant vada and bajji mixes : *Vada* and *bajji* are the unfermented traditional snack items. The instant *vada* and *bajji* mixes are generally made from flours of bengal gram, black gram, wheat and rice. Frying contributes to destruction of anti-nutritive factors, moisture evaporation and thereby, increases shelf stability. The bengal gram *dal*, black gram *dal* and raw rice are ground separately. The ground bengal gram is sieved in a BS 14 sieve and the rest of the items are sieved in BS 36. Then the ground flours are further dried separately in a cabinet drier at 80°C for two hours, cooled and stored in an airtight containers till they were used. The ground bengal gram *dal* is mixed with maida and the flours of black gram *dal* and rice. The salt, citric acid and sodium bicarbonate are added to it and mixed well.

Pulses as functional foods : Nutritional attributes associated with pulses can be a postive attribute for their use as functional foods. Pulses have a low glycaemic index (due to the low amylose content of pulses). Lower glucose and insulin responses have been reported after a meal due to low glycaemic index. Pulses are a source of soluble fibre which lowers blood cholesterol as well as blood glucose levels. Due to their low amylose content, pulses contain resistant starch which is believed to contribute to healthy bowel function.

Pulses are a healthy option for people with celiac disease. Compared to many of the other gluten-free grain alternatives, pulses are an excellent source of dietary fibre, protein, iron, and other minerals and vitamins. The pulses have two major advantages in the gluten-free market. One is that pulses are high in iron, as iron deficiency anemia is a common nutritional concern for celiacs, along with vitamin B deficiencies. Secondly, pulses help to good bowel movement due to good amount of soluble dietary fibres available in them.

Production of Tannin Acyl Hydrolase (Tannase) from pulse milling by-products: Tannase is produced from pigeon pea husk by solid state fermentation and is extremely important in various industries such as in the manufacture of instant tea, beer, fruit juice, coffee-flavour soft drink and grape wine. In addition, gallic acid, the product of hydrolytic cleavage of tannic acid, is a chemical used in pharmaceuticals. World over greater emphasis is given on secondary agriculture, which is the utilization of waste material for production of value added products which leads to high economic returns. With the advent of biotechnology, attempts have increasingly been made globally to make potential use of agro-industrial residues for value addition by production of enzymes, organic acids, bioactive secondary metabolites, single-cell protein, etc. Solid-state fermentation (SSF) involves the growth of microorganisms on moist solid substrates in the absence of free flowing water and is an alternative cultivation system for the production of value added products from microorganisms, especially enzymes or secondary metabolites. The maximum production of tannase was found to be 43 U/g/min at 35°C, incubated for 96 h. The tannin content in the substrate decreased during incubation period. The tannase prepared from Aspergillus niger had a specific activity of 45 U/g/min in immobilized form (Paranthaman *et al.*, 2009).

Snack foods : Pulse-based snack foods are popular in South-East Asia. New snack foods are being launched based on pulses, such as chickpea chips, *fafda, sev* of different diameters and tastes, *Alu Bhujia*, *Khatta Mitta Mixture*, *Masala Chana Dal* etc.

Pretreatment and development of machinery for pigeon pea milling at CIPHET

Experiments were conducted at Central Institute of Post Harvest Engineering & Technology (CIPHET) Ludhiana to develop and standardize the pre-milling treatments on pulse milling to increase *dal* recovery. The simple water spray and heat treatment was found better than soaking and sun drying method, it increasesd the protein content marginally, reduced cooking time (30%) and less of cotyledon losses as powder with husk. The pre-milling treatment using steaming on pigeon pea was also optimized. The recovery of *dal* soaked for 60 sec and steamed for 5 min at low pressure 0.25 kg/cm^2 was higher than the recovery of *dal* from pigeon pea washed only and steamed at the same pressure. The time of steaming was inversely proportional to the recovery of *dal*.

The effect of types of oil was evaluated with mustard and cotton seed oil for pretreatment of pigeon pea during milling and was optimized with RSM. The moisture content of 10.11 %, time of dehulling 12.3 sec and 0.30% mustard oil gave optimum dehulling efficiency while in case of cotton seed oil the optimum values of these parameters were 9.84%, 11.96 sec and 0.28%, respectively. The effect of moisture content on pitting of pigeon pea was also studied. The pitting efficiency increased up to 10% m

c. and decreased with increase in moisture content beyond 10%. It could be attributed to less friction between husk and carborumdum at higher moisture content. Therefore 10% moisture content was found optimum for pitting of pigeon pea grain.

The objective quality standards were developed for grade I and grade II *dal* in terms of physico-chemical properties like physical dimensions, bulk density, cotyledon percentage, germ and fat content.

Single feed composite pigeon pea mini *dal* mill.

Since *dal* milling practice for self consumption at village level is being done away with, the availability of brokens and husk for animal feed has also got reduced. Instead of *dal* milling individually, if it is done as a small scale industry at village level it still offers scope for employment generation and increased income to the farmers. To facilitate such operation at village level many institutions have made efforts in developing a small scale *dals* mill having about 100 kg/h capacity unit. The dal mills developed by CFTRI, CIAE, IIPR and PKV are depicted in figures 2 to 5.

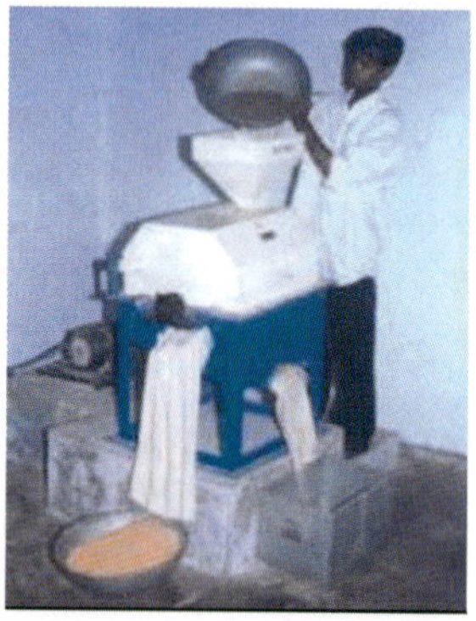

(Fig.2) CIAE Dal Mill

(Fig.3) CFTRI Dal Mill

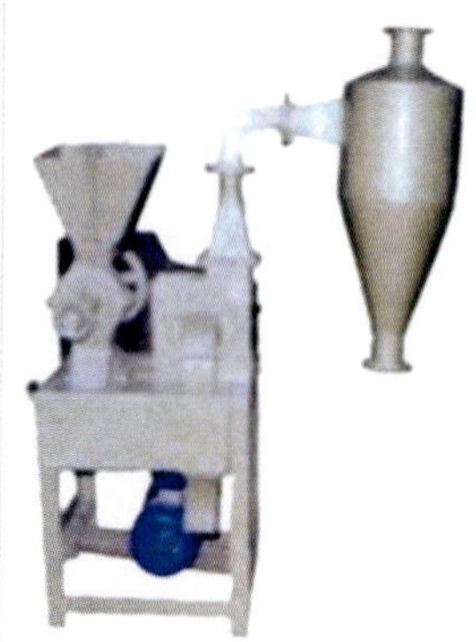

(Fig.4) IIPR Kanpur Dal Mill

(Fig.5) PKV Akola Dal Mill

Fig. 2-5. The models of Dal machines developed by different Institutions

However, since all these mills have only one abrasive roller, removal of 100% husk is not physically possible. Hence, appearance of dal produced in such mills does not match with the dal produced in commercial mills though cooking quality and nutritional value is better for dal in these mills. Since removal of husk from dal is not possible in a single pass, it is essential that even in a small mill, provision needs to be made to create the similar effect of multi pass. The *dal* milling machine developed by CIPHET is based on abrasive dehulling of pigeon pea (Fig. 6). The machine consists of feed hopper, cylindrical emery, and concave of perforated mild steel. This machine has two specific features (emery made of three different grades of carborundum and stoppers inside the concave to increase the residence time inside the machine).

Fig.6. CIPHET Single Feed Multi Stage Dhal Mill

The Mini Dal Mill has the dimension: Length 1000 mm, width 555 mm, Height 1225 mm, abrasive circle surface perimeter is 1100 mm. It is driven by 3 Hp motor which rotates at 1580 rpm and dal mill shaft rotates at 615 rpm. The capacity of the mini *dal* mill is 100 kg/h.

Feed Hopper: A trapezoidal feed hopper of capacity 15 kg is fitted at the top of the machine. A feed gate is provided at the outlet of the machine to control the feed rate of grain. The treated pigeon pea grain is filled in the hopper. After running the machine, the feed gate is opened to feed the material between emery roller and concave.

Emery Roller: The emery roller is fitted inside the machine. It is a mild steel cylinder of 75 cm length on which emery is pasted. At the feed section Grade A carburendum followed by Grade B and at outlet section Grade C is pasted and screen is encircled for dehulling in the pigeon pea. The grades of emery pasted are 18, 24 and 32. First 25 cm of the roller is pasted with 18-grade emery, middle 25 cm with 24-grade emery and last 25 cm with 32-grade emery. The material is fed at the start of the roller where 18-grade emery is pasted. The diameter of the emery roller is 30 cm. Emery rotates at 900 rpm with a 3 hp three-phase motor. The power is transmitted from motor to roller through v-belt drive pulley. A dehusking roller with three different grades of carborundum has been fabricated. The grades of carborundum are: 24 (0.3-0.5 mm), 30 (0.22-0.49 mm) and 36 (0.16-0.25 mm) .

The market value of *dal* prepared from pigeon pea is based on the characteristics of the dehulled whole splits. At present two grades namely grade-II and grade-I dal are sold in Indian market though these are mainly for the miller or retailer. The consumer who buys the *dal* looks at it from his own perception. Some consumers may prefer to buy *dal*, which is shining due to polishing whereas others who know the disadvantages of polishing buy unpolished *dal*. Again, both types of consumers are hardly aware of the quality aspect of both grades of dal. CIPHET developed mill can produce commercial grade dal upto 73% whereas in single roller small capacity mill, a single pass produces less than 50% *dal*.

References

Anonymous 1992. Technical informations received from various centres of AICRP on Post Harvest Technology (ICAR).

Anonymous 2006. Economic Survey, National Informatics Centre, Ministry of Finance, Govt of India.

Choudhary, B.S. and S. C.Pathak 1989. Relative preference of calloso-bruchus chinensis (LINN) for different varieties of Bengal gram. Bulletin of Grain Technology. 27(3), 181-7.

Choudhary, B.S. R.R.Rawat and S. C.Pathak 1989. Relative preference of calloso bruchus chinensis (LINN) for different varieties of green gram. Bulletin of Grain Technology. 27(2) : 107- 12.

DMI 2002. Estimation of Marketable Surplus & Post Harvest Losses of Foodgrains in India. Directorate of Marketing & Inspection (DMI), Nagpur, Department of Agriculture and Co-Operation, Ministry of Agriculture, Government of India.

Grover, P.C. and Deepak Kumar 1985. Some engineering properties of pulses. Proc. On ISAE, SJC 1985, vol. 3 : (2)-27-35.

Hand Book of Agriculture. ICAR. New Delhi.

Nimkar, P. M., and P. K. Chattopadhyay 2001. Some Physical Properties of Green Gram. J. Agric. Engng Res.80 (2), 183-189 doi:10.1006/jaer.2000.0664

Paranthaman, R., R. Vidyalakshmi and K. Alagusundaram 2009. Production on Tannin Acyl Hydrolase from Pulse Milling By-Products Using Solid State Fermentation, Academic Journal of Plant Sciences 2 (3): 124-127,

Sharma, S.K, RK Dubey and C.KTechchandani 1985. Engineering properties of black gram, soybean and green gram. Proc. of ISAE, SJC 3 : 181-185.

Shrivastava, AK, B.Panda and N.Darshan 1990. Comparative nutritive values of different cereals in quail diets. Indian J.of Animal Sci. 60(6), pp 720-724.

Singh, B.P.N P.C. Sah and A.K. Tikoo. 1980. Engineering properties of food materials (Tech. Bull. CIAE/ 80/15), PHTS, CIAE, Bhopal. 1980. (Edited and Illustrated Alam, A)

Srivastava, S J., D.P.Mishra and B.P.Khare 1988. Effect of insect infestation on biochemical composition of pigeon pea (Cajanas Cajan L) seeds stored in mud bin. Bulletin of Grain Technol. 26(2) : 120-125.

US Department of Agriculture, Agricultural Research Service. 2001. USDA Nutrient Database for Standard Reference, Release 14. Nutrient Data Laboratory Home Page, http://www.nal.usda.gov/fnic/foodcomp

□□□

Chapter - 13

Pulses Trade in India : Problems, Opportunities and Policy Issues

Harbir Singh and B Gangwar
Project Directorate for Farming Systems Research, Modipuram, Meerut - 250 110, Uttar Pradesh.

Pulse legumes are considered to be the best choice for intensification of predominant cereal-based systems. The studies conducted in different agro-climatic zones under AICRP on cropping systems have revealed that the productivity of existing cropping-systems can be enhanced by the inclusion of legumes during summer or as catch crop, or in intercropping systems (Gangwar *et al.*, 2009). Further, these food legumes also serve as a feed crop in many farming systems and fetch higher prices compared to cereals and are increasingly being grown to supplement farmers' incomes (Gouda *et al.*, 1999). Although pulse legumes have many desirable characteristics in terms of increasing system's productivity, raising nutritional standards and environmental benefits, in most countries of the world they are considered secondary crops. Globally, the harvested area under pulse crops is about one-tenth of the harvested area under all cereal crops. In view of the important and diverse role played by pulses in the farming systems and in the diets of the poor population, this chapter presents the global trade scenario in pulses followed by pulses production and trade opportunities in India. An attempt is also made to highlight the major concerns and to meet the future challenges in production and trade of pulses in India.

Global Production Scenario

Global production of pulses was little more than 55 million tonnes during the first half of the 1990s. Despite fluctuations, mainly due to aberrant weather, pulses production has increased to around 64 million tonnes in recent years. Developing countries account for about 76 percent of world production of pulses. Asia dominates with 50 percent of production, mostly concentrated in India which alone produces close to one-fourth (23%) of the global output of pulses, followed by Canada (Fig. 1). Brazil is the largest pulse producer in South America, while Nigeria dominates production in Africa.

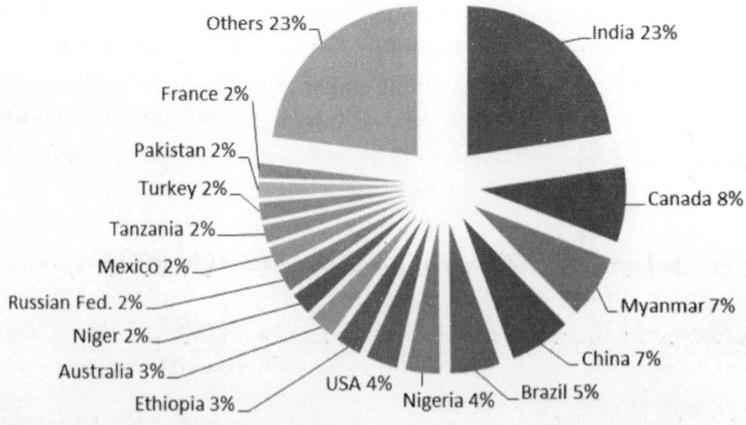

Fig.1: Global production of pulses, 2009

Source: FAOSTAT, 2009

Cultivation of pulses is not as widespread as that of cereals across the globe. Similarly, consumption of pulses as regular human food in significant amounts is concentrated only in few countries. Owing to these two characteristic features of pulses, their cultivation and trade is restricted to a few countries only. A closer look of Table 1 reveals that pulses productivity is the lowest (except that of Nigeria) in India. As against the world average of 836 kg/ha, India's pulses productivity level is less than 40 percent of the productivity in China and 29 percent lower than that of Myanmar which are India's neighbouring countries and figure among the top ten producers of the pulses in the world.

Table 1: Production of pulses in top ten producing countries, 2009

Top ten countries	Production ('000 tonnes)	Share in world production (%)	Yield (kg/ha)
India	14480	22.61	697
Canada	5185	8.10	1650
Myanmar	4406	6.88	1627
China	4331	6.76	762
Brazil	3514	5.49	906
Nigeria	2412	3.77	428
United States of America	2267	3.54	4436
Ethiopia	1840	2.87	1004
Australia	1806	2.82	1729
Niger	1571	2.45	840
Others	22234	34.7	n/a
World	64050	100.0	836

Source: FAOSTAT, 2009

Concentration of pulses production is much more revealing at the level of individual pulse crop in comparison to the aggregate pulses production. This is clearly reflected in the table 2 which provides information on production and productivity of four most important pulses in terms of output in top ten producer countries. Dry peas ranks first with about 18 percent share, and gram ranks second with about 13 percent share in total production of pulses in the world. Lentil and pigeon pea occupy third and fourth position with 5.6 and 5.4 percent share, respectively. India alone produces 71 percent of pigeon pea (*tur*) and 64 percent of global output of chick pea (*gram*). Its share in lentil and dry peas is 30 and 7.5 percent, respectively. Except dry peas, where Canada and France are major producers, India's share in pulses production, even among top ten countries is significantly higher than that of rest of the countries. Except Canada, Myanmar, Pakistan and Turkey where production of chick pea, pigeon pea and lentil varies from 0.5 to less than one million tonnes, production of these three pulses in other countries is very small.

Table 2: Production of major pulse crops in top ten countries

Chick-peas (*gram*)			Pigeon pea (*tur*)			Lentil (*masur*)			Dry peas		
Country	Prodn. ('000 ton)	Yield (kg/ha)	Country	Prodn. ('000 ton)	Yield (kg/ha)	Country	Prodn. ('000 ton)	Yield (kg/ha)	Country	Prodn. ('000 ton)	Yield (kg/ha)
India	5124	792	India	2300	674	India	985	693	Canada	2276	1863
Turkey	623	959	Myanmar	484	940	Canada	612	1097	France	1650	4668
Pakistan	549	573	Kenya	99	547	Turkey	548	1198	China	1353	1488
Iran	307	407	Uganda	83	1000	Nepal	152	830	Russian Fed.	1187	1620
Mexico	238	1598	Malawi	79	642	Syria	142	1073	India	800	1263
Myanmar	223	1098	Tanzania	49	716	USA	138	1321	Ukraine	540	1761
Ethiopia	146	868	Nepal	25	894	China	136	1513	Germany	426	3118
Australia	143	923	Dominican Republic	13	1000	Iran	121	495	USA	348	2313
Iraq	100	557	Congo	5	625	Australia	119	867	Australia	329	935
Canada	92	1080	Haiti	3	401	Bangladesh	118	758	UK	274	3613
World	7993	771	World	3152	708	World	3265	869	World	1755	10731

Source: FAOSTAT, 2006

Global Trade Scenario

About 15 percent of total production of pulses is traded globally and total trade volume in pulses hovers around 9 million tons. Disaggregate trade figures show that more than one-third of total pulses trade is accounted for by dry peas and another 31 percent is by dry beans. The size of trade in chick pea is around 0.82 million tonnes while lentil trade varies around 1 million ton.

India is the largest importer with close to 20 percent share in world import of pulses. Pakistan and Egypt are next with each accounting for 4.3 percent share in global imports. Fourth and fifth place is held by Bangladesh and USA (Table 3). Canada and Australia remain the top two largest exporters of pulses followed by China, Myanmar, France, the United States and Turkey. Canada alone accounts for about one-third of world exports of pulses which consists mainly of dry peas and lentils. Canada's major markets include India, Bangladesh, and Pakistan, the Middle East, North Africa, and Latin America.

Table 3: Average trade in pulses during 2001-2004 and major exporters and importers

Pulses	Trade volume ('000 tonnes)	Share of each pulse (%)	Major exporter	Share (%)	Major importer	Share (%)
Dry peas	2968	34.0	Canada	25.1	India	19.1
Dry beans	2721	31.0	Australia	10.0	Egypt	4.3
Lentil	1090	12.0	Myanmar	11.1	Pakistan	4.3
Chick pea	822	9.0	China	9.1	Bangladesh	3.9
Broad-beans	561	6.0	France	9.0	USA	2.7
			USA	6.4		
Total pulses	8730	100.0	World	100.0	World	100.0

Source: FAOSTAT, 2006

Trends in Pulses Production in India

Pulses in India are grown under a wide range of agro-climate conditions. Though in recent years, India's share in global pulse production has increased to 27-28 per cent owing to record harvest in 2010-11 (18.24 million tonnes) traditionally, the pulses production hovered around 13.5 to 15 million ton during the last decade. To alleviate protein-energy malnutrition in India, a minimum of 50 gm of pulses per capita per day is required in addition to other sources of protein such as cereals, milk, meat, egg, etc. To make up this shortfall about 27 million tonnes of pulses are required by 2015 (TMOP, 2000).

The annual domestic demand for pulses is around 18-19 million tones, leaving a gap of around 3 million tons for meeting domestic requirements in the past. This is for the first time that pulses production in the country has reached close to be adequate for meeting the domestic demand. Among various factors contributing to a record production of pulses during 2010-11, was the adequate availability of quality seeds of high yielding varieties and price incentives (e.g., 30% increase in case of pigeon pea) in the form of minimum support prices (MSP) which played a significant role in moving closer to achieve self-sufficiency in pulses.

The long term trend in area, production and yield of pulses is depicted in Fig. 2. It clearly shows that though productivity of pulses has been fluctuating around 600 kg/ha, it shows an increasing trend, and the latest data show pulse productivity touching a record 697 kg/ha in 2010-11. However, the present productivity growth is not sufficient enough to meet the diversified pulses demand of the rising population. As discussed earlier, pulses productivity in India (697 kg/ha) is the lowest, except that of Nigeria (428 kg/ha).

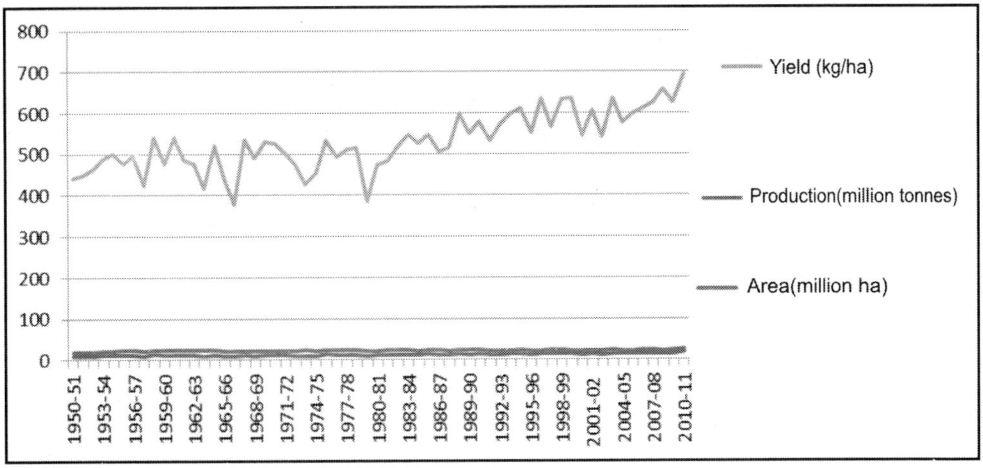

Fig. 2: Area, production and yield of pulses in India, 1950-51 to 2010-11

Source: Agricultural Statistics at a Glance, 2011, Government of India.

Trade Opportunities for India

Exports

Though India imports large amount of pulses from the rest of the world, at the same time, it is interesting to note that India is also the largest processor of pulses, because the pulses exporting nations such as Myanmar, Canada and Australia, do not have adequate pulses processing facility. The reason is that these countries do not have much domestic consumption of pulses as food and therefore, they did not attempt to develop domestic processing industry. Due to this India re-exports a considerable amount of pulses. Pulses export has increased from 0.34 lakh tonnes in 1992 to 1.61 lakh tons in 2001-02, reaching a record 4.4 lakh tons in 2005-06 (Fig. 3). However, there was no consistent increase in exports. Instead, there are sudden hikes which appeared during the years 2000, 2004 and 2005. Latest available data shows that quantity of pulses exports has declined to less than 1.0 lakh tonnes (0.99 lakh ton) in 2010-11 which may be attributed to the ban on pulses export owing to higher domestic prices. Export as a percentage to the pulses production has also increased consistently to one percent during the first ten years. After that the percentage has fluctuated between one and 4 per cent. To augment availability and check prices, the Central government extended the ban on export of pulses, except *kabuli chana*, for another year till March 31, 2012.

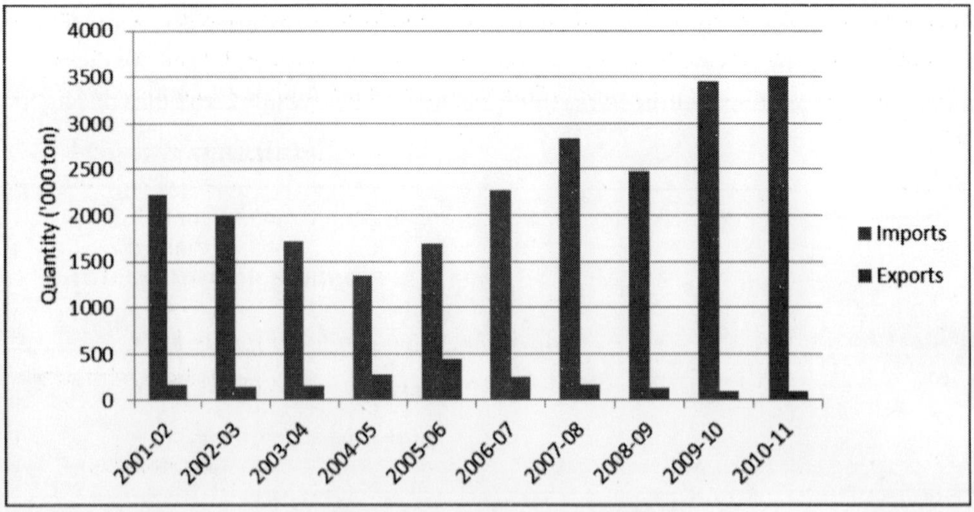

Fig. 3: India's imports and exports of pulses in quantitative terms: 2001-02 to 2010-11

Source: Agricultural Statistics at a Glance, 2011, Government of India.

Commodity-wise exports

Though export of pulses from India was kept under the category of open general license (OGL) in previous Export-Import policy of India, there is a ban on export of pulses, except *Kabuli Chana*, since last three years to meet domestic requirements and containing the domestic prices. India's exports of pulses in 2004-05, as may be seen from Table 4, made a phenomenal growth of 68 per cent over the previous year when the same reached a level of Rs. 566 crore as against Rs. 366 crore. Lentils *(masur)* continues to dominate exports, accounting for a share of 52.58 per cent of the total exports of pulses in 2004-05 followed by other varieties such as "other dried & shelled leguminous vegetables other than split" (28.2%), *tur* (8.1%), "chickpeas, dried & shelled" (6.3%), and "other dried and shelled leguminous vegetables split (3.03%). Variety wise export trend during the period reveals that barring "peas, dried & shelled, all other varieties showed exceptionally steep growth e.g. "chickpeas, dried & shelled" (565%), followed by "Lentils dried & shelled" (70.6%), *tur* (61.2%), and "Other dried & shelled leguminous vegetables other than split" (59.2%).

After 2004-05, the destination of India's exports has undergone a major change (Table 5). The share of Sri Lanka in pulses exports from India was the highest (17.24%) in 2004-05 but it declined to around half (9.35%) by 2010-11. During the same period, pulses exports to Pakistan increased more than nine-folds from 3.8% in 2004-05 to 31.7% in 2010-11. Similarly, Algeria, Turkey, Tunisia and Egypt have emerged a major destination countries for pulses exports from India in recent years.

Table 4: Commodity-wise India's exports of pulses in the major markets (Rs. crore) .

Market/Variety	2003-04	2004-05	% change
Bangladesh			
Lentils (dried & shelled)	120.37	153.64	27.64
Other dried & shelled leguminous vegetables other than split	2.07	4.51	117.87
Pigeon pea (*tur*)	1.07	2.10	96.26
Total (incl. others)	125.45	158.90	26.66
Sri Lanka			
Other dried & shelled leguminous vegetables other than split	4.66	13.38	187.12
Pigeon pea (*tur*)	0.91	6.52	616.48
Lentils (dried & shelled)	4.61	6.07	31.67
Other dried & shelled leguminous vegetables, split	0.28	5.51	82.14
Chickpeas, dried & shelled	1.56	2.55	63.46
Total (incl. others)	13.13	89.52	581.80
UAE			
Other dried & shelled leguminous vegetables other than split	10.22	22.39	119.08
Lentils (dried & shelled)	5.52	9.12	65.62
Pigeon pea (*tur*)	5.80	7.52	29.66
Chickpeas, dried & shelled	0.14	2.71	1835.7
Other dried & shelled leguminous vegetables, split	0.11	2.40	2081.8
Total (incl. others)	23.25	44.19	90.06
USA			
Other dried & shelled leguminous vegetables other than split	16.68	16.23	-2.70
Other dried & shelled leguminous vegetables, split	0.10	5.34	5240.0
Pigeon pea (*tur*)	6.60	7.53	14.09
Chickpeas, dried & shelled	0.51	1.02	100.0
Lentils (dried & shelled)	5.33	6.93	30.02
Total (incl. others)	33.52	37.34	11.40
Nepal			
Other dried & shelled leguminous vegetables other than split	18.87	23.01	21.94
Other dried & shelled leguminous vegetables, split	0.98	1.20	22.45
Pigeon pea (*tur*)	2.12	4.85	128.77
Lentils (dried & shelled)	9.58	4.18	-56.37
Total (incl. others)	31.89	30.83	-3.32
Total (All India)	**328.60**	**553.81**	**68.54**

Source: Adapted from Gandhi, 2006.

Table 5: Changes in the destination of India's export of pulses, 2004-05 to 2010-11

Country		Percentage share in total exports	
		2004-05	**2010-11**
1.	Pakistan	3.81	31.70
2.	Algeria	3.19	13.20
3.	Turkey	0.84	11.72
4.	Sri Lanka	17.24	9.35
5.	UAE	8.06	7.68
6.	Tunisia	0.36	4.05
7.	Saudi Arabia	4.77	3.08
8.	Egypt	1.58	3.45
9.	Portugal	-	2.17
10.	Kuwait	3.56	1.87

Source: CMIE Report, September 2011.

Imports

Government of India has permitted unrestricted imports of pulses with low duties for about 20 years. Still, India remains the world's largest pulses importer. India imports its requirements from various countries, such as Myanmar (black gram & pigeon pea), Canada, Australia and various other countries (Yogesh and Sukhadeve, 2011). Historically, chickpeas and dun peas (field peas) have been the largest pulse imports, but in recent times, peas (green and yellow) have become prominent because of their low price and increasing availability. Imports of high quality (9 mm and above) *kabuli* chickpea have also shown a marked increase in recent years (USDA, 2001). For many pulses, large shares of import, including desi chickpeas, pigeon peas, mungbeans, and kidney beans, come from Myanmar. Import firms favour neighbouring countries (Myanmar, China) because these offer many varieties with qualities similar to those produced in India as well as at reasonable prices, low freight rates, and relatively fast delivery. Canada and Australia are major suppliers of dry peas and *kabuli* chickpeas to the Indian market, each supplying about one-third of India's pea imports. In fact, Canada is the world's largest producer and exporter of yellow and green field peas (Pulse Canada, 2007). Despite lower per capita consumption of pulses over the past 30 years, consumption is still increasing in volume and value, and pulses are still in high demand among Indians as they are a major source of protein and used in many traditional Indian dishes.

Australia is a major supplier of chickpeas and low-priced dun peas (field peas). Most *kabuli* chickpeas come from Mexico, Australia, Canada, Turkey and Iran. Nepal and Syria account for the largest shares of Indian lentil imports. Import of pulses generally increased over the period between 1991 and 2010 (Fig. 3). Volume of import has increased from 3.13 lakh tonnes to 34.48 lakh tonnes during the same period. Negative annual change in import of pulses has appeared only for 8 out of 17 years in the above period but the changes are generally marginal. Though the quantum of imports dipped during the mid 2000, it reached a record high level of 3.5 million tonnes in 2010-11 in spite of bumper harvest. In value terms, the import of pulses increased continuously in most of the years, except during 2002-03 to 2004-05. Thereafter, the value of imports has risen consistently and reached around Rs.9,700 crores in 2009-10. It is worth noting that though India imported a record 3.5 million ton of pulses in 2010-11, the value of imports was lower (Rs.7,000 crores) in comparison to that in 2009-10 (Fig. 4). This may be attributed to the highest production in the country which helped in meeting the pulse demand for human consumption, while the low-cost imports helped in meeting the feed demand. Thus, India's export of pulses is a meagre 4.2 percent only of the total imports of the country, and the gap between value of exports and imports has widened in recent years which is a major cause of concern. It is quite evident that despite being the world's largest producer of pulses, India also imports large quantities of pulses to meet her domestic demand.

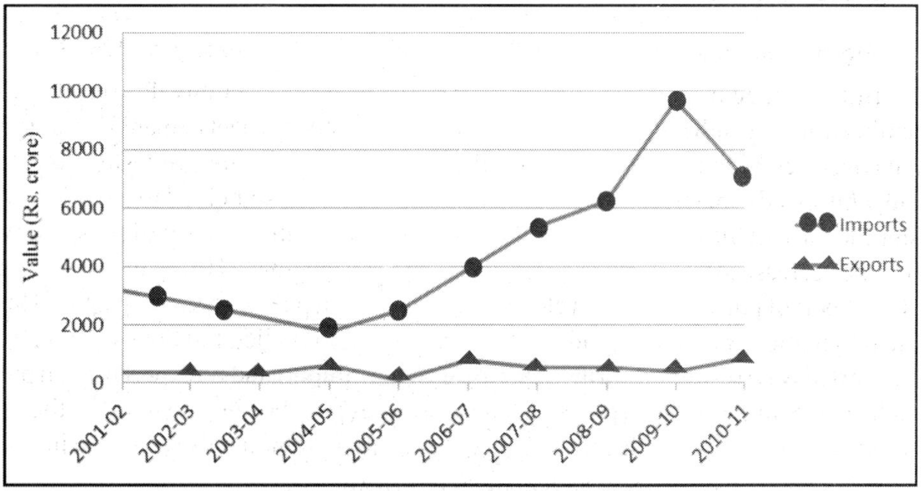

Fig. 4: India's imports and exports of pulses in value terms: 2001-02 to 2010-11

Source: Agricultural Statistics at a Glance, 2011, Government of India.

Trade Management Options

Global pulses scene indicates that production and trade in pulses are concentrated in a few countries. India is not only the largest producer but also the biggest importer of pulses. Annually, around 9 million tons of pulses are traded globally and India imports close to 2-3 million tons. Despite a ban on exports and free imports flowing in to the country, the wholesale price index of pulses has risen sharply in comparison to the price index of cereals (Fig. 5). This a matter of serious concern and calls for adoption of a broader strategy which focuses not only on higher price support and free flow of imports of pulses, but also provides technological backstopping for managing biotic and abiotic stress, achieving a breakthrough in pulse productivity on sustainable basis and providing effective institutional arrangements for marketing of pulses.

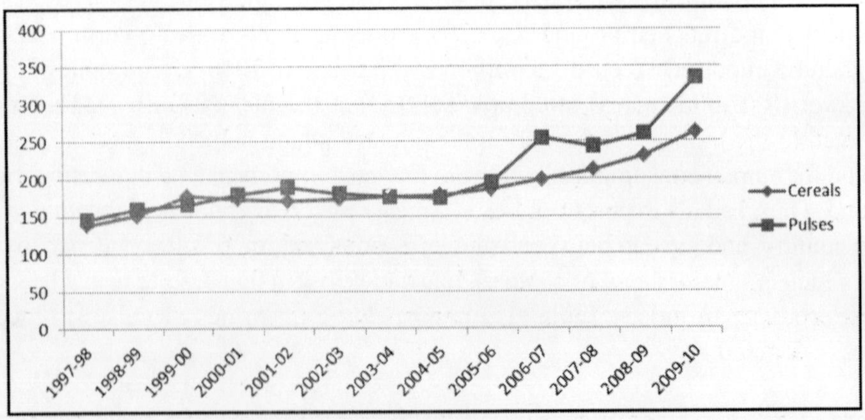

Fig. 5: Trend in wholesale price index of cereals and pulses (base year: 1993-94)

India's trade in pulses has impacted the country's economy. Particularly, the liberalized and subsidized import of pulses in India helped to meet demand-supply gaps, which occurred because of stagnation in the area under cultivation, very slow growth in yield, and speedy increase in population. These imports also helped to slow down the faster increase in the prices of different types of pulses items. Import of pulses slowed down the decrease in the net per capita availability of food grains. However, ban on export and re-export of pulses resulted in closure of some Indian pulses processing units. Thus, a more pragmatic approach to handle the pulse shortages is called for. For example, while the export may be restricted during scarcity/shortage, import and re-export of such pulses which are meant for feed purposes in other countries may be allowed to make the pulse processing units viable while meeting the demand for common pulses through imports.

Major concerns

Due to stagnant production and increasing population, it seems India's dependence on global market to meet the pulse demand would increase in future. The Indian Government has banned the export of pulses to meet the growing domestic demand, and also has given more attention towards providing price incentives to the growers in the form of higher MSP which has been raised upto 30 percent in some cases (e.g., pigeon pea). Thus, prices of pulses in India are expected to remain firm as there are prospects of minor decline in global production of pulses during the current year in comparison to 60 million tonnes harvested in the year 2010. However, demand projections shows doubling of demand for pulses (like chickpea and pigeon pea) in Africa and Asia (Rao *et al.*, 2010). Further, the scope for meeting increasing demand for pulses is limited. The acreage under pulse production in some countries like Canada and U.S. is on the decline. One of the reasons for the decline of production area in Canada, the largest producer of peas and lentils, is unsuitable weather conditions. In addition to these, there are other factors which will have bearing on the pulses production, consumption and trade. For example, as a result of increased competition for farm land use from other crops and government policy aimed at moving farmers towards producing more visible "food security" crops (*wheat and paddy*) and bio-energy crops (*corn*), pulses production may decline further. As production falls behind domestic needs, imports will rise. Imported pulses are not always cheaper and do not always meet the consumers' preferences for color, flavor and cooking characteristics as locally grown pulses do. Relying on imports also leaves countries vulnerable to supply shocks, which can periodically have a negative effect on the amount people buy and consume. Also, increase in income and urbanisation does not always have a positive impact on demand for traditional foods, which may include pulses. Initially, higher incomes allow people to buy more basic foods, stimulating demand. Once disposable income grows, eating habits change and new patterns evolve, typically resulting in some movement away from traditional foods, which may be an additional driver for inducing the declining trend (Akibode and Maredia, 2011).

Under such circumstances, India cannot and should not depend heavily on imports to meet the demand-supply gap in pulses. Instead, there is need to give more focus on basic and strategic research to address the biotic and abiotic constraints and to break the yield barrier in pulses. Further, there is huge scope to minimize the yield gap at experimental and field levels. The results of successful technology demonstrations by the Indian Institute of Pulses Research (IIPR), Kanpur shows that productivity in different pulse crops can be increased upto 50 percent using high yielding varieties along with improved package of practices. However, apart from availability of quality seeds of high yielding varieties, the strong technology back-up, remunerative support prices and effective government programmes for enhancing pulses production would go a long way in increasing pulses production in the country.

Action plan to meet the challenges

The Department of Agriculture & Cooperation (DAC), Ministry of Agriculture, Government of India has launched several programmes including National Food Security Mission (NFSM), Agricultural Technology Management Agency (ATMA), Rashtriya Krishi Vikas Yojana (RKVY). The RKVY aims at achieving 4% annual growth in the agriculture sector during the XI Plan period, by ensuring a holistic development of agriculture and allied sectors through increase in public investment, providing more flexibility and autonomy to states in the process of planning and executing agriculture sector schemes, ensuring the preparation of agriculture plans for the districts and the states based on agro-climatic conditions, availability of technology and natural resources, ensuring that the local needs/crops/priorities are better reflected in the agricultural plans of the states, reducing the yield gaps in important crops through focussed interventions and maximizing returns to the farmers in agriculture and allied sectors. The components for the RKVY cover among others the integrated development of major food crops such as wheat, paddy, coarse cereals, minor millets, pulses and oilseeds. The NFSM aims at increasing the production of rice, wheat and pulses by 10 million tons, 8 million tons and 2 million tons, respectively, by the end of the Eleventh Plan (2011-12). The planned efforts have borne fruits in the form of achieving target production of pulses even before the end of 11th Plan. The efforts for increasing pulse production got a boost in the form of Rs 300 crore being allocated in the current Budget (2011-12) under RKVY for integrated development of 60,000 villages of pulses crops during 2011-12. Since then, a set of guidelines has been prepared through appropriate interventions for effectively utilizing the allocated money in a manner that supplements the ongoing works on pulses development under NFSM and RKVY as also to build institutions that provide market linkages to the pulses growers for sustainable development of the crop.

Presently, to contain the inflationary trend in prices, export of pulses from India, except *Kabuli chana,* has been banned while duty free imports are permitted. Incentives through price policy, have also encouraged the farmers to grow more pulses. Under NFSM-pulses programme, Government of India has provided enough support for breeder and quality seed production as well as training to extension workers and farmers. During the last year alone, 240 extension personnel and 2628 farmers of different pulses growing states were trained on improved pulses production and protection technologies. All these combined efforts have played a major role in achieving the highest pulse production in the country. But keeping in view the rising population, higher incomes and scope for value added products of pulses, these effort need to be further strengthened not only to achieve self-sufficiency in pulses production but also to export value-added products of pulses. NAFED, set up under the Ministry of Agriculture, Government of India, is another government agency engaged in promoting direct export/import and procurement of pulses in the country. Its International Trade Division is involved in constant research and exploration work for this purpose. Regular efforts are made by the Division to identify new markets and new commodities. The Division has also been engaged to promote direct export business of pulses like lentils, gram, etc. It also undertakes import of pulses.

Many ICAR research institutes as well as various State Agricultural Universities (SAUs) are engaged in R&D of pulses in the country. Some of the major ICAR institutes include: Indian Institute of Pulses Research (Kanpur) with a major focus on basic and applied research in pulses such as chickpea, pigeon pea, *mung* bean, *urad* bean, lentil, etc. To boost production of pulses, it has developed a host of high yielding varieties in respect of these crops. Besides, Indian Agriculture Research Institute, New Delhi has also developed a large number of improved varieties of pulses. Similarly, Project

Directorate on cropping systems (PDCSR) now project directorate for farming systems (PDFSR) was activiely engaged in promoting and developing efficient alternate cropping systems involving pulses through on station and on farm experiments at Cultivator & fields under AICRP on CS. The R&D efforts of the ICAR and SAUs have paved the way for Green Revolution in pulses as a group of thirty-one Indian scientists have recently (2 November 2011) decoded the genome of pigeon pea (*Arhar*), the second most important pulse crop of India. This is the first plant genome sequenced entirely through a network of Indian institutions and it will provide highly valuable resource for varietal improvement of pigeon pea.

Policy and Institutional Issues

There are a number of key economic, political and technological factors that are reshaping pulse trade in India (Agbola, 2003). Keeping in view the stagnant production of pulses, the government has adopted two-pronged strategy for enhancing pulses production and meeting domestic requirements. While incentives are given in the form of assured prices or minimum support prices (MSP) for increasing production, the import of pulses are freely allowed for meeting overall domestic requirements. However, the mechanism of price support (MPS) has not been much effective in the past as the prices declared by the Government always remained lower than the ruling prices. A closer look at the production of pulses and their MSP reveals an interesting picture (Table 6).

Table 6: Minimum support prices (MSP) of pulses and their production in India

Year	Production	Minimum support prices (Rs/ *Quintal*)				
	(million tons)	Chickpea	Pigeonpea	Green gram	Black gram	Lentil
2000-01	11.08	1100	1200	1200	1200	1200
2001-02	13.37(20.7)	1200(9.1)	1320(10.0)	1320(10.0)	1320(10.0)	1300(8.3)
2002-03	11.13(-16.8)	1220(1.7)	1320(0.0)	1330(0.8)	1330(0.8)	1320(1.5)
2003-04	14.91(34.0)	1400(14.8)	1360(3.0)	1370(3.0)	1370(3.0)	1500(13.6)
2004-05	13.13(-11.9)	1425(1.8)	1390(2.2)	1410(2.9)	1410(2.9)	1525(1.7)
2005-06	13.39(2.0)	1435(0.7)	1400(0.7)	1520(7.8)	1520(7.8)	1535(0.7)
2006-07	14.2(6.0)	1445(0.7)	1410(0.7)	1520(0.0)	1520(0.0)	1545(0.7)
2007-08	14.76(3.9)	1600(10.7)	1550(9.9)	1700(11.8)	1700(11.8)	1700(10.0)
2008-09	14.57(-1.3)	1730(8.1)	2000(29.0)	2520(48.2)	2520(48.2)	1870(10.0)
2009-10	14.66(0.6)	1760(1.7)	2300(15.0)	2760(9.5)	2520(0.0)	1870(0.0)
2010-11	18.24(24.4)	2100(19.3)	3000(30.4)	3170(14.9)	2900(15.1)	2250(20.3)
2011-12	17.21(-5.6)	2800(33.3)	3200(6.7)	3500(10.4)	3300(13.8)	2800(24.4)

Source: CACP Report (different years). Figures in parentheses are the percentage changes over the previous year.

During the period 2000-01 to 2007-08, the MSP of pulses were raised only marginally (maximum 11.8%) and in some years for few crops it remained even constant. Therefore, it is not surprising that pulses production also stagnated during this period and it hovered around 14 million ton. For the first time in 2008-09, MSP for green gram, black gram and pigeon pea were enhanced substantially (29% to 48%). It is well known that prices have a lag effect. Therefore, total pulse production improved only marginally in 2009-10 (14.6 million ton). Since the policy support for ensuring higher prices for pulses growers continued, the country achieved a record pulse production of 18.09 million ton in 2010-11. Thus, it can be inferred that a proper mix of price incentives, infrastructure and technological back up will be the key elements of the strategy for further raising the production levels. Recognizing the importance of pulses in household livelihoods and nutritional security, the Government has taken a right step in setting up a National Mission on Pulses to further bolster domestic production in the country.

Future outlook

Though recent years have seen a record harvest of pulses, the overall historical trend has been one of a steady decline in the world pulse production and consumption. This trend has also been observed in India– one of the largest producer, consumer and also the largest importer of pulses. There is no doubt that higher and sustainable production of pulses would be needed not only to be self- sufficient in meeting domestic requirement but also for having sufficient surplus for export to the niche markets for higher realization from value-added products as well as foreign exchange earnings. This calls for a holistic approach addressing all the issues related to technology, policy, institutional set up and infrastructure facilities. More specifically, such an approach should be based on *inter alia* bringing more area under cultivation of pulses, intensifying public sector research and exploring new avenues for technological breakthrough (such as genome sequencing) for developing high yielding varieties for cultivation of pulses crops. Access to market, crop insurance and reasonable minimum support price to the farmers would help in making pulse production more profitable in the prevailing farming systems. Government's efforts towards enhancing pulse production through integrated programmes such as NFSM, RKVY and pulse villages hold the key to not only to increase pulse production in a sustainable way but also to meet domestic requirements and harness trade opportunities in pulses.

References

Akibode, Sitou and Maredia M 2011, Global and Regional Trends in Production, Trade and Consumption of Food Legume Crops. Report submitted to SPIA.

Agbola, Frank W. 2003, Pulse Policy Reform in India: Implications for Trade, Prices and Production, 1970-1999, Paper presented at the 47th Annual Conference of the Australian Agricultural and Resource Economics Society, Fremantle, Australia, 11-14 February, 2013.

Part - II
Cropwise Issues and Technologies

Gandhi, G. P 2006, Export prospects of pulses from India, Science Tech Entrepreneur, April 2006.

Gangwar B, NB Singh and Gangwar KS 2009, Crop intensification through legumes, In: Massod Ali, Sanjjev Gupta, PS Basu an Naimuddin (Eds.), Legumes for Ecological Sustainability, Indian Society of Pulses Research and Development, IIPR, Kanpur, India. p. 567

Gouda, Ali, Erskine, Halila, Johansen, Kusmenoglu, Mahmoud, Malik, Meskine, Rahman, Sapkota and Zong 1999, Trends in support for research and development of cool season food legumes in the developing countries, In: Linking research and marketing opportunities for pulses in the 21st century, 1999.

Pulse Canada 2007, Canada and the Global Pulse Industry, Portage Avenue, Minipeg, MB.

Rao, Parthasarathy, Birthal P.S., Bhagavatula S. and Bantilan M.C.S. 2010, Chickpea and Pigeonpea Economies in Asia: Facts, Trends and Outlook. Patancheru, Andhra Pradesh, India: International Crops Research Institute for the Semi-Arid Tropics. p. 76

TMOP (Technology Mission on Oilseeds and Pulses), 2000. Expert Committee Report onPulses, DARE, New Delhi.

USDA 2001, India Grain and Feed Shopping for Pulses, GAIN Report #IN1065, United States Department of Agriculture, USA.

Yogesh, Y B and Varsha S. Sukhadeve 2011, Emerging challenges and issues of pulses imports in India, Asian Journal of Management Research, 1 (2):539-549.

❑❑❑

Chapter - 14

Pigeonpea

N.D. Majumder and Deshraj
Indian Institute of Pulses Research (ICAR), Kalyanpur, Kanpur, Uttar Pradesh.

Of the total global pigeonpea production, over 90 percent of this is grown in India. The major pigeonpea growing states are Karnataka, Andhra Pradesh, Tamil Nadu, Maharashtra, Bihar, and Uttar Pradesh, contributing to 88.2% of the total pigeonpea basket and the same holds good over the years. Over the decades, whatever little increase in production level was observed i.e. from 1.75 mt (1970-1975) to 2.75 mt (2005-2009), it is neither uniform nor steady. It is mainly through the expansion of area from 2.52 m ha to a level of 3.58 m ha (2005-2009) without significant change in productivity which (1970- 1975) has remained almost stagnant around 700 kg/ha (Table 1) and is a matter of real concern (Srivastava, 1993).The gross geographical shifting of area (467 th. ha) took place mainly from traditional areas (Uttar Pradesh and Bihar) to new and less productive areas (Maharashtra, Karnataka, Gujarat, Andhra Pradesh, Orissa) with a net gain of area 837 th. ha in Central zone and South zone including expanded additional area (Table 2-3).

Pigeonpea grows well in the agro eco-region 4-13 mostly as rainfed crop and in marginal situations receiving natural precipitation ranging from 500 to 1600 mm extended over 90-210 days. It is grown as sole/mixed /intercrop/relay crop. The crop utilizes the nutrients from shallow to deep profile, grows well in sandy to loam soils with deep and active root system under different agro ecological situations (humid/sub-humid/semi arid/arid). Such diverse growing situations obviously, necessitate development of suitable varieties with different maturing periods, stature, canopy structure, flowering and biological behavior. Accordingly, varieties were developed with three distinct maturity group's viz., 90-150 (AER 4-8), 151-200 days (AER 9-12) and 180-260 days (AER 13). The gene- environment interaction is large, repeatable and variable in nature and predominantly played by environment. Obviously, under such situations, physiological modification will play a key role with peculiar phenotypic expressions. Since stabilization of environment is impossible and neither genotypic efficiency alone nor stability works, adaptability and that too specific adaptation would be more profitable.

Table 1 : Area, Production and Productivity in Pigeonpea in India

Year	Area (000'h)	Production (t)	Productivity (Kg/ha)
1970-75	2519	1747	695
1975-80	2645	1879	710
1980-85	3012	2254	746
1985-90	3350	2491	743
1990-95	3527	2341	663
1995-2000	3437	2444	710
2000-05	3470	2279	657
2005-09	3580	2748	710

Table 2 : Pigeonpea Area (000'ha), Production (000't) and Productivity (Kg/ha) in different states in India over the years

States	1970-75	1975-80	1980-85	1985-90	1990-95	1995-2000	2000-05	2005-09
				Area (000'ha)				
India	2519	2645	3012	3350	3527	3437	3470	3580
M.S.	569	657	726	828	1026	1030	1061	1121
U.P.	536	524	512	506	514	459	382	456
M.P.	512	500	513	453	411	366	312	446
Karnataka	280	319	352	461	415	454	534	581
A.P.	194	199	245	312	320	365	473	316
Gujarat	89	120	261	333	398	381	302	285
Orissa	52	60	110	144	163	142	131	133
Bihar	120	474	86	75	69	68	39	58
T.N.	84	77	105	124	105	87	51	39
Haryana	7	7	12	37	48	31	24	30
Rajasthan	30	38	31	22	24	30	20	20
Punjab	3	7	25	28	11	14	8	8
West Bengal	26	24	23	8	4	3	4	5
				Production (000't)				
India	1747	1879	2254	2491	2341	2444	2279	2748
U.P.	614	692	697	657	549	517	418	362
M.S	298	371	431	533	529	674	712	633
M.P.	403	322	407	451	366	298	232	238
Karnataka	136	197	166	183	141	206	228	401
Gujarat	39	63	194	209	326	323	197	313

Contd.

A.P.	45	38	55	71	94	125	198	301
Orissa	1	4	25	26	10	12	7	7
Bihar	97	67	93	88	77	81	48	26
T.N.	37	43	50	92	63	63	32	25
Haryana	4	5	10	38	49	32	23	32
West Bengal	21	23	20	8	3	2	3	4
Rajasthan	13	14	13	10	12	23	11	13
Punjab	1	4	25	26	10	12	7	7
			Productivity (kg/ha)					
India	695	710	746	743	663	710	657	710
U.P.	1143	1321	1357	1293	1073	1132	1089	961
M.S	519	565	601	637	515	653	672	565
M.P.	787	644	794	997	886	815	741	737
Karnataka	477	612	467	400	348	450	423	674
Gujarat	443	519	747	623	769	838	665	1098
A.P.	233	190	225	235	295	337	419	609
Orissa	529	477	692	769	770	628	635	744
Bihar	796	710	1080	1170	1133	1204	1223	448
T.N.	444	563	465	761	605	705	627	641
Haryana	560	807	862	1030	1002	977	928	1067
West Bengal	811	954	870	1006	695	702	829	800
Rajasthan	437	347	420	421	462	771	550	650
Punjab	453	575	970	905	914	851	877	875

Table 3 : Regional shifting of cultivating area of Pigeonpea

Particulars	Gain(CZ/SZ)	Loss(NEPZ/CZ)	Net gain
Area ('000 ha)	1304	467	837
Production('000 t)	863	398	465

'A' gain- MS (486), Kar. 266, A.P. 262, Guj. 193, Or. 78, Har. 19
Loss= UP 150, MP 219, TN 32, Bi. 66
Productivity (kg/ha) 613 (80's), 749 8 Ks)
Per capita availability 50 33

Regional disparity is very high with respect to production and productivity (> 1000 kg/ha in NEPZ vs.< 700 kg/ha in SZ) and highly subject to crop weather conditions. Increasing level of both area and productions noticed in Maharashtra, Karnataka, Gujarat and Andhra Pradesh while it decreased in Uttar Pradesh, Bihar, Rajasthan, Punjab, Orissa and West Bengal. Bengal, Rajasthan, and Madhya Pradesh dominate over the others states. But productivity remained stagnant, highly unstable and could not keep pace with the ever increasing population growth, wherein genetic potential and better management could be used partly, therefore, the major issues still remained a great concern.

Higher production could be achieved through the horizontal expansion of area with minimum interference of diseases, insect pest and other abiotic factors and vertical expansion utilizing genetic potential and heterosis breeding. On the contrary, in traditional pigeonpea growing areas, both area and production threatened due to blue bull damage while the availability of irrigation facility favoured wheat cultivation, replacing pigeonpea. Unfavorable crop weather conditions, policy matters, inputs and unavailability of inputs in time and other social economic conditions cause a great loss to the production. Despite the fact of higher productivity in U.P., Bihar, Haryana and West Bengal, pigeonpea growing area declined due to the above mentioned unfavorable conditions while the area increased remarkably in Maharashtra, Karnataka, Andhra Pradesh, Gujarat etc. at faster rate. During this transition phase of geographical shifting though high productivity area had to be sacrificed, it was partially compensated from large area expansion in Maharashtra, Gujarat, Karnataka, and Andhra Pradesh. High yielding varieties, favourable management practices and improved technological advancement could safeguard the production level as otherwise regional shifting would have brought down the productivity level upto 555 kg/ha.

Though the higher productivity was visible in some sectors but national productivity remained almost stagnant. It was not allowed to boost up due to decreasing act on other sectors. Undoubtably, production increased proportionally due to increased area but overall productivity remained stagnant because the increased productivity in some states was offset by an unusual decrease in some other states. Out of total pigeonpea area the maximum portion (65%) is represented by poor yielders (<700 kg. ha) contributing to 48% of production (Table 10). Another 33% area of moderate yielders contributes to 47% of national production. Thus, only 5% production is obtained from only 2% of high yielding (> 1300 kg/ha) potential area (Table 4). This high yield portion is insignificant in comparison to the remaining vast hostile area not allowing any quantum jump in production level.

Pigeonpea, as a rainfed crop, mostly depends on the onset of monsoon, the quantity of precipitation including its distribution. Various intrinsic factors combined with other factors such as environment and genetic- environment interaction accounted for 28% to 78% across the zone. Due to its long duration, this crop is obviously subject to various abiotic and biotic stresses, one or other stages of crop growth cycle, ultimately affecting the production.

Average productivity in UP & Bihar (>1000kg/ha) is higher than Tamil Nadu, Karnataka and Andhra Pradesh (<700 kg/ha). Also the frequency of favourable weather, year wise, during the last decade was ten out ten as evident in Uttar Pradesh and Bihar as compared to nil in southern states. Similarly, the number of high yielding potential states in Uttar Pradesh (45) and Bihar (30) is better than southern states (nil). With the availability of irrigation facility some pigeonpea growing areas have shifted to rice-wheat cultivation (Table 4).

Table 4 : State wise average yield (kg/ha), and potential districts

Area (000, ha)	Yield(kg/ha)	Frequency	Potential Districts
		North East Plain Zone	
Bihar (37)	1102	(10/10)	30
UP (378)	1037	(10/10)	45
Jharkhand (62)	769	(5/10)	3
Haryana (28)	996.0	(10/10)	7
Punjab (8)	892.0	(10/10)	8
Rajasthan (19)	618	(3/10)	13
		Central Zone	
Gujarat (290)	801	(7/10)	6
MP (314)	748	(5/10)	8
MS (1063)	700	(3/10)	4
Chhattishgarh (55)	472	(1/10)	-
Orissa (128)	703	(3/10)	3
T.N. (46)	643	(0/10)	-
Karnatka (538)	508	(0/10)	-
A.P. (469)	463	(0/10)	-

At some places, due to canal irrigation, pigeonpea cultivated areas turned into saline/alkaline areas unfit for cultivation. The disadvantage in pigeonpea cultivation is mainly poor shelf life, storage, marketing profitability, traders' dependence (milling) and high elasticity in requirement.

The biophysical stress cause serious damage and economic loss, *viz*. Blue bull *(Boselaphus tragocamelus),* Wilt (*Fusarium udum*), Sterility Mosaic Disease (viral disease vector *Aceria cajani),* Phytophthora blight *(Phytophthora drechsleri* f. sp. *cajani),* Alternaria blight, pod borer complex mainly *(Helicoverpa armigera),* podfly *(Melanagromyza obtusa),* and abiotic stresses like low temperature often exert additive effect with foggy weather and compel the farmer to discontinue the crop. However, considering the fact that pigeonpea with high remunerative values as pulse, vegetable, fuel, house hold use and enriching the soil due to heavy defoliation, finds an important place for human and soil health for sustainable agriculture.

Constraints: The constraints in pigeonpea production are as follows

Crop biology

- Nutrient reserve, poor pod formation due to intrinsic perennial Nature.
- Abscission promoting factor, Poor Harvest index

Genetic base

- Narrow (released variety), broad (Germplasm)

Ecological

- Rainfed (rains delayed-uneven), harsh crop weather conditions

Resources

- Marginality-farmers, land input, fertilizers (low, unbalanced)

Biological

- Disease (wilt, SMD, PSB)
- Insects (Pod borer) Pod fly, Maruca.
- Resurgence of new pest, Blue bull, Monkey

Production

- Quality seed
- Regional shifting of area, Fierce crop competition (Cereals, Cotton, sunflower).

Management

- Poor- crop management, monitoring, Transfer of technology, processing, Post harvest.

Socio-economic

- Poor land holding
- Economic status, cooperative system/risk, unstable, Price Policy

The intrinsic and extrinsic factors associated with pigeonpea are the bottlenecks mainly due to low photosynthesis and transpiration rate, poor sink potentials, poor remobilization of nutrients, more nutrient reserve for subsequent survivals, poor pod formation due to intrinsic perennial nature, abscission promoting factors and extremely variable phenology. The ecological constraints are rainfed situation, harsh crop weather and adverse soil conditions. The production constraints are inadequate seed supply of recently released varieties, reduction in pigeonpea cultivated area or shifting area across inter-zones and degraded fertility gradient. The genetic constraints are narrow genetic base of released varieties and unidentified / unutiiized physiologically efficient genotypes. The resource constraints are associated with marginality with respect to farmers, land and input, sub-optimum input application, non adoption of improved varieties and insufficient and unbalanced application of fertilizers. The crop also suffers due to biotic stress, insect pests, diseases and blue bull. The poor crop management, inadequate transfer of technology, poor post-harvest and storage conditions. Among the Socio-economic constraints, the poor and marginal economic status of farmers and lack of cooperative systems.

Future projections and yield gaps

With the medium fertility assumption of population growth and consumption of pulses @ 55 g/ capita/day, the requirement of pulses is 26.9 m.t. and the average seed, feed and wastage (@ 12.5% is equivalent to 3.4 m.t. Thus, the total pulse production is projected to be 29.43 m.t. by 2025 A.D (Table 5). As per the present production varying from 13.39 to 14.76 m.t. during 2006 to 2009 and the demand projection, from varying 1.22 to 2.79 m.ton. need to be enhanced upto 4.58 and 5.04 m.ton by 2022 and 2025 respectively. To meet the present demand, a total of 1.22 m. ton of pulses is being imported costing around of Rs. 3095 crores. Yield potentiality of early (1000-1500), medium (1500-2500) and long duration (2000-3500) was evident in both the experimental and field demonstration trial. However, realized yield in the farmers field is far below than the expected and potential yield and it can be increased reasonably with the combined effect of suitable varieties and improved technologies simultaneously minimizing the biotic and abiotic impacts. Yield gap across the zones, can be minimized further with the introduction of suitable photo-thermo insensitive varieties, thus, avoiding the transformation loss of bio-energy, meaningful use of phenology and appropriately directing the harvest index. The yield gap based on current production and demand accounts for 1.22 m.t. Within a couple of year (June 2008 to June 2010), prices of pulses (Pigeonpea) shot up to, more than double (Rs. 45/kg + Rs.100/kg) level and became unaffordable to many people even after cooling of prices.

Table 5 : Demand projection (m.t.)

Year	08-09 (Actual)	2012	2017	2022	2025
Pigeonpea	2.37	2.75	3.59	4.58	5.04
Total Pulses	14.13	17.02	21.72	26.43	29.43

Table 6 : Production and import of pulses scenario

Year	Production (m t)	Quantity ('000 tonnes)	Value (Rs. crore)
2000	13.42	252	358
2001	11.08	350	500
2002	13.37	2232	3163
2003	11.13	1995	2741
2004	14.91	1725	2288
2005	13.38	1312	1741
2006	13.39	1628	2347
2007	14.20	2271	3891
2008	14.76	2791	5278
2009	14.13	1222	3095

Major cropping system involving Pigeonpea

Short duration and extra short duration varieties play an important role in cropping system and intensive cultivation (Chauhan *et al.* 1999, Nam *et al.* 1993). The state wise major cropping systems involving pigeonpea are as follows:

Pigeonpea+ sorghum/maize: U.P., M.P., Maharashtra, Gujarat, A.P., Karnataka, T.N.

Pigeonpea + sorghum: U.P., M.P., Maharashtra.

Pigeonpea + groundnut: Maharashtra, Gujarat, U.P., A.P., T.N.

Pigeonpea + cotton: Maharashtra, A.P.

Pigeonpea+ mungbean/urdbean: U.P., Bihar, M.P., T.N., A.P., Gujarat, Haryana, Maharashtra.

Improved varieties

State-wise suitable improved varieties developed are given below:

Maharashtra: BMR736, BSMR 853, BDN 708, Vipula, Asha, BDN2, ICPL 87, C 11, AKT 8811, Maruthi

Karnataka: BRG 1, BRG 2, Maruthi, Asha, BSMR736, TS 3, Hy 3C, TTB 7, WRP 1, GC 11-39

Madhya Pradesh: JKM 7, JKM 189, JA 4, MA 3, Asha, ICPL 87, ICP 8863

Uttar Pradesh: UPAS 120, NDA 1, MAL 13, Bahar, Pusa 992, Amar, Azad, MA 6, ICPL 87

Andhra Pradesh: LRG 41, LRG 55 Maruthi, WRG 27, Laxmi (ICP 85063), Asha, MRG 66, Durga

Gujarat: GT 101, BDN 2, BSMR 736, 853, GTH 1 (Hybrid), Banas, GT 1, GT 100, ICPL 87

Bihar: Pusa 9, Sharad, BR 65, Bahar, NDA 2, NDA 1

Orissa: Asha, Laxmi, UPAS 120

Tamil Nadu: Asha, Maruthi, CO5, CO6, CORG 9701, Vamban 1, Vamban 2

The genotypes *viz.,* Pusa 992 in Punjab, GT 101, JKM 189 and Asha in Gujarat, BSMR 853 in Maharashtra and TTB 7, JKM 186, CO 6 and LRG 41 in South Zone, were found most compatible in intercropping systems.

Vegetable pigeonpea

Pigeonpea is used as vegetable (as a substitute of pea) at various places in South and Central zone. Pigeonpea when used as vegetable, Grain quality is important when large grain size, consumer preference, colour texture, higher sugar content, protein, fibre, and minimum level of anti nutritional factors such as phenolic compounds were important. Reserved food material as a phenomenal requirement for subsequent flushes and fruiting are important. Sowing time is very important so that it can fetches better price during the lean season. The physicochemical characters of a few varieties are given below (Table 7). With appropriate sowing time and good agronomic management, three harvests can be made for better profit.

Table 7 : Estimates of different contents (Protein, Phenol, green pod/seed weight, dry seed weight, sugar and fibre) in dual type of Pigeonpea grain.

Parameters	**At 50% Maturity weight**		**Dry Seed weight (gm)**	**Fibre content (%)**	Sugar content (%)	**Protein content (%)**	**Total Phenol content**
	Green pod (kg)	Seed (gm)					
Mean	4558	982	14.3	6.88	4.19	18.19	0.29
Range	3995-5619	96-1989	12.0-16.7	6.56-7.19	3.86-4.44	17.48-19.40	0.26-0.33
Promising Culture Rank (I-III)	BRG 1-2 (6770)	BSMR 846 (3127)	BRG1 (17.3)	BRG1 (7.59)	BRG 1 (4.80)	SKNP 9711 (20.86)	Hy3C (0.19)
	BRG 1 (6708)	BRG 1-2 (2265)	BRG1 (17.0)	BRG1 (7.48)	BRG1 (4.78)	BRG 1 (20.47)	BSMR 846 (0.19)
	BRG 2 (6319)	PTV 607 (2067)	Hy3C (17.0)	Hy3C (7.36)	BRG1-2 (4.72)	BRG 2 (20.11)	PTV 607 (0.22)

Figures in parentheses show the values for respective varieties.

Some of the important vegetable pigeonpea varieties are BRG 1, 2, Hy 3c, BSMR 846, SKNP 9711 and PTV 807. On an average, 4000-5600 kg/ha of green pod can be harvested. These varieties had large seed size (14-17g/100 seed) with higher content of fibre content (5-6%), sugar (3.8-4.4%), protein (17-19%) and less phenol content 0.26-0.33, thus, suitable for vegetable consumption. Among the varieties tested BRG 1 BRG 2, BSMR 846, HY 3C, SKNP 9711, besides being good yielders, also registered good seed quality characters. These varieties can be selectively grown depending on regional suitability. The physic-chemical characteristics, very often environment influenced by the environment, primarily, temperature, light intensity, soil moisture and fertility etc. Grain size and colour also play an important role in consumers preference. Selecting the variety to be grown varies from place to place.

Table 8 : Seed weight in dual purpose Pigeonpea (Location wise) entries

Entries	Bangalore	S.K. Nagar	Badnapur	Coimbatore	Rahuri	Mean
BRG 1	16.7	16.0	16.6	17.3	17.0	16.7
BRG 1-2	16.3	12.9	15.9	14.7	16.0	15.2
BRG 2	12.7	11.0	12.9	11.7	15.0	12.7
PTV 607	12.7	9.5	11.7	13.2	13.0	12.0
BSMR 846	15.3	13.6	14.1	14.8	14.0	14.4
Hy 3C	16.3	12.3	14.1	14.0	17.0	14.7
Mean	15.0	12.6	14.2	14.3	15.3	-

Recommended production technology

Pigeonpea was usually considered as a low-input marginal crop and therefore, not much attention was paid for development of improved production technology except for some nutritional requirement. Based on recent studies, the production technologies were strengthened and the most suited to different zones are highlighted in Table-9.

Some specific management strategies

Hybrid Pigeonpea : For hybrid pigeonpea, row spacing of 75-90 cm was found effective and application of 20-60-25-20 and 15 kg N, P, K, S and Zn per ha was found optimum. Improved varieties alone are not sufficient but in combination with agronomic management (Chauhan, Y.S., 1990) and mineral nutrition (Johnson *et al.*1990), production can get a significant boost.

Nutrient management : Application of *Rhizobium,* Phosphate solubilizing bacteria, PGPR and compost (2.5 t/ha) saved application of 50 percent fertilizers. A good number of beneficiary and efficient *Rhizobium* stains were identified of which CPR 9 performed consistently better. Sources of host-plant resistance to wilt, sterility mosaic disease and tolerance to pod fly were also identified. Foliar application of urea, KCI and seed soaking with $CaCl_2$ and K_2HPO_4 reduced the adverse affect due to stress' in pigeonpea. Seed treatment with micronutrients such as zinc, boron, molybdenum and iron was as found as effective as soil application. Zinc and boron in the form of $ZnSO_4$ and Borax helped in increasing harvest index by 35 to 39%, respectively. Applications of 20 kg Sulphur along with 100 kg DAP/ha enhanced the productivity of pigeonpea. Addition of pigeonpea, as a sole crop or as an intercrop in the cropping system, not only helped in soil N build up, but also made more of phosphorus reserves available for subsequent crops.

Table 9 : Recommended package of practices for different zones

Planting time	Seed rate (kg/ha) & spacing (cm)	Seed inoculation	Fertilizer	Irrigation	Weed management	Cropping system
North west plain zone : Punjab, Haryana, Delhi, Western Uttar Pradesh and north West Rajasthan						
2nd fortnight of May to 1st fortnight of June.	18-20 60x20	*Rhizobium* culture @ one packet for 10 kg seed.	N:P:K:S: ZnSO4@ 20:40:20:20:25 kg/ha.	Pre sowing and in absence of rains : at flowering and pod development stage.	Two hand weeding at 25 and 45 days after sowing (DAS) or application of Pendimethalin @ 1.25 kg a.i./ha followed by one hand weeding at 45-60 DAS.	1. Pigeonpea-wheat sequential cropping. 2. Intercropping of mungbean/ urdbean with pigeonpea (1:1 row ratio)
North east plain zone : Eastern part of Uttar Pradesh, Bihar, West Bengal, Assam & Jharkhand						
Early: 1st fort night of June. **Late:** 1st fortnight of July **Pre-*rabi*:** 1st fortnight of Sept.	18-20 60 x 20 12-15 75 x 25 25-30 30 x 10	-do-	-do-	**Early** : Pre sowing **Pre-*rabi*:** At 45 days after sowing (DAS) and 100-110 DAS in absence of rains.	-do-	**Early** : Pigeonpea-wheat sequential cropping. **Late:** Pigeonpea sorghum mungbean/ urdbean/ Sesamum by pairing pigeonpea rows at 40/8- cm and planting one rows of intercrop.
Central zone : Gujarat, Maharashtra, M.P., Rajasthan & Chhattisgarh.						
1st fortnight of July onset of monsoon	15-18 60 x 20	-do-	N:P:S @ 20:40:20 k g/ha	In absence of rains at flowering and pod development stage.	-do-	Pigeonpea + Groundnut(2:4) Pigeonpea = Soybean (2:4), Pigeonpea+ Sorghum (2:1) and pigeonpea+ urdbean (1:1)
Southern zone : A.P., Karnataka & Orissa						
2nd fortnight of June/on set of monsoon.	-do-	-do-	-do-	-do-	-do-	Pigeonpea= mungbean/ urdbean /cowpea (1:1), Pigeonpea+ groundnut (4:2).

In pigeonpea-wheat system, pigeonpea grown in rotation with wheat as sole pigeonpea, pigeonpea + mungbean, Pigeonpea + cowpea did not benefit much from residual fertility. Pigeonpea, being a long duration crop, utilized most of the fixed N, hence could not be called a soil ameliorator. But its beneficial effect on biomass production and grain yield of wheat has been reported on account of its fertilization through noticeable amount of leaf fall.

Regarding management of crop residues and rain water harvesting, progressive farmers take advantage of run-off and other mode of water harvest and use the water for life saving irrigation for successful drought management. Residues are also used for nutrient management as organic source as well as for the preparation of vermicompost and are beneficial for efficient residue management (Rao *et. al.* 1983 Malik 1994)

Rain water harvest : Zone wise varieties were identified for both intercropping and delayed cropping in case of aberrant rainfall conditions. So, under such situation farmers may select suitable varieties. In sorghum+pigeonpea intercropping system, the growth period of both the crops fall exactly at the same time as a result of which tall sorghum adversely affects the growth of the associated pigeonpea. Further, higher amounts of N applied to sorghum + pigeonpea system result in enhansed growth and yield of sorghum and consequently, the growth and yield of pigeonpea is adversely affected Therefore, moderate quantity of N (25-50 kg N/ha) should be applied to sorghum+pigeonpea system under rainfed condition to achieve higher yield and N use efficiency. Research work on red and black soils revealed that there was non- significant transfer of N from pigeonpea to sorghum, therefore, suggested applying N and P to sorghum in sorghum/ pigeonpea system under rainfed condition.

In India, about 80% of pigeonpea is grown under intercropping with different crop combination. Pigeonpea intercropped with short duration pulses/ legume is most popular combination in all the pigeonpea growing areas. Among them, pigeonpea+ groundnut system is the most prevalent where groundnut is characterized by rapid canopy coverage of the ground and uses the resources more efficiently. It was emphasized that both the crops as sole and intercrop responded upto 90 kg P_2O_5/ha.

System based agronomical practices for different zones

The impact of different agronomical practices in different zones are highlighted as below:

Agronomic Practices	Impact
Plant Population	
• Optimum plant density of pigeonpea group genotypes (early maturing) 25-33 plants/m^2. • 22 plants /m^2 (*rabi*), wider spacing 11 plants/ m^2 (long maturity group).	• Optimum plant population significantly contributed towards the yield and critical solar energy harvest, conversion and higher production per unit of land.
Intercropping	
• Pigeonpea +Sorghum/ Pearl millet/Seasmum (2:1 row ratio) and (2:4 row ratio)	• Bonus yield of intercrops, highly remunerative as compared to sole crop of pigeonpea both under optimal and sub-optimal fertility levels.
• Pigeonpea+Mungbean/Urdbean/Soybean/ Maize/ Groundnut	• Reduced wilt incidence (12-21%) than sole crop of pigeonpea.
• Pigeonpea + sorghum/Sesamum in 2:1 (row ratio) and in mixed additive system.	• Reduced pod borer incidence 27% than sole pigeonpea damage 49% with the help of bird perches (Sorghum, maize)
• Pigeonpea+rice (extra early) while pigeonpea	• Reduce the grain yield of Pigeonpea upto such

Contd.

(long duration)+ rice (early maturing- cv Narendra 118) was viable (LER 1:2).	extant that it was found productive and remunerative.
• Pigeonpea + maize (2:1) supplemented with FYM @ 5 tonnes per hectare.	• Higher pigeonpea equivalent yield (PEY) net return, return /Re invested and LER.
Noh-West Plain Zone (NWPZ):	
• Pigeonpea (Paras, Pusa 992, AL 1455) + Mungbean/ Urdbean/Soybean/Groundnut	• Bonus yield of intercrops
Central Zone (CZ):	
• Pigeonpea (Asha, JKM7, 189,GT 101, BSMR 736, 853) + Mungbean/ Urdbean/Soybean/ Groundnut	
Southern Zone (SZ):	
• Pigeonpea(TTB 7, JKM 186, Co6, LRG 41)+ Mungbean/ Urdbean/ Urdbean/ Soybean/ Groundnut/Fieldbean	• Spray (2% urea) at 15 and 45 DAH of intercrop is beneficial
• Inter Cropping of Pigeonpea + Sorghum or Sesamum in 2:1(row ratio) row ratio and in mixed additive system has not been effective in reducing the incidence of wilt disease. However intercropping of pigeonpea with sorghum was most productive	• Reduced yield of pigeonpea but higher gross monitory return alongwith additional harvest fodder sorghum
Ridge and Furrow raised beds planting	
• Pigeonpea (ridges)+Rice (Furrow)	• Ensured optimum plant population and higher grain yield as a bonus without any adverse effect on grain yield on pigeonpea.
• Pigeonpea (Ridges)+Intercrops (side of furrow)	
• Pigeonpea (Ridges)+ Urdbean 2:2 (raised beds)	
Integrated Nutrient management	
• 40 kg N, 22 kg P_2O_5/ha	• Grain yield advantage as recorded 177 kg /ha.
• 20 kg/ha Sulphur or through Gypsum & SSP	• Rabi pigeonpea (Sharad), yield increase 25% over control
• Sodium molybdate @ 1.0 kg/ha and $ZnSO_4$@ 15 kg/ha	• Yield advantage 18-20 kg/ha respectively with NPKS (20-18-27-20 kg/ha) as compared to without sulphur application.
• At optimum P fertility 40 kg P 20 kg/ha S per ha (also increases P availability utilizing Rock phosphate).	• No response of Mb in *rabi* Pigeonpea.
• Pigeonpea required fertilizer dose of 20:18:27: 20:25 kg/ha N:P:K:S:Zn, 20:18:27:20:10 kg/ha N:P:K:S Zn	• Application of $ZnSO_4$ with RDF is beneficial.
• Soil application @ 4 g/kg of seed	
• Sodium Molybdenum @ 1kg/ha chelated iron @ 1 kg/ha.	
• RDF + Seed treatment borax (4 g/kg of seed) and molybdate (4 g/kg seed)	
• Seed treatment with micronutrients such as	

Contd.

zinc, boron, molybdenum and iron has been found as effective as that of soil application.	
• Application of borax, molybdenum and iron (deficient soil) through soil application, foliar application and seed treatment however, rates will vary depending on extent of deficiency in the soil.	
• Soil ameliorathan through lime in low pH soil supplement with molybdenum.	
• Vermicompost (2.5 t/ha) + organic fertilizer (75% RDF) + biofertilizer (Rhizobium + PSB+ PGPR)	
Weed management	
• Raised bed system (2.7 m width) of planting and one hand weeding at 30 DAS.	• Yield advantage (23%) over flat system and sufficient control of seasonal weeds.
• Pre-emergence application of pendimethalin (10-1.5 kg/ha) or Benthiocarb, Metalachlor Alachlor @ 2.0 kg/ha hand weeding 25-30 DAS optional second weeding (45-60 DAS)	
• Pre-emergence spray of Pendimethalin @ 0.75 kg/ha followed by post emergence spray of Paraquot for weeds in pigeonpea.	
Insect Pest management	
• Pigeonpea +Sorghum (1:2) + optimum dose of fertilizer application than sole pigeonpea with sub-optimum dose of fertilizer <50%.	• Recorded lower pod borer incidence 20%.
• Pigeonpea + Sorghum (1:2) + RDF+ chemical control (20%).	• Lower wilt incidence (11%)
Cultural Practices	
• GT-1, VRG 17, TTB 7 and VBN 2 were recorded as efficient user of Phosphorus in P- deficient soil.	• Efficient use of phosphorus through P- tolerant varieties for higher yield.
• Under delayed planting, upto first week of Aug., use of suitable genotypes such as Asha , JKM 189, BSMR 853(2), TTB 7, LRG 41 30, 38, WRP 1 recommended.	• Use of suitable varieties for contingent planning.
Drought Management	
• Spray of 2% KCl for minimizing the adverse effect of terminal drought.	• Drought management with physiological and agronomic practices.
• Foliar application of urea, KCl and seed soaking with $CaCl_2$ and K_2HPO_4 in reducing the adverse effect of water stress at the terminal growth of pigeonpea.	
• Cultural mulch and water spray to minimize the adverse effect of terminal drought.	
Hybrid Pigeonpea	
• For hybrid GTH-1 plant density of 75 x 20-30 cm. and application of 20-60-25-20 and 15 kg/ha NPKS and Zn was optimum.	• Improved technology through agronomic management.

Pigeonpea varieties for inter cropping : The varieties of pigeon pea most suited for different zones are as follows:

Pigeonpea varieties (main crops)	Associated Crop	Zone
Paras AL 1455, Pusa 992	Urdbean, mungbean	North West Plain Zone
Asha (ICPL 87119) JKM 7, GT 101, BSMR 853, JKM 189, BSMR 736	Urdbean, mungbean, soybean, groundnut maize, sorghum	Central Zone
TTB 7, JKM 186 CO 6, LRG 41, WRG 27	Urdbean, mungbean, groundnut, ragi	South Zone
NDA1, Pusa 9	Urdbean, mungbean, maize	North East Plain Zone

Pigeon pea on bunds

Farmers practice bund-cropping of pigeonpea in order to meet their own requirement in many areas in Bihar, Orissa, Chhattisgarh and Bundelkhand of Uttar Pradesh (Photo plate). This bund cropping is generally not accounted for area estimation and production.

Photo : Periferal cropping of Pigeonpea

Pigeonpea varieties identified for delayed planting

For late sown (20 days): JKM 189, Asha, BSMR 853 in Central zone.

For late sown (30-35 days): TTB 7, LRG 30, LRG 38, LRG 41, WRP 1 in Southern zone.

For late sown (30-35 days): up to late August, Azad in North-east plain zone

Transplanting of Pigeonpea

Apprehension of crop failure due to delayed monsoon can be managed or avoided by transplanting of pigeonpea. Before the onset of monsoon the seedling can be ready, thus saving 30 to 35 days of seedling stage and crop duration is also reduced. It gives higher yields with minimum plant population, and is also effective for agronomic management. This needs ensured watering at transplanting stage. This technology has a limitation that it cannot be adopted at large scale due to lack of irrigation facilities.

Some management strategies

Weed management

Pre-emergent application of Pendimethalin (0.75 kg/ha) followed by one hand weeding at 50 DAS gave higher weed control efficiency. However, pre emergence spray of pendimethalin @ 0.75 kg/ha followed by post emergence spray of paraquat @ 2 l/ha (6 weeks after sowing-WAS) was as efficient as that of weed free situation. Similarly, application of post-emergent herbicide, Imazethapyr (75g/ha) at 15-20 DAS followed by paraquat (0.48 kg/ha) at 6- 8 WAS was found promising for efficient weed management.

Water management

Water spray at the terminal stage was found quite effective in minimizing the adverse effect of terminal drought. However cultural mulch was equally effective.

Integrated pest management

In pigeonpea the average crop loss due to insect-pests accounts for about 30-50% (out of 2.77 mt total productions the loss accounts 1.1 mt) result this 15,540 crores of loss in monetary terms. The major insect pests and diseases with their spatial distribution are as follows:

Key insect pests	Problem areas
Pigeonpea-early and medium: *Maruca vitrata, Helicoverpa armigera, Melanagromyza obtusa, Mylabris pustulata, Lampeds (Exelastis atomosa), Bruchid*	NEPZ, NWPZ, CZ
Pigeonpea-late: *H. armigera, M. obtusa,* Pod bugs, *Mylabris pustulata,* Bruchids	NEPZ, NWPZ, CZ, SZ
Diseases and pathogen	
Wilt (*Fusarium udum*)	NEPZ, CZ, SZ
Sterility mosaic (*Virus*)	NEPZ, SZ
Phytophthora blight (*Phytophthora drechsleri f.Sp. cajani*)	NEPZ, NWPZ
Alternaria blight (*Alternaria alternate*)	NEPZ (post rainy)

Wilt and sterility mosaic diseases have been reported from all the zones but incidence is very low in Central zone. Phytophthora blight is more severe in CZ and NWPZ and in some parts of NEPZ but is almost absent in SZ. For other regions, specific diseases of pigeonpea are also listed. The management practices of pigeonpea diseases (zone wise) are also mentioned to combat the diseases. Efficient disease management and integrated pest management are beneficial, cost effective and environment friendly (Reddy *et. al.* 1990, Reed and Lateef 1990).

Management of nematodes

Nematodes parasitic on pulse crops attack roots, causing stunting and yield reduction. Nematode-infected plants are not only weakened, but their root systems are more susceptible to secondary infections by fungi or bacteria. There are mainly four types of nematodes, which parasitize pulse crops and cause economic losses. Root-knot nematodes are predominant and widespread throughout the pulse growing regions. Mainly two species *viz. Meloidogyne incognita* and *M. javanica* are infecting almost all the pulses. Other important ones are reniform nematodes (*Rotylenchulus reniformis*), which also infect almost all the pulse crops. These are semi-endoparasites. Third important nematodes are pigeonpea cyst nematodes (*Heterodera cajani*) which are mainly pests of *kharif* pulses like pigeonpea, etc. Fourth important are root-lesion nematodes, *Pratylenchus thornei,* which infect mainly chickpea and field pea. This nematode is migratory endoparasites, which cause lesions on the roots.

Nematodes not only damage the plants directly but also interact with other soil-borne microorganisms and cause more damage to pulse crops. They also interact with different soil-borne fungi and cause disease complexes. In pigeonpea, nematodes have been found to increase the severity of wilt and other soil-borne diseases.

Various IPM modules (incorporation of chemical and non- chemical approaches) have been developed in which different control tactics are used in combination to suppress pest below economic threshold level. These tactics also include judicious use of pesticides, based on economic threshold level and regular scouting.

Integrated pest management

Before sowing

- Deep summer ploughing
- Selection of disease resistant varieties as per recommendation for the particular location
- Application of *neem* seed powder @ 50 kg/ha or *neem* cake @ 500 kg / ha in nematode infested soils.
- Balanced use of fertilizer, including K to ensure pest tolerance in crop.

At the time of sowing

- Sowing on ridges
- Intercropping with sorghum, sesamum, finger millet, etc. as per ratio recommended for the particular location
- Seed treatment with carbendazim @ 1g/kg seed or *Trichoderma* (10 g/kg seed) + carboxin (1g / kg seed) for disease management.

- Seed treatment with *Beauveria bassiana* @ 10 g/kg or imidacloprid @ 5g/kg seed for early stage insect pest management.
- Seed coating with carbosulfan 25DS @ 3 g/kg seed or with *neem* seed powder @ 1g/kg seed
- In nematode infested soils, soil treatment with phorate 10G @ 1 to 1.5 kg a.i./ha.
- Crop rotation with non-host crops such as mustard, onion, garlic, sesamum, wheat, rice, maize, finger millet, etc.

Standing crop

- Regular monitoring of the crop for the occurrence of diseases and pests
- Use of pheromone trap and /or light trap @ 4-5/ha for insect monitoring
- When the trapped moth number reaches 4-5 moths per trap for 3-4 nights, start the first spray of insecticides

The Issues

- Low and unstable productivity
- Terminal moisture stress
- Heavy incidence of *Helicoverpa armigera*

Management strategies

- Seed production and timely supply to farmers in attractive bags treated with Fungicide (carbendazim @ 1 g a.i./kg)
- Improved production package that should include: Sulphur, Zinc, PSB, *Rhizobium,* Urea /DAP spray, Ridge/furrow planting, *Trichoderma* for wilt prone areas, IPM and other crop protection measures
- Under mechanization, sowing machines, Ridge maker, Sprayers, Sprinklers, Threshers and Mini *Dal* mill are considered important.
- For proper storage the 500 kg capacity storage bins for farmers need to be encouraged.
- The policy related strategies requiring due attention are: Revision of statistics through remote sensing (FASAL of SAC-ISRO), Price policy measures, Loans, Incentive for seed producers, Procurement/MSP and, formulation of State *Tur* Board as in Karnataka.
- Promotion of short duration pigoenpea Varieties for hilly areas (Uttarakhand, J&K, Himachal Pradesh, NEH states): VLA 1, Pusa 998, AL 881.

Future strategies

To obtain a significant and rapid increase both in production and productivity the emphasis should be given on the following.

- Development of an effective and economical I PM programme is very important.
- Validation of IPM modules and their demonstration in wider areas.
- The genetic material which has been identified to be tolerant to pod borer needs to be for pyramid resistance genes to pod borer and subsequently evaluated for pod borer resistance and their multiplication
- Development of suitable genotypes for intercropping systems.
- Installation of bio-control laboratories for mass production of NPV in major pigeonpea producing states.
- Supply of quality bio-pesticides
- Education of farmers to develop their competence
- Farmers should procure and safely store their own *neem* seed during the summer months.
- Proper training of extension staff at all levels on the IPM concept.
- Various IPM components should be made available at the farm level.
- Refinement of the technology as per the farmers' requirement.
- Involvement of women should be encouraged.

References

Ae, N., Arihara, J., Okada, K., Yoshihara, T., and Johansen, C. 1990. Phosphorous uptake by pigeonpea and its role in cropping systems of the Indian subcontinent. Science 248: 477-480.

Chauhan, Y.S. 1990. Pigeonpea: optimum agronomic management. In: The pigeonpea (Nene, Y. L., Hall, S.D., and Sheila, V.K. eds.) Wallingford, U.K.: CAB International. pp 257-278.

Chauhan, Y.S., Atukorala, W.D., Perera, K.D.A., Joseph, K.D.S.M., Saxena, K.B., and Johansen, C. 1999. Potential of extra-short-duration pigeonpea in the short rainy season of a tropical bimodal rainfall environment. Experimental Agriculture 35:87-100.

Johansen, C. 1990. Pigeonpea: mineral nutrition. In: "The Pigeonpea" (Nene, Y.L., Hall, S.D., and Sheila, V.K, eds.) Wallingford, U.K.: CAB International. pp 209-231.

Kumar Rao, J.V.D.K, Dart, P.J., and Sastry, P.V.S.S. 1983. Residual effect of pigeonpea (Cajanus cajan (L.) Millsp.) of different maturity groups. Plant and Soil. 99: 255-266.

Malik, R.P.S. 1994. Public policy and sustainable agricultural development - a natural resource accounting approach. Indian Journal of Agricultural Economics 49 (4): 569-578.

Nam, N.H., Chauhan, Y.S., and Johansen, C. 1993. Comparison of extra short duration pigeonpea with short season legumes under rainfed conditions on Alfisol. Experimental Agriculture 29: 307-316.

Reddy, M.V., Sharma, S.B., and Nene, Y.L. 1990. Pigeonpea: Disease Management. In: "The Pigeonpea" (Nene, Y.L., Hall, S.D., and Sheila, V.K. eds.), Wallingford, UK: CAB International, and Patancheru, Andhra Pradesh, India: ICRISAT. pp 303-348.

Reed, W., and Lateef, S.S. 1990. Pigeonpea: pest management. In: "The Pigeonpea" (Nene, Y.L., Hall, S.D, and Sheila, V.K., eds.) Wallingford, UK: CAB International, and Patancheru, Andhra Pradesh, India: ICRISAT pp. 349-374.

Srivastava, S.K 1993. Problems and prospects of boosting productivity of pigeonpea. International Pigeonpea Newsletter 18:22.

❑❑❑

Chapter - 15

Mungbean and Urdbean

B.B. Singh, G.P. Dixit & P.K. Katiyar
AICRP on MULLaRP, Indian Institute of Pulses Research, Kanpur-24, Uttar Pradesh

Mungbean (*Vigna radiata* L Wilczek) is considered to be originated from *Vigna sublobata*. It is also known as green gram, an important short duration grain legume with wide adaptability, low input requirements, and the ability to improve soil fertility by fixing atmospheric nitrogen. Mungbean is well suited to a large number of cropping system and constitutes an important source of cereal based diets. Mungbean is used for different purpose. The major portion is utilized in making *dal*, curries, soup, sweets and snacks. The food values of mungbean lie in its high and easily digestible protein. Mungbean is grown throughout Asia, Australia, West Indies, South and North America, Tropical and subtropical Africa. However, India alone accounts for 65% of the world acreage and 54% of the world production. It is widely cultivated in India in different seasons; however, the maximum area is under *kharif* cultivation. Extremely plastic morphology, phenology and short maturity duration (60 days) makes the crop ideal as a catch crop, inter crop or relay cropping. It is grown as sole relay crop in rice fallows during *rabi* season in A.P., Tamilnadu, Karnataka and Orissa and sole catch crop during spring/summer season in Uttar Pradesh, Bihar, West Bengal, Jharkhand, Punjab, Haryana & Rajasthan (Table 1).

The mungbean seeds contain approximately 25-28% protein, 1.0-1.5% oil, 3.5–4.5% fiber, 4.5–5.5% ash and 62–65% carbohydrates on dry weight basis. Amino acid analysis indicates that as with most grain legume crops, the concentration of sulphur containing amino acids methionine and cystine, are small. Methionine concentration is larger in urdbean than in mungbean. Lysine values are comparatively large that is why the protein of mungbean and urdbean are an excellent complement to rice in terms of balanced human nutrition.

Urdbean [*Vigna mungo* (L.) Hepper] is another important crop of the '*Vigna*' group. Archeological studies have shown that it was cultivated in the country as far back as 2200 B.C and considered to be originated from *Vigna silvestris*. It is a short day plant. Flowering is both photoperiod and temperature sensitive. However, there is a range from completely photoperiod insensitive types to strongly photoperiod sensitive types.

Table 1 : Statewise Area, production and Productivity of Mungbean

States	1970-71	2010-11	1970-71	2010-11	1970-71	2010-11
	Area (000 ha.)		Production (000 t)		Yield kg/ha.	
Andhra Pradesh	496.6	378.0	149.2	166	300	439
Bihar	103.0	175.0	36.2	117.0	351	669
Gujrat	156.6 *	209.0#	132.8 *	114#	848 *	545#
Chattisgarh	14.1 **	16.1#	3.5 **	4.3#	248 **	290#
Haryana	22.5	17.7#	9.4	7.6#	418	429#
Karnataka	80.5	404	18.9	111.0	235	275
M.P.	230.6	99	62.5	35.0	271	353
Maharastra	434.5	580.0	97.4	380	224	655
Orissa	322.4	268.0	192.6	79.0	597	293
Punjab	6.90	11.7#	3.9	8.4#	565	718#
Rajasthan	249.70	1050	90.7	652	363	621
Tamilnadu	60.10	197	11.6	72	193	368
Uttar Pradesh	13.20	78	3.8	45	288	577
West Bengal	17.40	2.20	7.7	1.4	443	624

*** Data of 1983-84 ** Data of 2000-01 #Data of 2007-08**

Based on seed colour and other characteristics, urdbean has been grouped under two main types viz, Var. *mungo*-with large black seeds and early maturity and Var. *viridis* with small greenish seeds and late maturity. The green seeded types are locally known as 'Katikalhia' urd and generally grown as mixed crop with sorghum, pigeonpea and cotton. In India like mungbean, urdbean is used for different purposes. The major portion is utilized in making *dal*, curries, soup, sweets and snacks. In South India, the most popular *idli* and *dosa* are prepared by mixing rice with urdbean flour. Urdbean contains trypsin and other growth inhibitors but these are denatured by heating.

Urdbean is the third most important pulse crop of India cultivated over a wide range of agro-climatic situations. The major urdbean growing states of the country are Maharashtra, Tamilnadu, Andhra Pradesh, Uttar Pradesh, Madhya Pradesh, Rajasthan and Karnataka (Table 2). Urdbean is also grown in all the seasons but maximum area coverage is underk *kharif* season mostly as intercrop with sorghum, pearl millet, maize, cotton, castor, pigeonpea etc. When intercropped with tall cereals or pigeonpea it smothers weed flora appreciably (20-45%) and, consequently, minimize the cost incurred on weed control. Development of short duration, photo thermo insensitive and disease resistant varieties has led to its cultivation as a sole or intercrop during spring season in North India and as a sole relay crop during winter season in the rice fallows of the coastal peninsula.

Table 2 : State-wise Area, Production and Productivity of Urdbean

States	1970-71	2010-11	1970-71	2010-11	1970-71	2010-11
	Area (000 ha.)		Production (000 t)		Yield (kg/ha)	
Andhra Pradesh	190.3	461.0	90.2	252.0	474	547
Bihar	108.8	24.30#	53.0	19.4#	487	798#
Chattisgarh	121.9 *	113.50#	34.1*	34.4#	280*	282#
Gujrat	65.4 **	104.0#	29.90 **	71.0#	457**	683#
Haryana	9.9	2.00#	3.9#	1.0#	394	500#
H.P.	28.2	12.30#	9.8#	6.3#	348	512#
J & K	18.3	15.40#	7.0#	6.6#	383	429#
Jharkhand	40.0*	97.50#	27.9#	61.0#	698#	625#
Karnataka	67.8	128.0	13.6	45.0	201	352
M.P.	627.8	592.0	156.4	231.0	249	390
Maharastra	456.7	485.0	114.2	326.0	250	672
Orissa	382.3***	149.0	126.1***	46.4	330***	345
Punjab	25.9	3.5#	11.8	1.6#	456	457#
Rajasthan	55.	128.0	27.7	94.0	504	737
Tamilnadu	97.3	319.0	24.4	127.0	251	398
Uttar Pradesh	151.7	546.0	51.5	365.0	340	668
West Bengal	174.6	53.0	73.6	35.0	422	659

*** Data of 2000-01 ** Data of 1983-84 *** Data of 1979-80 #Data of 2007-08**

Substantial reduction in crop duration coupled with photo – thermo insensitivity, early maturity and resistance to key diseases are the major research achievements in mungbean and urdbean. This has increased the per day productivity, besides making it suitable for introduction in new niches and diversification of the existing cropping systems. In the last 3 decades, mungbean has registered phenomenal increase both in area and production. Though there is lot of year to year fluctuations in acreage and total production because of the complete dependence of *kharif* season crop on rainfall, (Fig.1). The area has increased from 1.99 million ha in 1964 – 65 to 3.55 million ha in 2010-11 and it became third largest pulse crop in the country. The production has increased from 0.60 million tonnes to 1.80 million tonnes during the same period. The increase in area was 38%. This expansion in area has resulted due to development of relatively photo and thermo insensitive varieties which allowed a greater flexibility in their planting dates. Further the shortening of duration paved the way for its inclusion in a wide range of intensive cropping systems. The area under winter mungbean under rice fallows and spring/summer mungbean has increased from 5.56 lakh ha in 2001-02 to 6.04 lakh ha in 2008-09. There is ample scope for further increase in area to the tune of 2-3 million ha under rice – wheat and rice – rice cropping systems, which however, requires enhanced seed supply of improved varieties.

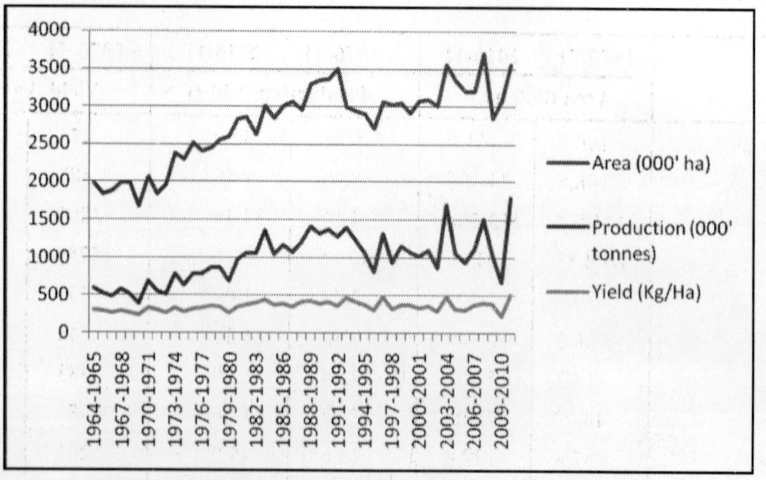

Fig. 1 : Trends in Area, Production and Productivity of Mungbean in India

In the last three decades, urdbean has registered progressive increase both in area and production (Fig. 2). The area has increased from 2.09 million ha in 1971 – 75 to 3.26 million ha in 2010 – 11. Similarly, the production has increased from 0.64 to 1.76 million ton during the period. There was positive growth of 3.05% annually in its production during 1971-2011 mainly because of positive growth in area (1.53%) and productivity (1.48%). This occurred mainly due to the development and adoption of new varieties and improved production technology.

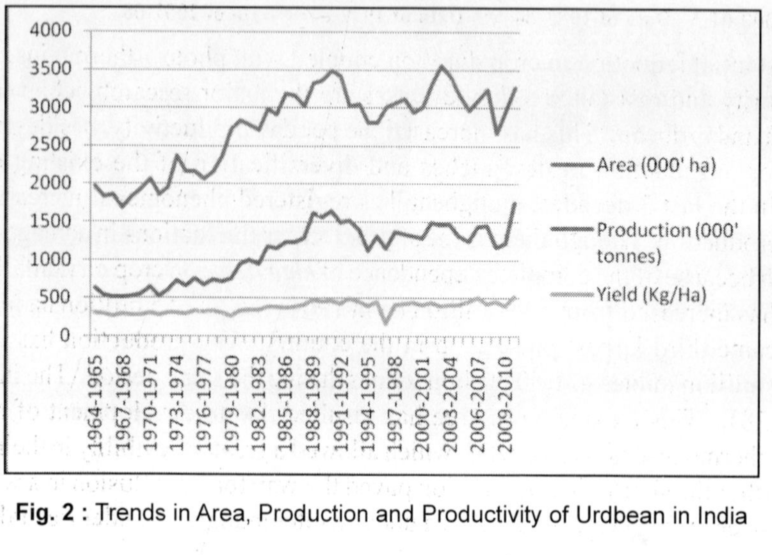

Fig. 2 : Trends in Area, Production and Productivity of Urdbean in India

Considering the realistic approach of making available 55 g / capita / day of pulses to the medium population projection, the total pulse requirements for consumption would be 26.9 million tonnes. In addition, it is assumed that nearly 12.5 per cent of gross production is used for seed, feed and wastage (5% for seed, 5% for feed and 2.5% as wastage). Taking into account these factors, a goal of producing 30.3 million tonnes of pulses by 2020 AD could be fixed. Considering the total pulses production of 12.05 million tonnes during the base year 1991-92 and 14.1 million tonnes in 1994-95 the target for 2020 AD is more than double. Mungbean and urdbean offers good opportunity to increase its area from 3 m ha to 4 m ha and productivity from 400 kg / ha to 800 kg / ha through development of efficient and stable plant types and production technologies.

Constraints

Mungbean and Urdbean are grown mostly during rainy season, however development of short duration and disease resistant varieties led its cultivation during spring season in almost all parts of country and during winter (rice fallows) in peninsular India. The major constraints for slow growth rate of mungbean production and productivity in these crops are characterized as under:

- Insect pests during *kharif*/spring seasons cause major losses to Mungbean and urdbean. During the vegetative stage, defoliators like, hairy caterpillars, semilooper and til caterpillar are common. Activity of thrips starts at the bud stage and poses serious problems when the crop attains peak flowering resulting in heavy flower drop.
- Mungbean yellow mosaic in *kharif* and spring in northern, and powdery mildew during winter season in southern part of the country cause considerable losses to these crops.
- Lack of high yielding varieties of efficient plant type for different situations which can provide 1.5 - 2.0 tonnes seed yield to compete with other crops. To serve as a commercial crop, the plant type should have determinate, photo insensitive, early maturing with high harvest index and reasonable seed yield.
- Pre harvest sprouting in rainy season is a major concern which restricts their intercropping with other companion crops such as maize, sorghum pigeonpea etc. which are being grown at the onset of monsoon.
- Lack of on farm storage facilities and its vulnerability to store grain pest result in considerable post harvest losses. Mungbean is most preferred host of bruchids and strong resistance against the insect is not available in germplasm.
- Low seed replacement rate of improved varieties has been major concern of these crops. In spite of best efforts presently the seed replacement rate is below 10 per cent. The agencies responsible for providing foundation and certified seed failed measurably in spite of great demand of seed. It is assumed that availability of quality seed uplift about 20% production with the available varieties.

Varietal development

The initial phase of varietal development had been aimed at improving locally adapted but genetically variable populations mainly by the methods of pure line and mass selections with major emphasis on traits other than yield. This resulted in the release of large number of pure lines, some of which are still cultivated in certain parts of the country. The first variety of mungbean Type 1 developed in the year 1936. The breeding work on Mungbean and Urdbean got momentum with the sanction of an adhoc research schemes on pulses by the Indian Council of Agricultural Research (ICAR) founded in 1929 and many varieties were developed during this period. Establishment of All India Co-ordinated Pulses Improvement Project (AICPIP) in 1967 provided an opportunity to access of improved germplasm to pulse breeders and to test their improved breeding lines in multi location evaluation across the country. As a result more than 130 varieties of mungbean and 88 varieties of urdbean were developed. During the last two decades, major emphasis has been to breed short duration, photo and thermo-insensitive varieties coupled with resistance to biotic stresses *viz.,* yellow mosaic virus and powdery mildew, which contributed significantly to the national production.

Impact of improved varieties

To demonstrate the production potential of newly released high yielding varieties, 1133 FLDs in mungbean were conducted in different states. On the basis of overall mean, improved genotypes recorded grain yield of 781 kg/ha, which was 22.08 per cent higher than the local varieties (Fig. 3).

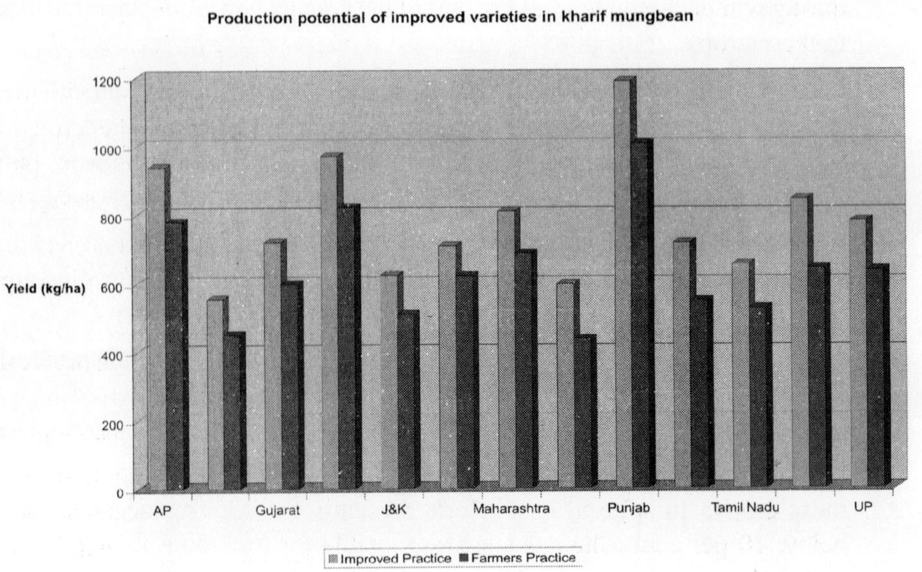

Fig 3 : Production potential of improved varieties in *kharif* mungbean

In *rabi* mungbean, 197 demonstrations were conducted and overall increase in yield was recorded to be 36.91 per cent more than the local varieties (Fig. 4).

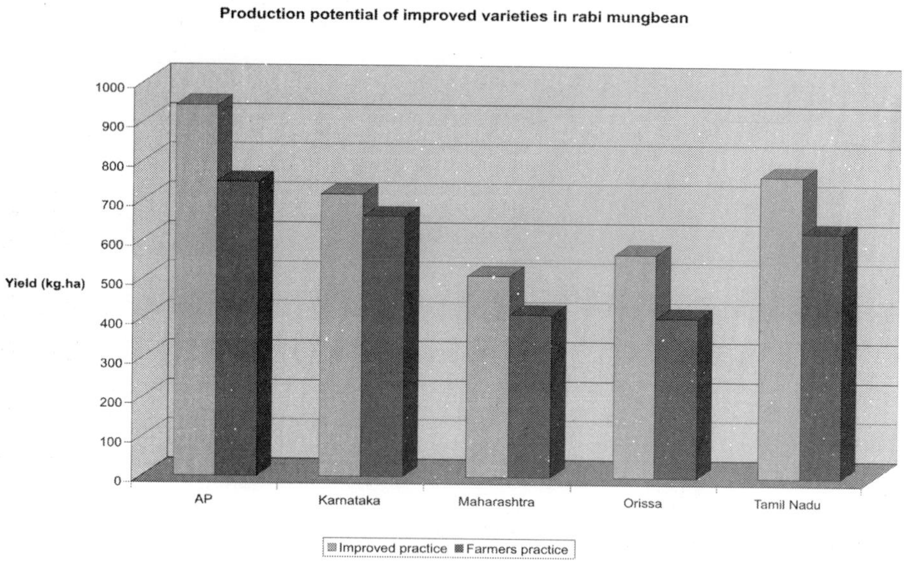

Fig 4 : Production potential of improved varieties in *Rabi* mungbean

In Spring/summer mungbean, 448 FLDs were conducted and overall increase in yield was observed to be 24.15 per cent more than the local varieties being grown by farmers (Fig. 5).

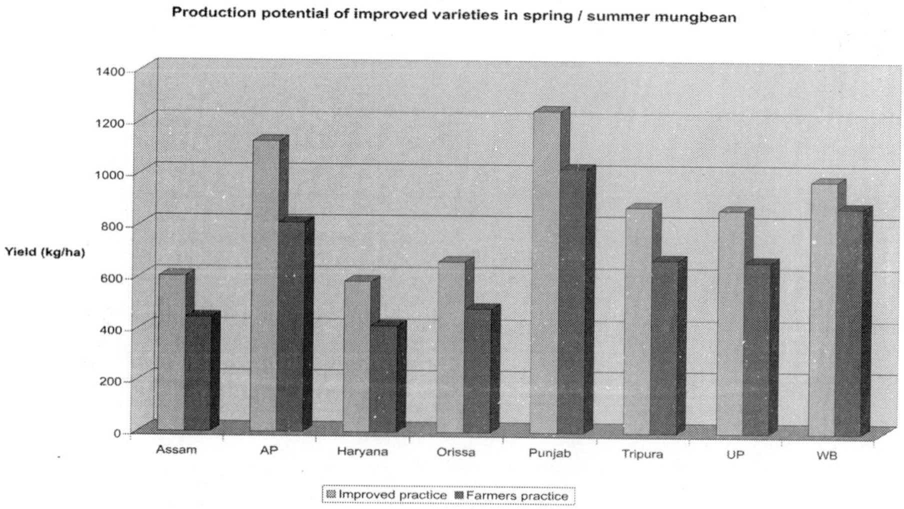

Fig 5 : Production potential of improved varieties in *spring/summer* mungbean

To demonstrate the potential of newly released high yielding varieties, 770 FLDs in urdbean were conducted in different states. On the basis of overall mean, improved genotypes recorded grain yield of 770 kg/ha which was 22.98 per cent higher than the local varieties (Fig. 6).

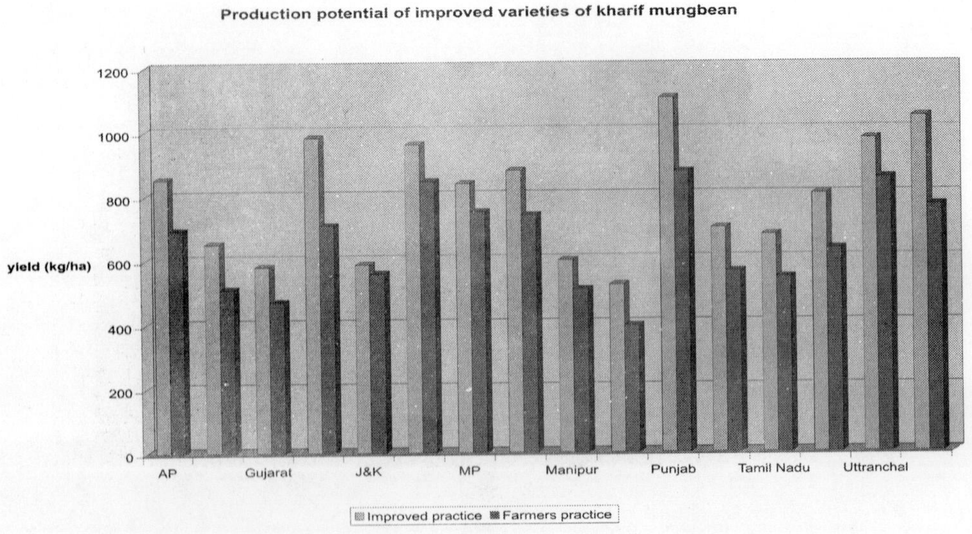

Fig 6 : Production potential of improved varieties in *Kharif* mungbean

In *rabi* urdbean, 227 demonstrations were conducted and overall increase in yield was recorded to be 25.59 per cent more than local varieties (Fig. 7).

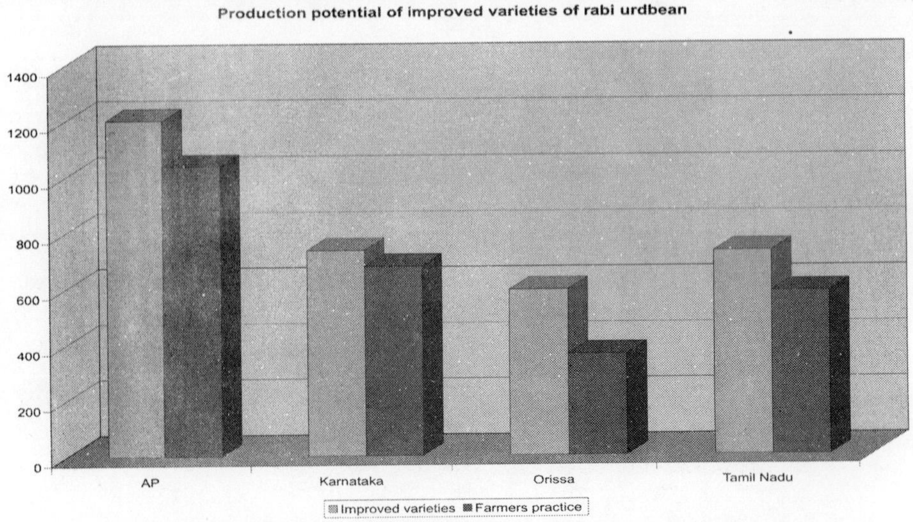

Fig 7 : Production potential of improved varieties in *Rabi uradbean*

In spring urdbean, 170 FLDs were conducted and overall increase in yield was observed to be 26.20 per cent more than the local varieties being grown by farmers (Fig. 8).

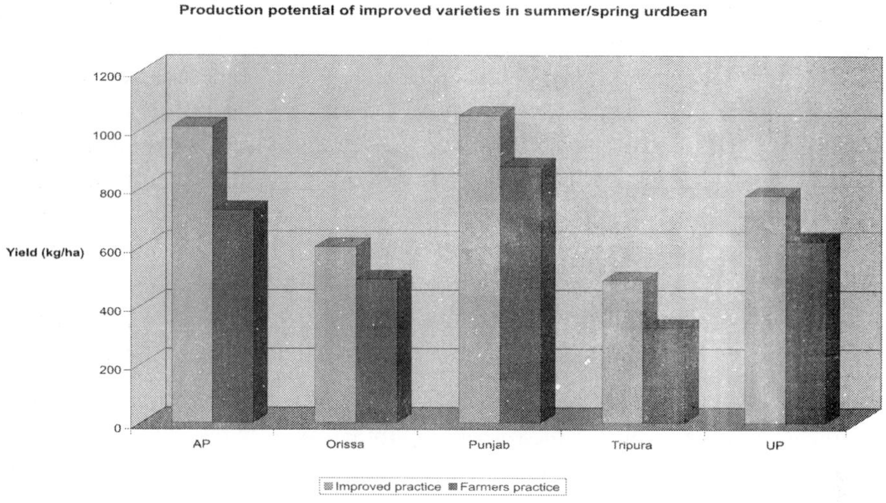

Fig 8 : Production potential of improved varieties in *summer/spring uradbean*

Agronomic Management : Based on the research accomplishments and experiences, the following agronomical techniques are suggested for realizing optimum yield potential of mungbean and urdbean under agro-ecological conditions of India.

Selection of variety : In mungbean, for spring/summer planting, varieties maturing in 60-70 days are preferred, whereas in *kharif* season long duration varieties of mungbean and urdbean maturing in 70-80 days could also be successfully cultivated. Since rainy season is most congenial for spread of yellow mosaic virus diseases, selection of varieties resistant to yellow mosaic virus is of paramount importance. In intercropping systems early maturing variety is more compatible. The recommended varieties for different seasons and agro-ecological regions are presented in Table 3.

Table 3 : Important varieties of Mungbean and Urdbean for specific traits

Trait	Crop	Varieties
Short duration for spring/summer cultivation (60-65 days)	Mungbean	Samrat, Pusa Vishal, OUM 11-5, Meha, TMB 37, Pant Moong 5, SML 668,HUM1, HUM 16, IPM 2-3.
	Urdbean	WBU 109, PDU 1, Pant U 31, KU 300, KU 92-1
Large seed size (> 4g/100 seeds)	Mungbean	Pusa Vishal, Pant Moong 5, SML 668, HUM 16, TMB 37
Resistant to PM for *rabi* season.	Mungbean	TM 96-2, Vamban 2, Vamban 4, TARM 18, BPMR 145
	Urdbean	LBG 625, LBG 709, LBG 645, Vamban 4, LBG 623, WBG 26
MYMV resistant	Mungbean	Samrat, Meha, Pant Moong 4, MH 2-15, Narendra Moong 1, Pant Moong 6, HUM 1, HUM 12, IPM 2-3
	Urdbean	WBU 109, Pant U 30, Pant U 31, Pant U 40, Azad U 1, Sekhar 3, Sekhar 2, Azad U 2, Uttara, Narendra U 1, Mash 1008, IPU 02-43

Seed treatment : Seed treatments include use of *rhizobium*, PSB, VAM or treating the seed with some chemicals for better nodulations, crop establishment and yield. Seed inoculation with specific *rhizobium* culture should be done before sowing of the crop. The efficiency of the leguminous crop to trap atmospheric nitrogen depends upon the degree and extent of nodulations in the roots which in turn is dependent upon the population of the r*hizobium*. It is therefore important to ensure desired population of these bacteria in the soil. Since the natural populations of the bacteria are often inadequate in the soil, seed inoculation with appropriate *rhizobial* culture is essential. Under the all India coordinated pulses improvement project (AICPIP) soil samples were collected from 506 places and it was observed that out of these, 245 samples were low and 153 medium in *rhizobium* populations, thus 79% of the total area had low-medium status.

Realizing the low *rhizobium* status in Indian soils where pulses are grown, the inoculation of seeds with appropriate *rhizobium* strain is must to take full yield advantage. Inoculation of seeds with *rhizobium* culture which cost Rs 8-10/ha increases production by 10-15%.

Fungicides like captan, thiram, bavistin etc. are compatible with culture. Seeds should always be treated with fungicides 4-5 days before culture treatment. Inoculation of seeds during summer season is more important as the population of the native bacteria declines drastically due to high temperature.

Sowing time : The time of sowing varies according to growing season. During kharif season the ideal time of sowing of Mungbean and urdbean is the last week of July in high rainfall zones of U.P., Bihar, Jharkhand, Chattisgarh and West Bengal. Timely planting with the onset of monsoon, is vital in rainfed areas where double cropping is followed. However, in the event of late onset of monsoon or failure of timely sown crop, it could be sown successfully by July end. In sub mountaineous areas of Himanchal Pradesh and Jammu and Kashmir, the second fort night of July is the optimum time of planting.

During spring/summer season sowing of mungbean/urdbean should be done as soon as the field is vacated by the preceding crops like potato, mustard, pea, cotton, sugarcane etc. The ideal time of planting of 70-80 days duration varieties during spring is the first fortnight of March.

Summer planting of mungbean after harvest of the wheat is quite prevalent in Uttar Pradesh and Bihar. The time of planting for summer crop is from 25 March to first week of April. The late planted crop may suffer heavy loss due to pre monsoon rains coinciding with the maturity of the crop. The spring and summer mungbean, thus, can be planted (i) after the harvest of *rabi* crops viz., potato, tobacco, pulses, oilseeds and wheat and before the planting of major *kharif* crops such as maize, sorghum, millets and rice, and (ii) in rice fallows as sole and intercropping systems rather than mixed cropping system.

Seed rate, spacing and sowing: Growing season, cropping system and seed size are the major factors for deciding the seed rate. As a sole crop during *kharif* season, 20-

25 kg seed/ha is required to maintain the desired plant stand. But during summer season, the seed rate should be increased to about 30-35 kg/ha as the plant does not have much vegetative growth.

As an intercrop with cereals, millets, pigeonpea and cotton during *kharif* season, the seed rate depends upon the area available to the crop. Generally one row of mungbean/ urdbean is sown between two rows of the main crop spaced at 45 cm keeping ratio of the component crop as 1:1 or two rows of main crop paired at 30 cm covering 60 cm interpaired space where one or two rows of mungbean or urdbean are adjusted. Thus, maintaining the ratio as 2:1 or 2:2 when intercropped with spring sugarcane, one or two rows of the mungbean/urdbean are adjusted between two rows of sugarcane planted at 75-90 cm distance.

Only healthy and pure seeds, taken from reliable seed stores/ agencies treated with fungicides and *rhizobia*l culture, should be used. During spring season, phorate/ aldicarb granules @ 10 kg/ha should be dropped in seed furrows to avoid insect-pest incidence.

Cropping system : There is enough scope for adoption of twin approaches for meeting the growth targets viz., increasing the area through multiple and inter cropping systems without much competition with other crops, and increasing the yield per unit area and time in the unexploited niches holding more promise and economic viability. Since the area under pulses has been shrinking, the non traditional area and cropping systems appeared to be the potent tools for triggering pulses production in the country. A significant increase in the area and production of mungbean and Urdbean has been possible through introduction of early synchronous maturing varieties in more desirable ecosystems they have covered a large chunk of spring/summer fallows in northern zone and rabi sowing in southern zone. Non traditional areas have been referred here in the context of time, space and horizontal spread in area where pulses were not grown earlier.

Development of short duration, photo and thermo insensitive disease resistance, with different architecture plants having high yielding capacity, new cropping systems with high productivity and monetary returns have been developed which have become popular with the farmer.

During *kharif* season, Mungbean or Urdbean is generally intercropped with maize, sorghum, pearlmillet, pigeonpea or cotton both under irrigated and unirrigated conditions. Intercropping of pigeonpea and Mungbean/Urdbean in 1:1 row ratio is quite prominent in the state of Uttar Pradesh, Bihar, Maharashtra, Gujarat and Tamil Nadu while spring sugarcane + Mungbean / Urdbean in 1:2 row ratio is popular among farmers in the state of U.P. and North Bihar and spring sunflower + Mungbean / Urdbean in 2:6 row ratio in the state of Punjab, Haryana and Bihar. In West Bengal it is intercropped with *tossa* jute (mid February to March sowing)

For sustaining productivity of major cropping systems prevalent in irrigated areas of north India, introduction of short duration, high yielding and yellow mosaic virus resistance varieties of mungbean as catch crop during the period of mid March to mid June, has proved quite beneficial.

In monocropped rainfed areas, early maturing varieties of Mungbean/Urdbean sown with the onset of monsoon offers good scope for double cropping. Mungbean/urdbean-mustard, Mungbean/urdbean-safflower, Mungbean/urdbean-linseed and mungbean (green manure) – wheat, are successful crop rotations of rainfed areas, if appropriate moisture conservation practices are adopted. In irrigated areas following cropping systems are prevalent.

Maize – wheat – mungbean/urdbean

Rice – wheat – mungbean

Rice – mungbean/urdbean

Rice - rice – mungbean/urdbean

Pigeonpea + mungbean/urdbean – wheat – mungbean

Cotton – mungbean

Maize – toria – mungbean/urdbean

Maize – potato – mungbean/urdbean

Maize – potato – sugarcane + mungbean/urdbean

Maize (cobs) – peas (pods) – mungbean/urdbean

Urdbean-wheat – mungbean

Arhar – wheat – mungbean

Urdbean – sarson – mungbean,

Maize - potato - wheat - mungbean,

Maize – potato – potato - mungbean/urdbean

Urdbean/mungbean – potato – wheat

Urdbean/mungbean – wheat – mungbean

Rice – potato – jute + urdbean

Nutrient management

Major nutrients: Like other crops, the requirement of nitrogen of pulse crops is quite high but in the field it is not realized because of their intrinsic capacity of fixing the atmospheric nitrogen thereby meeting 80-90% of the total nitrogen requirement of the crops by this mechanism. Yet several experimental data have revealed that 10-20 kg N/ha as starter dose is essential at the time of sowing.

After the nitrogen the second important element is phosphorus. Multi-locational experiments were conducted on farmer's field under AICPIP to determine the response of urdbean and mungbean crop to applied phosphorus. Based upon results of 173 trials on mungbean and 177 on urdbean it was found that the mean response to each kg of phosphorus applied at 13 kg P/ha was 1.6 and 3.0 kg grain, respectively.

Deficiency of sulphur is now common in several parts of the country where pulses are grown due to its removal from the soil. The deficiencies of sulphur in pulses are more deleterious than cereals because of higher sulphur needs in the latter. Mungbean and urdbean showed significant response up to 20 kg S/ha in multilocational trials under AICPIP. The mean response to this level of sulphur application was 166 and 194 kg/ha. Different sources of sulphur i.e., gypsum, pyrite and single super phosphate were almost identical in their efficacy. From the above studies it was concluded that mungbean and urdbean require 10 kg N, 30 kg P_2O_5, 20 kg K_2O and 20 kg sulphur for their optimum yield.

Micronutrients: Experiments conducted on different pulse crops on the use of micronutrients showed that application of Zn, Bo, Mo and Fe under deficient conditions is beneficial in increasing the yield but neither all the micronutrients nor their use in urdbean and mungbean have been tried for yield increase. Application of molybdenum favourably affects the yield of mungbean. Molybdenum pelleting of mungbean seeds @ 2, 4, 6 g/kg seed revealed response upto 4 g/ka seed during summer season under Kanpur conditions.

Nutrient economy in cropping systems: In sequential cropping involving mungbean or Urdbean, N economy to the extent of 30-40 kg N/ha to succeeding crop has been observed. In mungbean-maize cropping systems, the N economy to maize was 28 kg/ha. On the basis of soil test mungbean add 13 kg N/ha into the soil. Similarly in mungbean-rice system where mungbean crop was used as green manure after picking of pods, an economy of 85.2 kg N/ha was made in rice as compared to fallow-rice system. Mungbean grown in monsoon season benefited the following wheat and economized nitrogen by 37- 41 kg N/ha. Similarly in pulse-pulse cropping systems, pigeonpea-mungbean benefited the succeeding crop to the extent of 40 kg N/ha. In Mungbean-maize systems maize responded to higher doses of nitrogen when grown after phosphate fertilized mungbean than control. Further, incorporations of mungbean residue after picking of pods gave an economy of 40 kg N/ha to succeeding maize crop. In mungbean-maize sequence, maize following mungbean fertilized with phosphate showed better response to higher levels of N as compared to unfertilized crops. Incorporation of mungbean straw after picking of pods resulted in an economy of about 23 kg N/ha in succeeding maize. It indicates that benefit of legume effect could be increased by proper management of summer grown legumes. However, mungbean may not respond to applied P when raised after a cereal fertilized with recommended level of P.

Very little work has been done on fertility management in intercropping involving these crops. The knowledge of the period of the maximum demand of the nutrients of the component crop is essential in order to minimize the competition effect of nutrients.

In sorghum + mungbean intercropping system sorghum has two peak period of demand for nitrogen i.e., 40-50 days after sowing and 70-80 days after sowing whereas mungbean has its peak period demand for nitrogen i.e., 30-45 and 45-50 days after sowing. The demand of N at second peak period (45-50 days) is met by N fixation. Hence, the two crops did not suffer on account of competition for N. Generally the nutrient demand increased in intercropping system when planting is done under additive series mainly due to increase in plant populations per unit area. Higher uptake of major nutrients in pigeonpea + mungbean compared with sole pigeonpea has been reported.

In intercropping systems involving cereal and legume component nitrogen could be placed near the rows of cereal crops to avoid excess N availability to pulse crop which may otherwise affect the nodulations and nitrogen fixation adversely.

It is very unfortunate that under fertility management in intercropping systems, the much attention has been paid on management of N, very little on P and rare on micronutrients. The available information revealed that the recommended P level for pulse component is usually adequate to meet the requirement of both the components if soil is medium in available P. Additional P is required where soil is low in available P or if the non pulse crop is more responsive to applied P. In pigeonpea + mungbean intercropping systems, the application of 80-90 kg P_2O_5/ha gave favourable response. In legume-legume intercropping systems both crop should get adequate fertilizer, otherwise they will suffer possibly because of the similar requirement of nutrients.

Integrated Nutrient Management: Integration of chemical, organic and biological sources of nutrients and their efficient management have been found effective in not only sustaining the productivity and soil health but also in substituting a part of mineral fertilizers in pulse and pulse based cropping systems. The organic manures, green manures, crop residue and biofertilizer (Rhizobium, PSB and VAM) besides the mineral fertilizer are the major components of the integrated nutrient management. The inclusion of all the components in integrated nutrient management systems not only increases the yield but also improves the physical, chemical and biological properties of the soil. Incorporation of crop residue of mungbean in rice-wheat systems not only added 100 kg N/ha to the soil but also maintain high availability of N during various growth phases of rice. The residue management studies of *kharif* mungbean indicated a gain in yield of *rabi* crops to the extent of 15.6%. Further incorporations of the mungbean residue provided 50 percent NPK requirement of *rabi* sorghum and safflower. *Rhizobium* inoculation has sufficient beneficial effect on legume crops. This culture helps in better colonization of bacteria in roots. Under All India Coordinated Research Programme on MULLaRP crops, it was found that application of 20 kg P_2O_5/ha in mungbean through DAP along with phosphate solubilizing bacteria gave significantly higher grain yield and it was at par with application for 40 kg P_2O_5 through rock phosphate alongwith PSB. Dual inoculation of mungbean seed with *Vesicular arbuscular mycorrihza* and *Rhizobium* with 50% of recommended N recorded maximum grain yield in comparison to 100% of RDF or single inoculations. Mycorrhizal infection of green gram positively influenced rhizobial colonization for better nodulations and nitrogen fixation.

Water Management: Water is a prime natural resource and important input for assured agricultural production. Since long pulses are considered to be the crop which required no irrigation but recent researches have amply proved that water management in pulses is as much important as it is in any other crop. Since the inception of the AICPIP a lot of work on the aspect of water management has been done like determination of critical period of water requirement and measurement techniques for their rational use in irrigation management. Mungbean responded well to water application specially at their critical stage of growth (branching and podding). *Kharif* planted mungbean does not need irrigation as they mature by the cessation of monsoon but for spring/ summer crop an assured irrigation supply is a prerequisite for its cultivation. After harvest of *rabi* crops, a pre plant irrigation is necessary to prepare land and ensure adequate moisture for germination of seeds. Therefore, the first irrigation should be given 25 days after sowing and the subsequent irrigation at 7 days interval for good yield of summer crop, 3-4 irrigations are required.

When pulses are taken into sequential cropping, water management practices as recommended for pulse crop, are followed as the residual moisture from preceeding crop is not expected to influence the water requirement of succeeding crop. In summer mungbean crop, effect of kaoline spray was found to be beneficial in maximizing the water use efficiency. Yield of summer mungbean was 1.33 tonnes/ha with three kaolin spray at 15, 30 and 45 DAS with irrigation at 15 days interval compared with 0.67 tonnes from irrigation at 15 day interval without kaolin.

One important consideration in application of water is to its depth. Since pulses do not tolerate water logging the watering should be shallow and if there is a water logging situation it should be drained immediately. Water logging creates anaerobic conditions for *rhizobial* bacteria, hence, the bacteria residing in the roots will be killed and the whole systems of symbiosis will be destroyed.

Weed Management : Weed management research in pulse took place from 1970 onwards. A detailed review suggested that majority of the research work carried out in weed control in pulses in India restricted to the screening of the herbicides. The herbicides existed during that period were tested against different weed flora in pulse crops. Simultaneously the work started on ascertaining critical period of crop-weed competition, assessing potential damage caused by different classes of weed flora, weed suppressing ability of short grain legumes and crop weed competition.

The use of pendimethalin, oxyflurofen, alachlor, fluchloralin, oxidiazon etc. was spread to some extent to control the weeds in pulse crops grown under different cropping systems. The studies revealed that uncontrolled weeds can reduce the yield of mungbean and urdbean to the extent of 30-80%. Similarly some work on crop-weed competition, residual effect of herbicides on subsequent pulse crop and integration of herbicides with cultural practices was strengthened. Rotational practices adopted to control some of weeds are based on the principles that some weeds are associated with certain crop plants. Rotation of crop plants is able to break the life cycle of the weed plants. Water tolerant weed flora reduced in rice-mungbean cropping systems only due to adoption of mungbean cropping system. Studies made in Cavery delta region of Tamil Nadu Rice Research Institute, *E. colonum* and *E. crusgalli,* the major weeds associated with rice–rice–urdbean cropping systems were effectively controlled due to urdbean crop. While studying nutsedge infestation in different crop sequences at Kanpur, it was reported that in *kharif* season, rice–chickpea–mungbean sequence recorded lowest nutsedge shoot, tuber population and dry matter.

Intercropping and cultivar competitiveness is the cheapest method of weed control. Inclusion of mungbean between the rows of the pigeonpea resulted in 16.1% weed suppression. Integration of one hand weeding with fluchloralin 1.0 to 1.25 kg ha^{-1} or pendimethalin 1.0 kg ha^{-1} followed by one hand weeding at 25 days after sowing reduced weed density and weed dry matter production.

Integrated management of drought

- Water stress decreased pollen germination resulting in poor pod set. Pollen germination exhibited positive correlation with relative humidity under non-irrigated condition.
- High temperature stress (+10 ^{0}C above normal) at early crop growth stage (0-15 days after sowing) caused an irreversible loss in biomass production and grain yield in urdbean when subjected to normal temperature during later growth stages.
- Use of VAM fungi (*vesicular arbuscular mycorrhiza*) along with rhizobium culture with different mungbean varieties was found better under moisture stress conditions as the mycelium extracts moisture beyond the root extracting zone. The late maturing varieties were found better than early maturing type.

Disease management : Mungbean and Urdbean are attacked by 38 fungal, 12 viral, 3 mycoplasma like organisms and 2 bacterial diseases. Fortunately, a few of them like mungbean yellow mosaic virus (MYMV), cercospora leaf spots (C. canescens, C. cruenta) and powdery mildew (Erysiphe polygeni) are of considerable economic importance.

Mungbean Yellow Mosaic Virus : Mungbean yellow mosaic virus (MYMV) is the most destructive and widely distributed disease of mungbean and urdbean. The disease is caused by mungbean yellow mosaic virus; transmitted by whitefly. It occurs in almost all the states of India wherever these crops are grown but attains serious proportions in Uttar Pradesh, Uttaranchal, Madhya Pradesh, Chhattisgarh, Bihar, Jharkhand, Orissa, Punjab, Haryana, Delhi, Rajasthan and Andhra Pradesh. The disease causes heavy yield loss from 10 to 100%, depending upon the crop stage at which the plants get infected.

Management : Cultivation of resistant varieties is the most economical method which should be supported by control of vector by using chemicals or adopting cultural practices. Since MYMV develops more severe symptoms and causes heavy yield losses in early infected plants it is suggested to avoid the early infections by cultural means or vector control.

Host plant resistance : Cultivation of resistant varieties is the most economical and practical method of disease management. Varieties/lines like Pant Moong 2, Pant moong 3, Pant Moong 4, Pant Moong 6, PDM139, Meha,IPM 02-3,IPM02-14,Sattya,Narendra Moong1,TM96-2,PKV AKM4,HUM1 in mungbean have exhibited broad-based resistance against the disease and could be utilized for breeding MYMV resistant varieties. In urdbean, DPU 88-1, DPU 88-31, DU 3, Mash 1-1, NP 16, NP 19, NP 21, NDU 88-8, UG 135, UG 218, Pant U 19, Pant U 30, K 66-110, KU 305, Uttara, Pant U 31, Pant U 40, WBU 109 and KU 96-3.

Integrated disease management (IDM) : Integration of the components like resistant varieties, cultural practices and judicious use of chemicals may reduce vector population and disease incidence. IDM schedule for MYMV of Mungbean and Urdbean are given in Table 4.

Table 4 : Integrated management schedule of MYMV in Mungbean and urdbean

Resistant/tolerant varieties	Cultural method	Chemical control
Mungbean: Pant Moong 2, Pant Moong 3, Pant Moong 4, Pant Moong 6, PDM139, Meha, IPM 02-3, IPM02-14,Sattya, Narendra Moong1, TM96-2. PKV AKM4, HUM1	Inter/mixed cropping of mungbean and urdbean with non host crops like sorghum pearl millet and maize.	Apply phorate, disulfoton granules @ 1-2 kg a.i./ha in furrows during sowing.
Urdbean: IPU 94-1 (Uttara), Pant U 19, Pant U 30, Pant U 31, Pant U 40, KU 96-3, WBU-109 UG 218, Vamban 1, TU 94-2 Narendra Urd 1, WBU 108, LBG 17 VB 3, WBG 26, KU 92-1 KU 300.	Destroy affected plants as soon as they are noticed in the field.Adjustment of sowing dates to avoid the disease	To control whiteflies, spray metasystox, malathion (0.1%), dimethoate (0.02%), monocrotophos (0.05%) at the initiation of disease and subsequently at 10-15 days interval as per disease pressure.

Cercospora leaf spots (*C. Cruenta (Mycosphaerella cruenta), C. canescens*): Characterisitic symptoms are formation of leaf spots with brown to greyish centres and reddish border. Severe damage from CLS has been reported from Indonesia, India, Bangladesh, Taiwan, Philippines, Thailand and West Malaysia. In India, the disease occurs in most of the mungbean and urdbean growing states particularly Punjab, Haryana, Uttar Pradesh, Bihar, Madhya Pradesh, Rajasthan, Himachal Pradesh, Orissa, Assam and Maharashtra. Yield loss to the tune of 50% occurs in severely disease fields.

Management

Host plant resistance : Although few sources of resistance to CLS have been reported in mungbean and urdbean, these have not been effectively utilized in resistance breeding programme probably due to their low level of resistance.

Cultural practices : Use of seeds from disease free crop may help in reducing the seed borne inoculum of the pathogen. Field sanitation, crop rotation, destruction of infected crop debris and avoiding collateral hosts in the vicinity of the crop may help in reducing the disease incidence.

Adjustment of date of sowing in Mungbean and urdbean so that the crop may escape heavy rains has been recommended to control the disease. Delayed sowing reduces the disease severity. The early sown crop of Mungbean and Urdbean in 2nd week of July at 20 cm row spacing contracts more disease as compared to delayed sowing with wider spacing of 40 cm. Mulching reduces primary inoculums of the pathogen thereby decreases disease incidence.

Chemical control : Mancozeb, carbendazim, copper oxychloride and benomyl are reported to reduce the disease incidence considerably. Two sprays of carbendazim (300 g/ha) at 30 and 45 days after sowing (DAS) significantly controlled the disease and increased the yield in mungbean/urdbean, however, highest cost: benefit ratio of 1:5.5 with one spray of carbendazim (0.025%) at 30 DAS with yield increase by 250 kg/ha, thiophanate-methyl (0.025%) has been reported to be another effective fungicide.

Powdery mildew (*Erysiphe polygoni*)**:** The disease is very wide spread and occurs wherever Mungbean and Urdbean are grown particularly in the Indian subcontinent and South-East Asian countries. In India, the disease occurs in almost all the states causing serious damage to these crops especially in winter season in the states of Maharashtra, Andhra Pradesh, Tamil Nadu and, Orissa. Yield losses up-to 21% have been reported in a severely infected crop.

Management

Host plant resistance : Resistant cultivars such as TARM 1, TARM 18, TM96-2, Vamba 2, Vaban 5, KKM 3, COGG912 in Mungbean, and WBG 26, AKU 4, LBG 625, LBG 648, LBG 402, TU 94-2, AKU 15 in Urdbean, have been developed for cultivation in winter season.

Chemical control : Several fungicides such as thiovit, elosal, karathane, calixin, bavistin, benlate, topsin M, microsulf, cosan, bayleton, wet sulf, sulphur dust and indofil Z-78, have been found effective in controlling the disease under field conditions.

Biological control : Preliminary studies have indicated hyperparasitism of fungus, Ampelomyces quisqualis ces. on powdery mildew fungus on urdbean. The possibility of its use as biological agent for powdery mildew control on wide scale may be explored.

Nematode Management : The important nematode pests of mungbean include root knot and reniform nematodes (Meloidogyne incognita and M. javanica), Cyst nematode (Heterodera cajani) and reniform nematode (Rotylenchulus reniformis). Wherever Mungbean/Urdbean is grown in the field infested with nematodes, the yield losses may occur. The avoidable losses due to root knot nematode have been estimated up to 19% while 23% respectively in mungbean. The population level of 1000 juvenile per kg is considered to be the damaging level in case of root knot and reniform nematodes.

The most discernible symptoms in case of root knot nematodes are the formation of the galls on the roots that can be observed by uprooting the plants. The aerial symptoms are stunted growth, patches of stunted plants and yellowish foliage. In case of cyst nematode infection, the white cysts can be observed on the roots after 35 days of sowing. The plants remain stunted and week. In this case also, the patches are formed in the field. The reniform nematode infection can be observed by examining the roots of the plants. The roots become dirty due to sticking of the soil particles on the egg masses on the root surface, which are not detached even after washing the roots.

Management : Most of the screening has been carried out against root knot nematode in mungbean. About 500 accessions of mungbean have been screened against root knot nematodes and are being utilized in developing resistan varieties. Few cultvars have also been found resistant. These are LGG460, LGG-461, Ganga 6, Pusa 9531, Gang 1 in mungbean and TPU-3, WBU 105, KU 92-1, TPU-9, TU 95-1, UG218 in urdbean.

For managing the nematodes in *kharif* Mungbean and urdbean, use of summer heat is very useful. The *kharif* crop is taken immediately after the summer. Deep ploughing during summer can help reduce the nematode to non-damaging levels. Taking non-host crop after susceptible crop can also help in reducing the nematode population. Use of carbofuran 3G or phorate 10G @ 1.5 kg a.i./ha can reduce the nematode population and increase the yield.

Insect Pest Management : The pest spectra of mungbean and urdbean are almost same. Insect pest problem is relatively much less in summer crop due to hot and dry weather conditions as compared to rainy season crop. Both these crops are attacked by a large number of insect species numbering 110 or even more. But 32 species are considered of major importance in one area or the other.

Hairy caterpillars : There are number of species of hairy caterpillars which infest these crops depending upon locality. Some of them are:

i) Bihar hairy caterpillar –*Spilosoma* oblique – Central and north India

ii) Red hairy caterpillar – *Amsacta albistriga*- South India

iii) Red hairy caterpillar – *A. moorei* – Central and north India

iv) Sunhemp moth – *Utethesia pulchella* – North India

These hairy caterpillars are widely distributed pest in India and occur almost in all the regions of the country. Though these are sporadic pest, their seriousness as polyphagous pest have been well recognized. These are recorded from groundnut, sesamum, sunflowr, cowpea, linseed, mustard, cotton, castor, cabbage, cauliflower and other crops.

The eggs are laid in masses on leaves and after hatching from the eggs, the young larvae feed gregariously on the same or adjoining leaves especially under the surface by scrapping the chlorophyll. The skeletonized leaves can easily be detected even from a distant place. When larvae grow they disperse and feed individually by devouring leaves, flowers and growing points. When large number of caterpillars infest the crop, only the base stem of plants remains resulting in heavy and often total yield loss.

Galeruid beetle (*Madurasia obscurella)* **:** This is one of the important pest in northern and southern India. Adult beetle feed on root hairs and rhizobial nodules and also on foliage and make small circular holes in the leaves. The damage is mostly restricted to seedling stage of the crop. Damage to rhizobial nodule have been reported to be as high as 25-60%. Due to infestation at early stage and on root the plant growth in checked and there is heavy reduction in yield.

Stemfly (*Ophiomyia phaseoli)* **:** This insect is reported to cause damage at seedling stage of the plant. There are number of species of Agromyzids infesting Mungbean/ urdbean.

On hatching the maggots mines the leaves, tunnel underneath the epidermis of the leaves till they reach one of the veins which lead to midrib and finally to leaf stalk and stem. Due to severe infestation the leaves turn brown and breakdown. The attacked plants bear less pods which are mostly empty or else the seed are very small.

Whitefly (*Bemesia Tabaci*)**:** This is not only one of the occasional and serious pests of Mungbean and urdbean but is a potential vector of mungbean yellow mosaic virus and reported to cause 30-70% yield loss. Both nymph and adult sick the cell sap while remaining at the underside of the leaves, lowering the vitality of plant, which give sickly appearance to the plants. Under heavy infestation development of shooty mould on the leaves is also common.

Thrips –*Megalurothrips distalis*

Caliothrips indicus : It is one of the major insect pests of Mungbean and urdbean in northern India. Adult and nymph suck the sap from flower and bud resulting in their drop, devitalization of plants and low yield realization. When crop is young they damage tender leaves resulting in silvery appearance and upward bending of leaf margins. Summer season crop suffer severe damage due to thrips than rainy season crop. The thrips infestation is more in drier conditions.

Integrated Pest Management

Monitoring

- Visual inspection at weekly intervals starting from vegetative stages for incidence of pest and diseases.
- Hairy cater pillars at early stage, feed gregariously on leaves. Such dusers clusters are visible from a distance.
- Monitoring of thrips at flowering (35-40 DAS) stage.

Economics threshold level (ETL)

For thrips, 4-5 thrips per flower.

Cultural practices

- Early sowing during spring (by 15th March) for reduced incidence of MYMV and less infestation of thrips.
- Intercropping with taller crops like sorghum reduces MYMV during *kharif.*
- Deep ploughing helps in burying the larvae/pupae too deep in the soil to merge and also expose them to predators, parasites or adverse weather conditions (soil solarization).
- Timely irrigation in summer/spring crop reduces thrips damage as their multiplication is very rapid under drier conditions.

Host Plant Resistance: Growing MYMV resistant varieties as given.

Mechanical practices

- Hand picking and destruction of egg masses and first/second larvae of hairy caterpillar.
- Removal of egg masses of Hemipteran bugs (*Nezara viridula* and *Clavigralla gibbosa*) and keeping them away from the cropped area. These egg masses are often parasitized by *Gryon* sp. which helps in their multiplication and subsequent large scale parasitization of the eggs of the bugs.

Chemical control

- Soil application of phorate or carbofuran granules at 1.0 kg a.i./ha at the time of sowing of summer crop of mungbean/urdbean for protection against thrips and white flies and for *Kharif* crop against stem fly, white fly flee beetle and jassids. Also reduces incidence of MYMV.
- Foliar spray (2-3) of insecticides such as dimethoate (0.03%) or monocrotophos (0.04%) for control of thrips, white fly, jassids, stem fly and other leaf eating insects, if needed.

Perspective

There is considerable scope for bringing additional area under Mungbean and urdbean. The expansion of cereal–cereal rotation posses a serious threat to the sustainability of its production system as decline in total factor productivity (total output/ total inputs) is being realized. Being short duration and having nitrogen fixing ability, mungbean and urdbean fits well in multiple cropping systems. The area under these crops has increased in recent years due to introduction in summer season in most parts of the country and under rice fallows in peninsular India during winter & summer season. There is further scope for increasing area under Mungbean and urdbean under the following niches areas (Table 5).

For example rice-wheat cropping system is covering about 10.7 m ha. Mungbean is grown only in about 0.3 m ha in region including both *kharif* and spring season indicating tremendous scope for bringing additional area under Mungbean and urdbean as spring crop. Short duration (60 days) varieties are available which can be very well fitted in between wheat and rice during summer season. Similarly additional area under Mungbean and urdbean can brought in southern and coastal India as winter crop in prevailing area under rice – rice cropping system in 2.7 m ha. Inclusion of mungbean in between rice wheat and rice – rice systems brings not only quantitative change in production base for long term sustainability but also helps protect the environment from the risk associated with high input agriculture.

Table 5 : Possible new niches for Mungbean and Urdbean

Cropping system	Possible niche areas	Expected area (M. ha)
Maize-potato/ Mustard – Mungbean /urdbean	Punjab, Haryana, W.UP.	1.0
Spring Sugarcane + Mungbean /urdbean	East Uttar Pradesh, Bihar, West Bengal	0.15
Rice – Mungbean /urdbean	Orrisa, Part of Karnataka, Tamilnadu, Andhra Pradesh	0.35
Rice– Mungbean/ urdbean	Coastal area of Andhra Pradesh, Karnataka, Tamilnadu	0.35
Rice – wheat – mungbean	Western Uttar Pradesh, Haryana, Punjab	0.10
Maize – Rajmash - mungbean	Central & Eastern Uttar Pradesh, North Bihar	0.07

Research approach : Varietal development is a continuous process targeting the prevailing cropping system and environment. With the emerging cropping systems and changing scenario of pest and diseases, varieties with additional set of alleles with following traits are required:

Multiple resistances : Genetic amelioration of crop plants has contributed significantly in developing varieties with desired traits including disease-pest resistance. Varieties developed in past with resistance to single stress may not be a viable solution under the changing scenario as incidence of diseases and insect-pests is often not predictable. Varieties having resistance to more than one stress provide greater insurance. Combining resistance to major diseases in desirable background will help in consolidating the yield.

Development of New Plant type : The presently available plan type in mungbean and Urdbean are largely photo thermo sensitive, indeterminate growth habit with low harvest index and low grain yield. To remain a commercially competitive crop Mungbean/ urdbean has to fit in gaps in high input cereal– cereal cropping systems. For this, plant type that are determinate, photo thermo insensitive, early maturing, high yielding (1.5-2.0 tones/ha) with high harvest index and resistant to lodging are to be developed. A plant type with thick stem with two–three erect branches, short internodes, having small and thick leaves needs to be developed. Good seedling vigour. A clear-cut distinction between vegetative and reproductive phases and high harvest index will be essential component of the plant type. The proposed plant types for different situations are described in table 6.

Table 6 : Proposed plant types for various situations and seasons

Agro-ecol. regions	Season	Major traits	Resistant to
Northern India	*Kharif* season	90-100 days maturity distinct vegetative and reproductive phases, high harvest index, lodging and preharvest sprouting resistant	MYMV, CLS & Rhizoctonia blight, Thrips
Northern India	Post rainy (20 Aug–10 Sept.)	60 days maturity, good seedling vigour, synchronous podding and high harvest index and photo thermo insensitive	MYMV & Thrips
Central India	*Kharif*	60 days maturity, good seedling vigour, synchronous podding, large seed	MYMV
Southern & Coastal India	*Kharif*	70-75 days maturity, distinct vegetative & reproductive phase, lodging and pre harvest sprouting resistant	CLS & Rhizoctonia blight
Southern & Coastal India	Winter	70-75 days maturity, good seedling vigour, photo thermo insensitive, synchronous podding	Powdery mildew

Broadening the genetic base : Limited variability has been exploited in mungbean and Urdbean varietal development programme. The genetic base of the available varieties is narrow. In Mungbean, T 44 has been a frequently/widely used ancestor whereas T 9 in Urdbean has the maximum share in the pedigree of released varieties. There is good scope to utilize wild and cultivated *Vigna* species to incorporate novel characters and broaden the genetic base.

Wild relatives of *Vigna* i.e. *V. radiata* var. *sublobata* and *V. mungo* var. *silvestris* were found resistant to MYMV and bruchids. The gene introgression of mungbean has resulted into mung derivatives which have shown potential for yield contributing traits and disease resistance. Moreover, these derivatives facilitate further genetic enhancement in mungbean. The efforts made in this direction led to development of improved cultivars of mungbean IPM 99-125 and IPM 02-3. Further, number of new sources are being examined and utilized for improvement programmes more extensively.

References

Ahlawat, I.P.S, Gangwar, B. and Gangaiah, B. 2004. Potential pulse based cropping system for new niches. In: Pulses in New Perspective (Eds., M Ali, BB Singh, Shiv Kumar and Vishwa Dhar). Indian Society of Pulses Research and Development, Kanpur. pp 260-270.

Ali, M. 2004. Role of pulses in crop diversification in India. In: Role of legumes in crop diversification and poverty reduction in Asia. (Eds. CLL Gowda and S Pande). Proceeding of the Joint CLAN Steering Meeting, 10-12 November 2003, International Crop Research Institute foe the Semi-Arid Tropics, India. pp 42-56.

Ali, M. 2004. Role of pulses in crop diversification. In: Pulses in New Perspective (Eds., M Ali, BB Singh, Shiv Kumar and Vishwa Dhar). Indian Society of Pulses Research and Development, Kanpur. pp 245-259.

Bandyopadhyay, S.K. and De, R. 1986. Nitrogen relationship and residual effects of intercropping sorghum with legumes. Journal of Agricultural Sciences. (Cambridge) 107: 629-632.

Dixit, G.P. and Masood, Ali. 2007. Technology transfer through Frontline Demonstrations on Pulse Crops, IIPR Kanpur.

Katiyar, P.K. and Dixit, G.P. 2007. Ancestral relationship of greengram advance breeding lines developed in India. Indian Journal of Agricultural Sciences 77 (9): 579-82.

Katiyar, P.K., Dixit, G.P. Singh, B.B. and Gupta, Sanjeev. 2009. Urdbean varieties in India. AICRP on MULLaRP, IIPR KanpuRego, T.J. and Rao, V.N. 2000. Long –term effects of grain legumes on rainy season sorghum productivity in a semi-arid tropical Vertisol. Experimental Agriculture 36: 205-221.

Katiyar, P.K., Dixit, G.P. and Singh, B.B. 2008. Genetic base of advanced urdbean breeding lines developed in India as revealed by pedigree analysis. Indian J. Genet. 68 (3): 324-26.

Katiyar, P.K., Dixit, G.P. and Singh, B.B. 2008. Morphological characterization of green gram varieties and their application for distinctness, uniformity and stability testing. Indian J. Agricultural Sciences. 78(5): 439-44

Katiyar, P.K., Dixit, G.P. and Singh, B.B. 2008. Morphological characterization of green gram varieties and their application for distinctness, uniformity and stability testing. Indian J. Agricultural Sciences 78(5): 439-44

Roy Sharma, R.P. and Singh, Ambika. 1969. Residual effect of legumes in winter legumes-maize rotataion. Indian Journal of Agricultural Sciences 36: 530-539.

Singh, B.B., Katiyar, P.K. and Dixit, G.P. 2006. National Test Guidelines for the conduct of tests for Distinctness, Uniformity and Stability in Mungbean and Urdbean. AICRP on MULLaRP, IIPR, Kanpur

Singh, B.B., Katiyar, P.K. and Dixit, G.P. 2007. Urdbean. In: National test guidelines for conduct of tests for distinctness, uniformity and stability (eds. Dinesh Kumar and A.B. Mandal), pp. 131-136, Directorate of Seed Research, Mau (U.P).

Singh, B.B., and Dixit, G.P. 2007 Achievements in research on MULLaRP crops In: SOUVENIR. National Symposium on Legumes for ecological sustainability: Emerging Challenges and Opportunities, 3-5 Nov. 2007, IIPR Kanpur (U.P.).

Singh, B.B., Dixit, G.P., Katiyar, P.K and Pratap, Aditya. 2009. Mungbean varieties in India. AICRP on MULLaRP, IIPR Kanpur (U.P).

Singh, B.B., Katiyar, P.K., and Dixit, G.P. 2007. Mungbean. In: National test guidelines for conduct of tests for distinctness, uniformity and stability (eds. Dinesh kumar and A.B. Mandal), pp. 123-130, Directorate of Seed Research, Mau (U.P).

Singh, Y, Chaudhary, D.C, Singh, S.P, Bhardwaj, A.K., and Singh, D. 1996. Sustainability of rice (Oryza sativa) wheat (Triticum aestivum) sequential cropping through introduction of legumes crops and green manure crop in the system. Indian Journal of Agronomy 41: 510-514.

Tauro, P., and Khurana, A.L. 1986. Problem of prospects of growth, extension and promotion of biofertilisers. In: Proceeding of the FAI Seminar. Fertilizer Association of India pp. 1-4.

Verma, K.P., and Warsi, A.S. 1999. Evaluation of suitable rice-based cropping systems for central plain zone of Uttar Pradesh. Indian Journal of Agronomy 44: 444-7.

❑❑❑

Chapter - 16

Chickpea

N.P. Singh, Shiv Sewak and M.A. Iquebal
All India Coordinated Research Project on Chickpea Indian Institute of Pulses Research, Kanpur 208 024, Uttar Pradesh

Chickpea, *Cicer arietinum* L, is the second most important pulse crop in the world. In India, it occupies prime position in terms of area, production and economic value and as an important constituent vegetarian diet. Chickpea is traditionally grown in many parts of the world covering Asia, Africa, Europe and North and South America. But the bulk of it is produced and consumed in South Asia and increasingly in Middle East as well as in some Mediterranean countries. Recently, chickpea cultivation started in Australia, Canada and USA mainly for export to south Asia. The cultivated species, *Cicer arietinum* L. (2n=16 and n=8), is characterized by possessing two distinct types-small and brown seeded known as *desi* type, and bold and white seeded known as *kabuli* type. The former is mostly cultivated in Indian sub-continent. Chickpea (*Cicer arietinum* L.) is also known by several common names *viz*., Bengal gram, Gram, *Chhola*, Hommes and Garbenzobean.

Chickpea provides 2-3 times more protein than many cereals, constituting a major component of the agrarian diet of the Indian sub-continent. Chickpea grains provide about 18-24% protein, 4-10% fat, 52-70.9% carbohydrate, 10-23% fibre, minerals and vitamins. Among essential amino acids lysine, methionine, threonine, valine, isolucine and leucine are major components of seed protein. It also contains considerable amount of vitamins such as B_1 and B_2, ascorbic acid (vitamin C) and niacin also. Chickpea, therefore, plays an important role in human nutrition.

Being a leguminous crop, chickpea fixes atmospheric nitrogen through bacteria present in their roots, which in turn improves the soil fertility. Thus, the crops sown following chickpea are also benefited. Due to the tap root system, it opens up the soil and heavy leaf drop increases the organic matter in the soil. This crop also has the inherent quality to trap the soil moisture from lower strata of the soil, therefore, they are considered drought tolerant, hence, grown as rainfed crop. In spite of the variable uses of chickpea in our diet and its importance in maintaining the soil health, the increase in production not maintained the pace of population growth.

The nutritional quality of chickpea is adjudged by related content of protein, carbohydrate, lipids, minerals, vitamins and snit-nutritional factor. The cultivars, growing season, soil and climate conditions and management practices influence the nutritional quality considerably. The average chemical composition of whole seeds is given in (Table 1). The seed coat contains most of the non-digestible carbohydrates and relatively higher proportion of calcium.

Table: 1. Average chemical composition of chickpea whole seeds

Constituents	Content per 100 g of seeds
Protein (g)	21.5
Fat (g)	4.5
Total carbohydrate (g)	63.0
Crude fibre (g)	8.0
Ash (g)	2.7
Minerals	
Calcium (mg)	200.0
Phosphorus (mg)	260.0
Iron (mg)	6.0
Magnesium (mg)	130.0
Potassium (mg)	800.0
Vitamins	
Thiamine (mg)	0.34
Riboflavin (mg)	0.22
Niacin (mg)	2.25
Pyridoxine (mg)	0.55
Carotene (mg)	0.12
Ascorbic acid (mg)	4.08

Source: Salunkhe and Kadam (1989)

Present scenario

Global scenario: Globally, chickpea is cultivated on about 11.56 million ha adding 8.78 million tonnes of grain to the global food basket with productivity of 760 kg per ha (FAOSTAT, 2008). As many as 42 countries grow chickpea but only a dozen of countries *viz*., India, Pakistan, Syria, Iran, Turkey, Australia, Ethiopia, Myanmar, Mexico, Canada, Spain and Bangladesh, contribute 94.6% and 94.5% to global area and production, respectively (Table 2). India dominates the global chickpea supply as it has the distinction of being the largest producer as well as consumer in the world and accounts for over 65.2% and 65.4% of area and production, respectively.

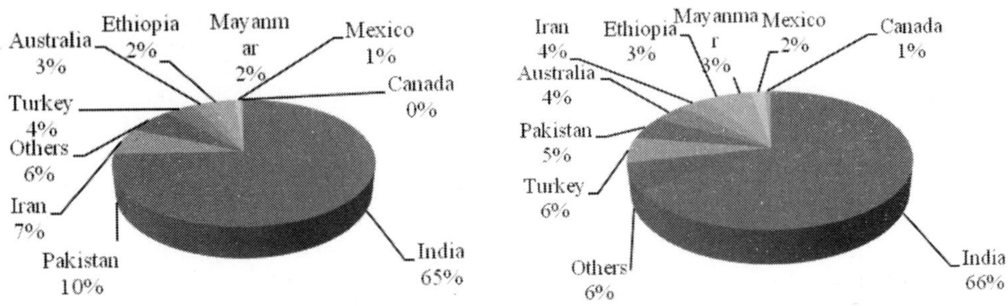

Fig.1: Distribution of Chickpea Area and Production (%) over different countries

Recently Australia has taken up its cultivation on a large scale touching 0.30 million ha area with production of 0.38 million tonnes and yield level of 1268 kg per ha from an almost insignificant level during early 80s. Similarly, in Canada, area and production of the crop have registered spectacular growth during this decade. As per FAO report (2008), Mexico has registered highest productivity (1809 kg/ha) followed by Australia (1268 kg/ha) and Ethiopia (1265 kg/ha).

Table 2: Global chickpea (Area, Production and Yield) scenario during 2008

Continent/Country	Area (Million hac.)	Production (Million tonnes)	Yield (Kg/hac.)
Continent World	11.56	8.78	760
Asia	10.47	7.56	722
Africa	0.53	0.45	859
America	0.20	0.32	1574
Europe Country	0.06	0.07	1136
India	7.54	5.75	763
Pakistan	1.11	0.47	429
Iran	0.79	0.32	405
Turkey	0.49	0.52	1065
Australia	0.30	0.38	1268
Ethiopia	0.23	0.29	1265
Myanmar	0.23	0.26	1156
Syria	0.09	0.05	585
Mexico	0.09	0.16	1809
Canada	0.04	0.07	1580
Spain	0.03	0.03	953
Bangladesh	0.009	0.007	766

Source: *FAOSTAT, 2008*

National scenario: As stated earlier, India is the largest producer of chickpea in the world with 65.24 (7.54 million ha) and 65.49 (5.75 million tonnes) per cent of the total area (11.56 m ha) and production (8.78 m t), respectively (FAOSTAT, 2008). In India, chickpea was cultivated in 8.74 million hectares with the production of 7.35 million tonnes of the grain during 2009-2010. India imported 1.74 lakh tonnes of chickpea from Canada, Australia, Turkey and Tanzania. However, at the same time the total export of *kabuli* chickpea was to the tune of 40,000 tonnes to USA, UK, UAE and Saudi Arabia during year 2009-10. Andhra Pradesh, Madhya Pradesh, Uttar Pradesh, Rajasthan, Maharashtra, Gujarat and Karnataka are the major chickpea producing states covering over 95 % area. During last five decades chickpea has registered significant increase (3.65 million tonnes in 1950-51 to 7.35 million tonnes in 2009-10) in production (1.69 % average annual growth rate for 1950-2010), which is primarily due to introduction of high yielding and disease (s) resistant varieties as well as adoption of improved production technologies. In this period, the area also increased slightly (7.57 million ha in 1950-51 to 8.74 million ha in 2008-09) with an annual average growth rate of 0.25%. However, the yield has increased steadily from 482 kg per ha in 1950-51 to 886 kg per ha in 2008-09 with an annual average growth rate of 2.38%. The north Indian states (Punjab, Haryana and Uttar Pradesh) have lost more than 82% area of chickpea cultivation which has caused about 28% reduction in production during 1975 to 2010. This reduction in area was because of increase in irrigation facilities causing a shift to wheat and non- availability of input responsive varieties of chickpea. This loss was compensated by increase in area (125%), production (46%) and productivity (24%) in central India (Maharashtra, Madhya Pradesh and Gujarat) and south India (Andhra Pradesh and Karnataka) during the same period (Fig. 2). Among major chickpea growing states of India, Andhra Pradesh has registered the highest productivity (1448 kg/ha) followed by Gujarat (977 kg/ha), Bihar (953 kg/ha), Maharashtra (825 kg/ha),

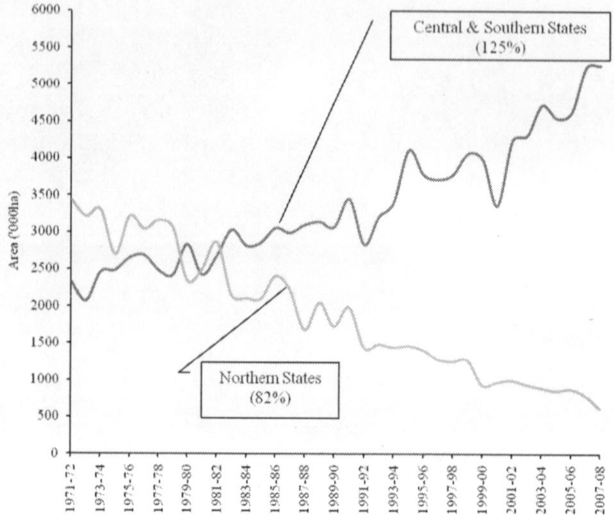

Fig. 2: Regional shift in chickpea area

Uttar Pradesh (745 kg/ha), Madhya Pradesh (726 kg/ha), Karnataka (607 kg/ha) and Rajasthan (466 kg/ha) when the national productivity was 761 kg/ha during 2007-08. The variation in productivity has clearly established that there is scope for improvement in average yield through increase in yield potential of the varieties or through better crop production and protection techniques. Potential exists for enhancement in productivity in states of Karnataka, Gujarat, Bihar, West Bengal etc. to enhance availability of chickpea at national level.

Future projections

The global chickpea scenario has witnessed rapid changes during the past two decades. Increasing production and profitability is threatened because of global warming, a direct consequence of climate change. The emergence of new pests and diseases, natural resource depletion and degradation, eroding farm profitability, strict regulation in food safety, trade competition *etc.* are formidable challenges. The country already has 1.2 billion people, which is predicted to cross 1.41 billion by 2025. In order to ensure self sufficiency, the pulse requirement in the country is projected at 29.44 million tons by the year 2025 which necessitates an annual growth rate of 4.02%. Presently, India import pulses to the tune of 1.5 to 2.0 million tones every year to meet the demand.

At present, chickpea appears to be in much better situation as compared to other pulses because of increase in production on account of both expansion of area in Central- and South India due to late termination of monsoon and increase in productivity due to development and spread of high yielding and disease resistant varieties in recent years. However, we can not afford to be complacent as a year with bad monsoon would change the entire scenario because more than 70% of chickpea is grown under rainfed conditions. In order to achieve the target of 29.44 million tons of pulses by the year 2025, chickpea has to contribute an additional 4 million tonnes production to the pulse basket of the country. This requires a paradigm shift in research, technology generation and dissemination, and commercialization along with capacity building in frontier areas of research.

The Strategy

To achieve the target of additional 4 million tonnes of chickpea by 2025, we need two pronged strategy as discussed below.

Area expansion approach : Scope for bringing additional area under chickpea is rather limited. Area under this crop can be increased in various inter-, relay or sequential cropping systems. In this context, rice fallows should get maximum attention. An additional 2 million ha area can be brought under rice fallows in the state of Chhattisgarh, Madhya Pradesh, Eastern Uttar Pradesh, Bihar, Jharkhand and West Bengal. Among innumerable cropping systems prevalent in Indo-Gangetic region, rice-wheat system is the most dominating one covering about 10.7 m ha area. Chickpea is grown in about 0.6 m ha area in this region accounting for 1.3% of the area only indicating good scope for bringing additional area under chickpea provided we develop input responsive varieties amenable to mechanical harvesting. Similarly cotton - chickpea sequential cropping, chickpea + mustard and chickpea + sugarcane can be rewarding in North West plains of India.

Approaches for increasing productivity : The varieties developed through breeding programmes require continuous efforts . The present day varieties of chickpea will require to be replaced with a new set of varieties for meeting the emerging challenges/ threats like emergence of races of *Fusarium* and *Ascochyta* blight. Besides, we need to develop resilient chickpea varieties having higher level of heat and drought tolerance for increasing productivity of chickpea in rainfed agro-ecosystem of central and peninsular India where >70% area of chickpea is under cultivation. There is need to develop pod borer resistant varieties through conventional plant breeding because of unavailability of resistant donor parents in chickpea gene pool. In order to achieve the target of additional 4 million tonnes by 2025, we need to increase the productivity level to 1200 kg/ha from the present level of 890 kg/ha. This should not be difficult as some states like AP has already average productivity of 1448 kg/ha.

Major cropping systems involving chickpea

Chickpea is grown on all types of soils *viz.*, heavy clay to light loam, except acidic, saline and sodic soils. Well drained sandy loam soil to deep loam soils of medium fertility are considered ideal for its cultivation. The alluvial soils of the Indo-Gangetic belt support a bumper crop of chickpea. Although chickpea is grown as a sole crop in many areas, its inclusion in various cropping systems *viz.*, mixed cropping including intercropping, sequential cropping etc. is also equally popular with farmers.

The successful growing of chickpea in multiple cropping systems is well documented in India. Howard (1974) was the first who documented mixed cropping system of chickpea in India. Thereafter, a flurry of reports was seen periodically. Growing of chickpea in mixed cropping system with cereals (Gupta 1955, Katyal 1963, Shyam 1963, Singh and Katyal, 1966), oilseeds (Gupta and Mathur 1964), sugarcane (Chowdhary 1953), forage crops, garden crops like coconut is also well documented (Anonymous 1952). With the availability of short duration varieties and expansion of irrigation facilities, there has been a manifold increase in mixed and sequential cropping systems involving chickpea in various ago-climatic regions of the country. Since chickpea is a traditional *rabi* season crop, most of the intercropping systems involve *rabi* crops such as wheat, *rabi* maize, mustard, safflower, linseed, *rabi* sorghum etc. The most popular cropping system involving chickpea are as follows:

Sequential cropping: Maize - chickpea, sorghum - chickpea, pearl millet - chickpea, cotton - chickpea, rice - chickpea, sunflower - chickpea and soybean – chickpea.

Intercropping: Chickpea + mustard, chickpea + linseed, chickpea + wheat, chickpea + barley, chickpea + safflower, chickpea + coriander

Expanding irrigation facilities, availability of short duration varieties and necessity to raise cropping intensity have resulted in evolving successful sequential or relay cropping systems involving chickpea. In fact, the sequential cropping systems with chickpea as component crops are more popular than intercropping or mixed cropping systems. In the Indo-Gangetic plains, chickpea follows various other crops such as pearl millet, sorghum, maize, short duration cotton, clusterbean, sesame or early rice in double cropping. In northeastern India, chickpea follows rice during monsoon. In northern plains, chickpea is grown after pearl millet. In central and southern India, chickpea in rotated with winter sorghum or wheat (Argikar 1970). Yadav (1996) listed major crop sequences involving chickpea in different agro-ecosystems as listed below.

Arid ecosystem

- Rainfed upland: Pearl millet - chickpea
- Irrigated upland: Sorghum - chickpea

Semi-arid

- Rainfed upland: Maize - chickpea
- Irrigated upland: Rice- chickpea

Sub-humid ecosystem

- Irrigated upland: Cotton - chickpea
- Irrigated lowland: Maize - chickpea

Humid ecosystem

- Lowland: Maize - chickpea

As chickpea is often grown on residual moisture under rainfed conditions, its time of sowing is of great importance in sequential cropping system to accommodate component crops. Adequate surface soil moisture at the time of sowing ensures better germination and uniform plant stand. The three major physical factors likely to influence the growth phenology and productivity are day length, temperature and water availability (Khanna-Chopra and Sinha 1987). Thus, these factors determine the optimum planting time of chickpea in cropping system. The preceding and succeeding crops of chickpea should be so selected as to provide optimum growing condition for chickpea. Pest management is also an important consideration in sequential cropping systems. Breaking the cycle of insect pest and pathogens through changing the component crops in systems, results in their reduced incidence. Like-wise, certain sequences tend to aggrevate the incidence of pest and disease. Chickpea following rice showed higher incidence of wet root not and collar rot disease of chickpea (Nene 1980). Further, wheat, oat and barley in rotation with chickpea, tend to enhance root knot nematode population in soil. Chickpea rotation still remains as an economic and effective way to prevent yield losses due to nematodes in chickpea. Rotating chickpea with non-legumes can easily control *Heterodera ciceri*. Incidence of *Meloidogyne artiellia* can also be controlled by altering chickpea with crops other than legumes, cereals and crucifers (Greco and Sharma 1990) In the long term crop rotation experiment, *Fusarium* wilt damage to safflower was much less where chickpea was grown as component crop as compared to continuous cultivation of sorghum and safflower in rotation (Kalpana Sastry *et al.* 1993).

Major constraints.

The major constraints of various chickpea based cropping systems are given below.

Intercropping systems

- Lack of input responsive cultivars
- Low and unstable yield
- Agronomic constraints

Sequential cropping

- Soil management
- Non-flexibility in sowing time
- Lower return
- Biotic stresses

Chickpea has been a traditional favourite with farmers in various cropping systems. Among intercropping system, chickpea + mustard / wheat / barley in north, chickpea + linseed in central, and chickpea + safflower / sorghum / coriander in peninsular India have been quite successful systems recording higher productivity and net returns. The major cropping systems involving chickpea in different states are given below:

Cropping systems	**Major states**
Rice-chickpea	Punjab, Haryana, Uttar Pradesh, Madhya Pradesh, Chhattisgarh, Bihar, Tarai region of Uttarakhand, West Bengal
Cotton-chickpea	Punjab, Haryana
Sorghum / maize-chickpea	Haryana, Uttar Pradesh, Rajasthan
Chickpea + mustard	Uttar Pradesh, Madhya Pradesh, Bihar, Rajasthan
Chickpea + safflower	Madhya Pradesh, Gujarat, Karnataka, Maharashtra
Chickpea + barley	Uttar Pradesh, Uttarakhand
Chickpea + linseed	Uttar Pradesh, Madhya Pradesh, Maharashtra
Chickpea + Safflower / coriander	Maharashtra, Karnataka, Tamil Nadu

Yield Gaps

Sustained efforts of the dedicated scientists have led to the development of many high yielding varieties of chickpea with resistance to diseases and pests and matching production and protection technologies. These technologies have been transferred to the farmers and end users through programmes like IVLPs, NATP, FLD and other institute sponsored on-farm demonstrations. Still there is wide gap in productivity (yield gap I) (Table 6). Data on FLDs and other on-farm demonstrations has revealed an average increase of 30% productivity of chickpea.

Since area under chickpea cultivation virtually declined during 1974-2004, gains in production were obviously the result of improvements in productivity. At the national level, a 31% increase in yield was observed in 2004 over 1974 (Table 3). However, intra-zonal increases in yield varied over the period. Mean yields of All India Coordinated Trials (2070 kg/ha) for *desi* type were reported for Central Zone while the North Western Plain Zone (NWPZ) reported the highest potential yield (1462 kg/ha) for *kabuli* type (Table 4).

Table 3: Chickpea production trends during 1971-2004 (AICRP-Chickpea Report, 2006)

Item	Year/Change	NEPZ	NWPZ	CZ	SZ	India
Area	1971-1974	383	4678	2137	226	7420
('000 ha)	2001-2004	314	1818	3236	786	6158
	Percentage change	-18	-61	51	248	-17
Yield	1971-1974	570	663	467	421	598
(kg/ha)	2001-2004	749	815	702	773	782
	Percentage change	31	23	50	84	31
Production	1971-1974	223	2883	1234	83	4433
('000 t)	2001-2004	243	1460	2531	596	4838
	Production percentage change	9	-49	105	614	9

NEPZ: North-East Plain Zone (Eastern UP, Bihar, and West Bengal); NPWZ : North-West Plain Zone (, Punjab, Haryana,Jammu, Himachal Pradesh, part of Rajasthan and part of Uttar Pradesh); CZ : Central Zone (Madhya Pradesh, Chhattisgarh, part of Rajasthan, Gujarat and Maharashtra); SZ : South Zone (Andhra Pradesh, Karnataka, Orissa and Tamil Nadu).

Yield gap analysis (YGA) shows that significant yield differences exist between desi and kabuli cultivars on research station. These differences range from 330 kg/ha in the NEPZ to 783 kg/ha in the CZ. Yield gap I, which is the gap between research station and on-farm trial yields, is highest in the SZ (30%) for desi variety and lowest (17%) in the NWPZ. Yield gap II, which is the gap between on-farm trials and zonal average yields, is very large in all zones, ranging from 64% in the NEZ to 148% in the CZ. Wider yield gap II is an indication that there is a large gap between on-farm demonstration yield and zonal average yield, which can be bridged by wider adoption of technology by farmers. The existing technology has the potential of doubling production at national level without increasing area under chickpea if farmers adopt recommended package of practices.

Table 4: Zone-wise yield gaps in chickpea production in India during 2000-2005. (From Annual Report of All India Coordinated Research Project (AICRP) on Chickpea, 2005).

Zone	Yield (kg/ha)						
	Research Station - desi	Research Station-kabuli	Difference Desi and Kabuli	On-farm trial	Zonal average	Yield gap I (%)	Yield gap II (%)
NEPZ	1491.2	1161.6	329.6	1230.0	749.0	21.2	64.2
NWPZ	1825.6	1461.8	363.8	1558.0	815.0	17.2	91.2
CZ	2069.6	1286.2	783.4	1745.0	702.0	18.6	148.6
SZ	1676.8	1316.8	360.0	1290.0	773.0	30.0	66.9

It has been realized that the outcome of the research on pulses has not been reached the growers to the extent required to make a significant breakthrough in the pulse production. The gap in transfer to technology has been highlighted time and again. Many technologies have proved to be beneficial to the growers but have not percolated well. To quote a few, fungal and bacterial antagonists against soil borne pathogens and NPV, a viral pathogen of gram pod borer, are menace to chickpea production, but have not yet been commercialized.

Linking production with trade is imperative for promoting chickpea cultivation. Development of export oriented chickpea like extra large seeded (>60 g/100 seeds) *Kabuli* types meeting international standards will receive priorities in future. It is well known that domestic demand for *kabuli* chickpea is increasing day by day compelling its import. To save the foreign valuable exchange, we have already developed 4 extra large seeded varieties (IPCK 2, MNK1, PKV 4-1 and Phule G 0517) that will fetch premium price in international market. Extra large seeded *kabuli* chickpea has great market in Europe.

Technologies to double the production

A) Improved cultivars

More than 75 high yielding varieties have been developed through collaborative efforts under the ambit of All India Coordinated Research Project on Chickpea by various State Agricultural Universities, Indian Institute of Pulses Research, Kanpur and Indian Agricultural Research Institute, New Delhi during last 20-22 years. International Crops Research Institute for the Semi-Arid Tropics, Patancheru has also contributed significantly in varietal development. These varieties possess wilt resistance, *Ascochyta* blight tolerance, early maturity, bold seeded *desi* and *kabuli* and varieties with restricted vegetative growth and suitable for mild salinity and moisture stress. List of varieties recommended for various states and agro-ecological conditions is given in Table 5.

(i) Disease resistant varieties : Development of disease resistant varieties has been a major activity in chickpea improvement. As a result, a number of wilt resistant and *Ascochyta* blight tolerant varieties have been developed. Some of these are given below:

- **Wilt Resistant:** KWR 108, H 82-2, DCP 92-3, GPF 2, JG 315, Vijay, JG 16 (SAKI 9516), JG 74, ICCV 10, JG 11, JG 322, Alok, GCP 101, Samrat, GCP 105, GNG 663, Radhey, Avrodhi, Pusa 372, Udai, Vijay, Vishal, Dharwad Pragati, Pusa Chamatkar, Pusa Kabuli 1003
- ***Ascochyta* blight Tolerant:** PBG 5, Samrat (GNG 469), PBG 1 and Pusa 372

Table 5 : Chickpea varieties released for different states of India

S. No.	State	Varieties
1.	Andhra Pradesh	Bharati (ICCV 10), JG 11, Kranti (ICCV 37), Phule G 95311 (K), Sweta (ICCV 2)
2.	Assam	JG 74, KWR 108, Radhey, Udai (KPG 59)
3.	Bihar	Gujarat Gram 4, KWR 108, Pusa 256, Pusa 372, Pusa 1003 (K), Udai (KPG 59), RAU 52
4.	Chhattisgarh	JG 16, JG 315, Jawahar Gram Kabuli 1 (JGK 1), KAK 2, Pusa 372, Pusa 391, Vaibhav, Vijay
5.	Delhi	Alok (KGD 1168), DCP 92-3, Karnal Chana 1 (CSG 8962), Pusa 372, Pusa Chamatkar (K) (BG 1053), RSG 888, Udai (KPG 59)
6.	Gujarat	Dharwad Pragati (BGD 72), Gujarat Gram 1, Gujarat Gram 2, JG 16 (SAKI 9516), Jawahar Gram Kabuli 1 (JGK 1), Pusa 372, Pusa 391, Vikas, Vijay, Vishal, Vishwas
7.	Haryana	Alok (KGD 1168), DCP 92-3, GPF 2, GNG 1581, Haryana Chana 1, Haryana Kabuli Chana1, Haryana Kabuli Chana 2, Karnal Chana 1, PBG 1, Pusa 329, Pusa 362, Pusa 372, Pusa 547, Pusa Chamatkar (K) (BG 1053), Rajas, RSG 888, RSG 963, Samrat (GNG 469), Udai (KPG 59), Vardan
8.	Jharkhand	Birsa Chana 1, Gujarat Gram 4, KWR 108, Pusa 256, Pusa 372, Pusa kabuli 1003, Udai (KPG 59)
9.	Karnataka	Annegeri 1, Bharati (ICCV 10), Chaffa, JG 11, Phule G 95311 (K), Sweta (ICCV 2)
10.	Madhya Pradesh	Dharwad Pragati, Gujarat Gram 1, JG 16 (SAKI 9516), JG 130, JG 218, JG 315, JG 322, JG 74, JGG 1, Jawahar Gram Kabuli 1 (JGK 1), Pusa 391, Vishwas (Phule G 5), Vijay, Vishal
11.	Maharashtra	Annegeri-1, Dharwad Pragati, Gujarat Gram 1, Jawahar Gram Kabuli 1 (JGK 1), JG 16 (SAKI 9516), , KAK 2 (K) (PKV 1), Pusa 372, Pusa 391, Sweta (ICCV 2), Phule G 12, Vihar (K), Vijay, Vikas, Virat (K), Vishal, Vishwas (Phule G 5),
12.	Punjab	Alok (KGD 1168), DCP 92-3, GPF 2, GNG 1581, Hare chhole 1, Haryana Chana 1, Karnal Chana 1, L 551, PBG 3, PBG 5, Pusa 256, Pusa 329, Pusa 362, Pusa 372, Pusa 547, Pusa Chamatkar (K) (BG 1053), Rajas, RSG 963, Samrat, Udai (KPG 59), Vardan
13.	Rajasthan	Abha, Abhar, Abhilasha, Akash, Ankur, Anvita, Aparna, Arpan,Arpita, Aruna, Asar, Asha, Alok (KGD 1168), DCP 92-3, GNG 146, GNG 663, GNG 1292 (K), GNG 1581, GPF 2, Harayana Chana 1, Karnal chana 1, PBG 1, Pusa 329, Pusa 362, Pusa 372, Pusa 547, Pusa Chamatkar (K) (BG 1053), RSG 44, RSG 888, RSG 963, Rajas, Samrat, Udai (KPG 59), Vardan
14.	Uttar Pradesh	Alok (KGD 1168), DCP 92-3, GPF 2, GNG 1581, Gujarat Gram 4, Haryana Chana 1, JG 16 (SAKI 9516), JG 315, Karnal Chana 1, KWR 108, Pant G 186, Pragati , Pusa 256, Pusa 372, Pusa 547, Pusa kabuli 1003 (BG 1003), Rajas, Sadabahar, Sadbhawana, Samrat (GNG 469), Surya, Udai (KPG 59), Vardan, Vishwas,
14.	West Bengal	Anuradha, JG 74, KWR 108, Mahamaya 1, Mahamaya 2, Pusa 256

ii) Bold Seeded : Adopting bold seeded varieties especially under irrigated conditions can increase the chickpea yield. Some of these high yielding and bold seeded varieties recommended are:

a) ***Desi* (>20g/100 seed weight):** Pusa 256, Samrat, Pusa 391, Pusa 362, KWR 108, Dharwad Pragati, Co 3, Co 4, Phule G 5, Radhey and Vishal

b) ***Kabuli* (>25 g/100 seed weight):** Jawahar Gram Kabuli 1 (JGK 1), KAK 2, Pusa Chamatkar (BG 1053), Phule G 95311, ICCV 2, Pusa Kabuli 1003 (BG 1003), Vihar, Haryana Kabuli Chana-1.

c) **Extra large seeded *kabuli* (>50g/100 seed weight):** PKV 4-1, Phule G 0517 Among these varieties Jawahar Kabuli Chana 1 (JGK 1), KAK 2 and Phule G 95311 have more than 30 g per 100 seed weight.

iii) Varieties for late sown conditions : In Punjab, Haryana, Uttar Pradesh, Bihar, parts of Chhatisgarh and Jharkhand states where rice fields are vacated quite late after harvest of rice or other rainy season crop, chickpea varieties like Udai (KPG 59), Pusa 372, RSG 963 and PBG 1 can be grown to harvest good yield. These varieties can be sown up to first week of December. Similarly, state releases like Pant G 186 and JG 74 have been recommended for cultivation under late sown condition in Uttar Pradesh and Madhya Pradesh, respectively.

iv) Varieties for specific situations : In areas where there is to high soil moisture or frequent winter rains or high water table and high fertility, chickpea crop attains more vegetative growth resulting in lodging and subsequent yield loss. The variety, DCP 92-3, can be grown successfully. Similarly, chickpea variety, Karnal Chana 1 (CSG 8962) can be grown successfully under mild salinity conditions of North-West plains. Recently, variety, RSG 888, has been recommended for cultivation in moisture stress conditions (drought prone areas) of Rajasthan, Punjab and Haryana. JG 14 is suited to late sown conditions in central India due to its high level of heat tolerance.

B) Improved management practices

(i) Improved method of sowing : Sowing depth decides the period of emergence. The period of emergence can be shortened with appropriate sowing depth according to soil types and moisture. Studies on loamy sand soil in northern India showed that seeding at 10 cm. depth was better than shallow seeding if the crop was raised without irrigation. In un-irrigated crop, mortality of seedling was high with 5 cm deep sowing. In areas where moisture content in top soil layer is low, seeds are placed deeper in moist zone. Since the crop is sown on conserved moisture, seed should be invariably dropped deep (8-10 cm.). In case soil moisture is sufficient for germination, a depth of 5-7 cm. is ideal for germination. Sowing is done in furrows through *pora* attached with *desi* plough seed drill. The sowing operation should preferably be done in the morning hours followed by planking.

(ii) Water management and Moisture conservation : Chickpea is capable of extracting water from a depth up to 150 cm (Singh and Bhusan 1979). However, a major proportion of the total water comes from the top 60 cm soil profile where active roots are concentrated. Some important measures for water management and moisture conservation in chickpea are : growing drought tolerant varieties, rain water harvesting through farm ponds and community reservoirs, promoting short duration varieties in drought prone areas, promoting micro irrigation systems and adoption of other moisture conservation practices.

Studies conducted under AICRP on Chickpea revealed that raised bed planting of chickpea was more economical than flat bed. Raised bed sowing is becoming popular in heavy soils. Flood irrigation in flat bed in heavy soils badly damages the crop due to water stagnation. In case chickpea is sown on raised beds, irrigation can be applied to crop under limited water availability condition without any damage to crop. Irrigation application to the crop sown on raised bed helped in saving about 30% of water. It also improved the seed size and grain yield of the crop in *kabuli* chickpea. During 2009-10, average yield of *kabuli* chickpea obtained from raised bed sowing was 1804 kg/ha.

(iii) Nutrient management : Total uptake of nutrients by chickpea crop has been estimated as varying from 60-200 kg N, 5-15 kg P and 60-170 kg/k per ha (Ahlawat 2000). Intensification of cropping systems has made soils deficient in nutrients leading to a reduction in availability, thus constraining the supply of essential nutrients to the plants. Consequently, deficiency of nutrients that started in 1950 for N has added six more till 1990 (Fe in 1960, P, Zn, K, S, Mn in 1970). The judicious use and balanced use of fertilizer is required not only for quantitative and qualitative increase in yield but also for vigorous growth of plants to enable them to overcome biotic and abiotic stresses. Imbalanced nutrition makes plant more vulnerable to insect pests and diseases, mainly, because of imbalanced distribution of assimilate due to restricted supply of nitrogen and photosynthates for developing reproductive parts.

In general, chickpea has shown response to application of nitrogen up to 15-20 kg per ha on coarse textured soils unless the soil is rich in organic matter. The response of applied nitrogen varies from 8.5 to 19.5 kg grains per kg nitrogen applied. However, response is poor under well-developed nodulation with efficient biological nitrogen fixation system when chickpea is grown after maize, sorghum or long duration rice and the planting of chickpea is delayed. Under such situations due to low soil N and short period for growth and development, the nitrogen dose should be increased by 25-50% under late sown conditions, as the crop responds well up to 40 kg N/ha.

Phosphorus is the most important nutrient for chickpea. Its role in proliferation and development of deeper root systems is well known. Beside beening a constituent of ADP and ATP, Phosphorus also increases protein content in the grains (Guhey *et al.*, 2000, Motsara 2002) reported that maximum P deficient samples were observed in soils of Maharashtra followed by U.P. Soils of Karnataka, Punjab and Tamilnadu were relatively less P deficient. Response of Phosphorus application in chickpea has been observed by several workers (Prabhakar and Sarat 1991). Experiments conducted under AICRP on Chickpea showed that grain yield of *kabuli* chickpea was 17-37% higher due to P application. The soils having less than 15 kg/ha available P showed responses up to 60 kg per ha of applied P, while locations having 15-22.5 kg per ha available P had shown response only up to 30 kg of applied P and no response of applied P was observed when available P was more than 22.5 kg/ha. Phosphorus response is more distinct under irrigated conditions than rainfed. In most of the cases, the response of 40-60 kg P_2O_5 per ha has been positive both at farmer's fields and research stations. The response to applied phosphorus is influenced by several factors such a method of application, source of phosphorus, profile moisture, which in turn is influenced by irrigation water. Phosphorus solubilizing bacteria (PSB) have been found effective in increasing the efficiency of applied phosphorus either with DAP or SSP in calcarious soil (Saad and Sharma 2001). Foliar application of DAP/Urea 2% solution has been found beneficial in increasing the seed yield of chickpea under rain fed conditions throughout chickpea growing areas.

Because of general the assumption about the higher available potassium in Indian soil, this element has rarely found place in fertilizer scheduling of chickpea. Chickpea growing regions have wide range of available K content in soil (Subba Rao and Srinivasa Rao, 1996). The response to K in chickpea is small and seldom significant. The response to applied K has been reported in laterite, red, coastal and deltic alluvial soils of India. Generally, the application of 20 kg K_2O per ha is recommended in deficient soils.

Being a pulse crop, requirement of sulphur in chickpea crop is higher basically because it is a constituent of predominant amino acids (methionine, cysteine and cystine). It is also a constituent of ferrodisiun – containing nitrogenase and also involved in metabolic activities of vitamins, thiamine and coenzyme. Due to increased cropping intensity under north Indian conditions and use of high analysis fertilizers, the deficiency of sulphur is being observed in most of the soils. Generally, soils of coarse texture and low organic matter have low S level. In sulphur deficient soils, the response to applied sulphur has been found in the range of 20-60 kg/ha (Aulakh and Pasricha 1986). Significant response up to 40 kg S per ha was also observed in central zone (Ali and Singh 1995). The respons to various sources of sulphur like gypsum, SSP, Pyrite and elemental sulphur was similar.

Deficiency of micro nutrients has become a common phenomenon in pulse growing regions. Among micro-nutrients, zinc deficiency is widely observed followed by boron and iron. Response to micro nutrients has been obtained only in deficient locations. Zinc and iron played an important role in collar rot, wilt and root rot in chickpea (Gupta 1999) and an application of 25 kg $ZnSO_4$ per ha at the time of sowing is recommended. Boron deficiency is an emerging problem in some states like Orissa, Bihar, Uttar Pradesh and Gujarat (Takkar 1996). Response to applied B is largely reported on calcareous and sandy soils. Molybdenum deficiency is reported in many areas particularly in black soils of Madhya Pradesh and Gujarat to the level of 18 and 10%, respectively. Significant impact of boron and molybdenum application has been observed in chickpea.

(iv) Weed management : Chickpea is a poor competitor to weeds because of slow growth rate and limited leaf area development at early stages of crop growth and establishment and they pose severe competition to chickpea crop under rainfed as well as irrigated conditions. Losses in seed yield due to weed was estimated to vary between

40 to 87% (Saxena *et al* 1976, Ahlawat *et al* 1981, Yadava *et al* 1983, Vaishya *et al* 1999). Weeds, besides posing above ground competition, are generally more hardy and drought tolerant and extract the scarce moisture and nutrients from soil. Depending upon the soil, cropping system, method of sowing, time of sowing and the weed control measures applied to previous crop, weed species differ. The dominant weed flora associated with chickpea crops are *Chenopodium album, Melilotus indica, Lathyrus aphaca, Medicago denticulate, Trigonella polycerata, Polygonum pblebilium, Asphodelus tenuifolius, Euphorbia dracunculoides* Lamk., *Anagallis arvensis* and *Trichodesma indicum*. The planting time has significant impact on weed dynamics. At Hisar, chickpea planting on 5th December (late) resulted into lower weed dry weight and weed density and significantly higher seed yield of chickpea (Table 6). The mean yield advantages were 197 and 71 kg per ha as compared to November 5 and November 20 plantings, respectively (Malik *et al* 1988). The choice of weed control methods depended on the farmers' resources, available technology, cost involved in implementation of that technology and the resultant increase in yield. Timely weeding is essential for achieving good yields. Since the initial four to eight weeks are most critical for weed competition (Ahlawat *et al* 1981), the first mechanical operation has been advocated three to four weeks after sowing, and the second one after six to eight weeks of sowing. However, in case of severe infestation, a third weeding may be needed around 9-12 weeks after sowing. However the late removal of weeds does not provide much relief to the crop.

Table 6 : Effect of planting dates on weed dry weight, density and yield of Chickpea

Planting Date	1985-86		1986-87		Yield (kg/hac)
	Weed dry weight (g/m²)	Weed density (no./m²)	Weed dry weight (g/m²)	Weed density (no./m²)	
November 5	63.4	2.9 (10.1)	74.9	3.0 (9.8)	1205
November 20	52.3	2.8 (8.2)	36.6	2.7 (4.7)	1331
December 5	33.7	1.7 (1.8)	31.4	1.7 (2.5)	1402
LSD	0.3	0.3	0.1	0.1	44

Although mechanical weeding provides efficient weed control but due to its limitations, chemical weed control through Fluchloralin, Pendimethalin, Prometryne and Trifluralin without having any phytotoxicity to the chickpea has been found beneficial. Prometryne @ 0.5 kg a.i. per ha (Ahlawat *et al* 1978) Alachlor @ 1.5 kg a.i. per ha as pre-emergence (Faroda and Singh 1981), and terbutryne @ 0.75-1.2 kg a.i. per ha (Kolar *et al* 1979, Dhingra *et al* 1982), applied pre-planting and Pendimethalin @ I kg a.i. per ha as pre-emergence (Yadav *et al* 1983, Vaisiya 1999) followed by one hand weeding at 45 days after sowing (Singh and Vaishya 2001) provide effective control of annual broad leaved and grassy weeds in chickpea crop. Weed infestation appreciably decreases when chickpea is intercropped with mustard or other crops. Similarly, close spacing at 25 cm also suppresses weed flora.

Storage : Lack of on-farm storage facilities and the vulnerability of chickpea to store grain pest results in considerable losses. As most of the farmers use farm produce as seed, the quality of seed is not up to the mark in most of the cases. Sowing of such damaged seeds is the main reason behind poor plant population of chickpea on farmers' fields. Chickpea is generally attacked by bruchids at the time of maturity. To avoid further development of these bruchids, it is recommended to fumigate the seed material with ALP @ 1-2 tablets per ton and EDB @ 30 ml per ton. Before storage chickpea should be cleaned and sun-dried to have low moisture (8% or below). The grains should be packed in muslin cloth bags upto 5 kg with proper stitching. If more quantity is to be stored, 30 kg packing in bag with straight streaming is recommended. When storing is to be done in bulk, they should be stored in iron containers and the top layer should be covered with 2.5 cm sand layer free from soil and dirt to avoid reinfestation and development of bruchids.

C) Integrated disease management

The major diseases are *Ascochyta* blight and wilt in North West Plain Zone (NWPZ) and wilt & dry root rot in Central Zone (CZ) and Southern Zone (SZ).

- **Wilt and root rot:** Seed should be treated with fungicides, *viz.*, Carbendazim + Thiram (1 g + 2 g/kg) at sowing. Seed dressing with bioagent *Gliocladium virens* @ 0.5 g/kg or *Trichodrma viride* @ 4 g/kg seed + fungicides Carboxin (Vitavex) @ 1 g/kg seed should be followed.The resistant varieties of chickpea (Zone Wise) are as follows.

North West Plain Zone (NWPZ)	:	GNG 146, PBG 1, Haryana Chana 1, GF 89-36, Pusa 362, Pusa 372, GNG 663, GNG 469, DCP 92-3, GNG 1581
North East Plain Zone (NEPZ)	:	Avarodhi, KWR 108, KPG 59, Pusa 1003
Central Zone (CZ)	:	JG 315, ICCV 10 (Bharti), Pusa 391, ICCV 2, Vijay, Vishwas, Phule G 5, JAKI 9218, Gujarat Gram 1, Pusa 244, JG 16
South Zone (SZ)	:	Annegiri 1, ICCV 10 (Bharti), ICCV 4, JG 11, ICCV 2

- **Ascochyta blight** : Seed treatment with Haxacap (3 g/kg seed), Carbendazim (2 g/kg seed) or Calixim-M (Tridemorph 2g/kg seed) is effective. Foliar sprays of Calixim-M (Tridemorph (0.1 %) or Hexacap (0.02 %) or Carbendazim (0.1 %) may be adopted. Disease resistant varieties, *viz.*, Gaurav, PBG 1, Haryana Chana 1 and GNG 469, are recommended NWPZ, where disease has generally occurred in severe form in some years.

- **Botrytis gray mould** : It is an important disease in sub-mountaneous (*terai*) regions where it appears sometimes when humidity is high. The resistant varieties against BGM are not available. It is not easily controllable.

D) Insect-pest management

Pod borer is the major pest. It requires monitoring at weekly intervals at flowering and podding stages for egg and larvae. Sex pheromone traps (5 traps/hac) should be put in January-March. Economic threshold limit may be 1 larva per 5 to 10 plants. The economic threshold level of important pests of chickpea as follows

Gram pod borer	:	2-3 eggs/plant or 2 early instar larvae/10 plants or 1 matured larva/10 plants
Cut worm (*Agrotis*)	:	1 larva/m^2 length
Semilooper	:	2 larvae/10 plants
Termite	:	5 plants/m^2
White grub	:	5 plants/m^2

- **Cultural practices:** Deep summer ploughing, clean cultivation, timely sowing (up to end of October), and intercropping with linseed, mustard, wheat, barley, coriander etc. minimizes damage by pod borers and nematodes.
- **Mechanical practices:** Collect and destroy the grown up larvae.
- **Biological control:** Conserve natural enemies, *viz*., *Campoletis*, ladybird beetles, stickbug, predatory wasps and spiders by using eco-friendly insecticides. Install bird perches to encourage insectivorous birds. Intercultivate nectar rich crop like coriander to support natural enemy population of *Compoletis chorideae*.
- Spray nuclear polyhedrosis virus (NPV) @ 250 larval equivalents (LE)/hac on noticing the first- or second-istar larvae or eggs of *Helocoverpa armigera*. Apply 3 sprays at weekly intervals. Use of *Bacillus thuringiensis* formulations, *viz*., Dipel, Delhin, Biobit etc. @ 1-1.5 kg/hac is useful.
- **Chemical control**: Selective pesticides, *viz*., Fenvalerate (0.01%), Monocrotophos (0.04%), and extract of Neem Seed Kernel (5%) or other Neem based formulations, *viz*., Nimbecidine, Repelin, Achook, Neemguard, etc., should be used as soon as economic threshold limit is crossed.

Seed and soil treatment with Chlorpyriphos may be adopted for control of termites and cutworm.

Integrated Pest Management (IPM) approach

Integrated pest management (IPM) utilizes all methods in a compatible manner. It should be on a community basis rather than on individual basis and needs group attention.

For effective management of *H. armigera,* all the farmers need to sow synchronously less susceptible varieties at optimum time and as far as possible, apply timely non-polluting or less polluting insecticides based on the eggs and larvae.

IPM has its utility under both the situations where insecticides have been indiscriminately used and where use of insecticides is not popular. The emphasis in the first case would be to reduce the amount of insecticides being used and rely more on the use of non-chemical methods, whereas in the second case, the focus would be to make pest control practice attractive so that it becomes popular and easily acceptable (Saxena and Duraimurugan 2008). Under both the situations, emphasis needs to be placed on cultural and bio-rational control.

Integration of intercropping along with imposition of bio intensive and chemo-intensive modules were synthesized and tested against *H. armigera* infesting chickpea for four consecutive years. Among different intercrops, chickpea +mustard (4:2) proved best as it recorded least mean pod damage (8.75%) and chickpea equivalent yield of 1672kg/ha as against 13.60% pod damage and 1402 kg/ha yield in sole crop. Among control modules tested, bio-intensive integrated pest management (BIPM) proved most efficacious recording 7.92% pod damage and yield of 1748 kg/ha as against 15.34% pod damage and yield of 1260 kg/ha in control. Considering economics of the treatments, chickpea intercropped with mustard with BIPM resulted in highest CBR (Table 7).

Table 7: Effect of IPM modules and intercrops on pod damage and yield of chickpea (2000-01 to 2003-04)

Treatment	Pod damage (%)	Chickpea equivalent yield (kg/ha)
Chickpea + mustard 4:2 row ratio	8.75	1672
Chickpea + linseed 2:1 row ratio	10.35	1573
Chickpea sole	13.60	1402
*Bio- intensive IPM	7.92	1748
# Chemo- intensive IPM	9.42	1684
Untreated control	15.34	1260

*BIPM = Seed treatment with *T. harzianum* + Vitavax (4g+1g/ Kg seed), first spray 250 LE HA NPV with adjuvant at ETL; Second spray NSKE 5%

CIPM= Seed treatment with Thiram + Carbendazim (2+1g/ Kg seed).

Based on the available information, implementation of IPM in chickpea is being suggested as follows:

Exploitation of host avoidance phenomenon

- Early sowing i,e,. by the end of October under rainfed situations to avoid peak period of *H. armigera* activity at podding stage.

- Sowing of short duration chickpea cutivars in November (under rice fallows) i.e., PDG 84-10.
- Inter/mixed farming. Growing chickpea with popular intercrops eg., mustard, linseed, wheat ,barley, etc.

Monitoring

- Use of sex pheromone traps.
- Counting of eggs and larvae on crop at reproductive stage during $10\text{-}16^{th}$ standard week (North India).

Thresholds

- One larva per meter row length at flowering/ podding.
- 5 to 6 moths per pheromone trap per day during post-winter months ie., March-April to be warming signal for close monitoring of eggs/ larvae population.

Plant products

- Use o f Neem Kernel Extract (5%) with 1% soap solution.
- Use of Neem based commercial formulations.

Bio-agents

- Application of HaNPV 250 LE/ha.

Eco-friendly pesticides

- Chlopyriphos have been found comparatively safer to *Campoletis chlorideae,* a potential larval parasite of *H. armigera.*
- Insecticides like methomy or such other insecticides should be used only when utmost necessary.
- Any particular insecticides should not be used continuously. It should be alternated with insecticides of varied group, like chlorinated hydrocarbon, organophosphate, carbamate and pyrethroids.

Augmentative/ biological control

- Installation of bird perches to encourage birds.
- Inter-cultivation of nectar rich plants like coriander

Critical input needs

A) Inadeguate supply.

The main issue is inadequate availability of quality seeds of improved varieties as detaled in (Table 8).

- Only 20% of the breeder seed is converted to certified seeds
- Seed multiplication ratio (SMR) in general is less than 10%.
- Seed indents of new varieties are low.
- Unrealistic seed indent.

Table 8: Total requirement of seed (state wise)

State	Area ('000ha)	Certified Seed requirement ('000 tonnes)	(SRR%)
Andhra Pradesh	630.0	10.08	20.0
Chhattisgarh	235.5	3.77	20.0
Gujarat	215.0	3.44	20.0
Haryana	107.0	1.71	20.0
Karnataka	605.0	9.68	20.0
Madhya Pradesh	2438.7	39.02	20.0
Maharashtra	1353.0	21.65	20.0
Rajasthan	1231.3	19.70	20.0
Uttar Pradesh	505.0	8.08	20.0
Total	7543.7	120.70	20.0

Strategies

- Advanced seed planning at state level with a rolling seed plan.
- Improving conversion of breeder to foundation to certified seeds.
- States need to be sensitized for giving indents of new varieties.
- Fungicide treatment should be made mandatory for all classes of seeds.
- SMR should be improved through proper crop management.
- Realistic indents should be given for breeder seed.
- Seed buffer of improved varieties may be maintained at State Seed Corporation level.

B) Inadequate and imbalanced use of nutrients

The state wise input requirement of some nutrients is given in table 9.

- Wide spread deficiency of sulphur. Out of 137 pulse growing districts, 87 districts show 20-60% S deficiency
- About 50% chickpea growing districts show zinc deficiency
- Severe boron deficiency on calcareous soils of Bihar.
- Sub-optimal or no application of fertilizers in chickpea.

Strategies

- Ensuring timely availability of S, Zn alongwith phosphatic fertilizers in the districts deficient in these nutrients.
- Promoting use of Bio-fertilizers (Rhizobium and PSB) as detailed in Table 10.
- Popularization of 2% foliar spray of urea/ DAP in rainfed areas
- Introducing the system of soil health card to ensure balance use of nutrients

Table 9: State wise input requirements

State	Area ('000 ha)	Sulphur* @ 20 kg/ha ('000 ton)	$ZnSO_4$** @15 kg/ha ('000 ton)	P_2O_5*** @ 40kg/ha ('000 ton)
Andhra Pradesh	630.0	3.15	1.89	12.60
Chhattisgarh	235.5	1.18	0.71	4.71
Gujarat	215.0	1.08	0.65	4.30
Haryana	107.0	0.54	0.32	2.14
Karnataka	605.0	3.03	1.82	12.10
Madhya Pradesh	2438.7	12.19	7.32	48.77
Maharashtra	1353.0	6.77	4.06	27.06
Rajasthan	1231.3	6.16	3.69	24.63
Uttar Pradesh	505.0	2.53	1.52	10.10
Total	7543.7	37.72	22.63	150.87

*To cover 25% area; ** To cover 20% area; *** To cover 10% area

Table 10: Total requirement of other inputs

Area ('000 ha)	Carbendazim* (tonnes)	Rhizobium** (tonnes)	Trichoderma** (tonnes)	HaNPV LE** 10^5 No
7543.7	1207.00	94.29	22.63	9429

*To cover 25% area; ** To cover 20% area

C) Water and land

Chickpea suffers heavily from soil moisture stress/drought as 75% is cultivated under rainfed conditions and there is very limited choice as far as availability of drought tolerant varieties are concerned.

Strategies

- Rain water harvesting through farm ponds and community reservoirs
- Promoting short duration varieties in drought prone areas
- Promoting micro irrigation systems
- Adoption of moisture conservation practices

Mechanisms for efficient utilization of resources

In order to utilize efficiently water, soil and nutrient resources, the following management practices may be used in chickpea based cropping systems (sequential and intercropping):

Integrated nutrient management

- 2% foliar spray of urea at flowering and podding stages increases yield by 15% in chickpea
- Basal application of sulphur @ 20 kg and zinc @15 kg/ha along with NPK improves yield by 20%
- Seed inoculation with *Rhizobium* and PSB ensures higher yield by 10%

Integrated soil and water management

Zero Tillage : This is one of most efficient moisture conservation technology available, particularly, in rice fallows. It also helps in saving time for sowing early crops. There are a number of good zero till drills available these days for different crops.

Furrow Irrigated Raised Bed Planting (FIRBS System) : Raised bed method of planting improves chickpea production over flat bed and has several advantages, e.g. heavy textured soil, better plant stand, saving in irrigation water and safe disposal of excess water. This also increases yield of chickpea to the tune of 10-15% over flat bed.

Laser Land Leveller : Normally chickpea is grown under rainfed situations with conserved moisture. However, it has been shown through various experiments that one irrigation at critical stage i.e. Flowering or pod formation stage improves productivity by 30%. If soil is properly leveled through laser leveller, a significant quantity of irrigation water may be saved.

Microirrigation (sprinkler and drip irrigation) : Micro-irrigation system has definite advantage over conventional irrigation (flood-irrigation) as it saves precious water and helps in avoiding water logging in poorly leveled area. It also promotes better growth and crop establishment. The real success of use of sprinkler system in chickpea may be seen in light sandy soils of Rajasthan (Bikaner, Jaisalmer, Barmer, Sikar and Jhunjhunu districts) and in heavy black soils of Western Maharashtra.

Seed priming : Soaking of seeds for 4-6 hours before sowing helps in better emergence, growth and crop establishment under rainfed conditions. The net advantage of seed primiming is 15-20% yield gain in chickpea.

Future thrust

- There is need to develop chickpea varieties specifically tailored for intercropping. These varieties need to have characteristics of erectness, less shyness against shedding and responsiveness to high inputs. These kind of varieties of chickpea may be intercropped with sugarcane and mustard in North West Plain Zone (Punjab, Haryana, Western UP and Northern Rajasthan).
- There is an urgent requirement to evolve multiple stress tolerant varieties lile more drought and heat tolerant varieties of chickpea which are also resistant to wilt and dry root rot for sequential cropping in rainfed conditions of central and peninsular India after sorghum, pearl millet, soybean and maize.
- In addition to above, more drought and heat tolerant varieties of chickpea which are also resistant to wilt and wet root rot, color rot for sequential cropping in rice fallows of eastern Uttar Pradesh, Bihar, West Bengal, Assam, Chhattisgarh and part of Madhya Pradesh are also required.
- Issue of weed management in intercropped chickpea need to be addressed on priority particularly, on appropriate post emergence weedicides.
- Integrated nutrient management in intercropped chickpea needs to be looked into with a major emphasis on micronutrients.

References

Agricultural Statistics at a Glance 2009. Directorate of Economics and Statistcs, Department of Agriculture and Cooperation, Ministry of Agriculture, Government of India.

Ahlawat, I. P. S. 2000. Bengal gram. In Techniques and Management of Field Crop Production. Agrobios, New Delhi, India. pp 317-335.

Ahlawat, I. P. S., Singh, A. and Saraf, C. S. 1978. Comparative efficiency of cultural and herbicidal methods of weed control in Bengal gram. Indian Journal of Weed Science, 10, 1-8.

Ahlawat, I. P. S., Singh, A. and Saraf, C. S. 1981. Effect of winter legumes on the N economy and productivity of succeeding cereals. Experimental Agriculture, 17 : 57-62.

Ali, M. and Singh, K. K. 1995. Dalhani Phaslon Mein Gandhak Ka Mahatwa (In Hindi) (Importance of sulphur in pulse crops). Technical Bulletin, Indian Institute of Pulses Research, Kanpur, India.

Anonymous. 1952. Intercrops in coconut gardens. Indian Coconut Committee Bulletin, 5 (12), 194.

Argikar, G. P. (1970). Gram. In Pulse Crops of India (Eds. P. Kachroo and M. Arif). ICAR, New Delhi. pp 54-135.

Aulakh, M. S. and Pasricha, N. S. 1986. Role of sulphur in the production of grain legumes. Fertilizer News, 31 (9), 31-35.

Chowdry, B. S. 1953. Sowing sugarcane with gram or peas. Allahabad Farmer, 27 : 94-96.

Dhingra, K. K., Sekhon, H. S. and Tripathi, N. P. 1982. Chemical control of weeds in chickpea. International Chickpea Newsletter, 6 : 14.

Faroda, A. S. and Singh, R. C. 1981. Weed control in pulses. Pulse Crops Newsletter, 1 : 50-51.

Food and Agriculture Organization of United Nations (www.fao.org)

Greco, N. and Sharma, S. B. 1990. Progress and problems in the management of nematode diseases. In Chickpea in the nineties, proceedings, Second International Workshop on Chickpea Improvement (4-8 December, 1989), ICRISAT, Patancheru, India. pp 135-137.

Guhey, A., Trivedi, A. K. and Khan, M. I. 2000. Change in sugars, amino acids and protein in chickpea as influenced by phosphorus levels. Advances in Plant Sciences, 13 (1) : 309-311.

Gupta, G. P. and Mathur, B. P. 1964. Mixed cropping studies in oilseeds. Indian Oilseeds Journal, 8 (3) : 206-213.

Gupta, M. L. 1955. Studies on wheat-gram mixture under irrigated and rainfed conditions in Kanpur. Agriculture and Animal Husbandry-Uttar Pradesh, 6 (6) : 33-34.

Gupta, O 1999. Effect of micronutrient on root pathogens of chikcpe (Cicer arietinum L.). Bhartiya Krishi Anusandhan Patrika. 14 (3-4) : 25-27.

Howard, A. 1974. Gram mixed cropping (Note). Indian Agriculture Journal, 9 : 308-310.

Kalpana Sastry, R., Rego, T. J. and Burford, J. R. 1993. Effect of cropping systems on wilt of safflower. In Proceedings, 3rd International Safflower Conference (Eds. Li Dajue and Han Yuanzhou) (June 14-18, 1993), Chinese Academy of Sciences, Beijing, China. pp 636-645.

Katyal, S. K. 1963. Studies on the relative utility of mixed cropping of wheat with lentil and kabuli gram over individual cropping under certain agronomic factors. M. Sc. Thesis, Punjab Agricultural Univeristy, Ludhiana, India.

Khanna-Chopra, R. and Sinha, S. K. 1987. Physiological aspects of growth and yield. In The Chickpea (Eds. M. C. Saxena and K. B. Singh), CAB International/ICARDA, Wallingford, UK. pp 163-189.

Kolar, J. S., Kang, P. S. and Sandhu, K. S. 1979. Studies on competition between some *rabi* crops and Chenopodium album. Indian Journal of Ecology, 6 : 61-67.

Malik, R. K., Beyon, R. S. and Bhan, V. M. 1988. Effect of herbicides and planting dates on weeds in chickpea. Indian Journal of Weed Science, 20 (2) : 75-81.

Motsara, M. R. (2002). Available nitrogen, phosphorus and potassium status on Indian soils as depicted by soil fertility maps. Fertilizer News, 47 (1) : 15-21.

Nene, Y. L. 1980. Pesticides of chickpea. In Proceedings, International Workshop on Chickpea Improvement, (February 28 – March 2, 1979) ICRISAT, Patancheru, India. pp 171-178.

Prabhakar, M. and Saraf, C. S. 1991. Influence of irrigation and phosphorus on growth yield and water use efficiency of chickpea (Cicer arietinum L.). Indian Journal of Agronomy, 36 (3) : 357-362.

Saad, A. A. and Sharma, H. M. 2001. Efficacy of PSB with phosphorus on the yield of chickpea in calcareous soil of north Bihar. In Proceedings, National Symposium on Pulses for Sustainable Agriculture and Nutritional Security (17-19 April, 2001), ISPRD, New Delhi, India. pp 42

Salunkhe, D. K. and Kadam, S. S. 1989. Hand Book of World Food Legume. Vol. III, CRC Publication, Press, Inc.Baca Raton, Florida. pp 1-31.

Saxena, Hem and Duraimurugan, P. 2008. Present scenario of integrated pest management in Pulses. National Conference of Pest Management Strategies for Food Security held at IGKV, Raipur during May 2-3, 2008. pp 9.

Saxena, M. C., Subramaniyam, K. K. and Yadava, D. S. (1976). Chemical and mechanical control of weeds in gram. Pantnagar Journal of Research, 1 : 112-116.

Shyam, G. 1963. Mixed cropping of wheat and gram. M. Sc. Thesis, university of Agra, India.

Singh, A. and Vaishya, R. D. 2001. Effect of weed management techniques and phosphorus levels on weed infestation and seed yield of late sown chickpea. Indian Journal of Pulses Research, 14 (2) : 119-121.

Singh, K. and Katyal, S. K. 1966. Effect of mixed cropping of wheat and gram with varying levels of nitrogen and phosphorus on yield. Journal of Research, Punjab Agricultural University, 3 (4) : 364-367.

Subbarao, A. and Srinivasarao, Ch. 1996. Potassium status and crop responses to K in different agro-ecological regions of India. In IPI Research Topics No. 20. International Potash Institute, Basel, Switzerland. pp 1-57.

Takkar, P. N. 1996. Micronutrient research and sustainable agricultural productivity in India. Journal of Indian Society of Soil Science, 44 : 562-581

Vaishya, R. D., Rai, O. P. and Sing S. S. 1999. Weed control in chickpea with pre-emergence herbicides in eastern Uttar Pradesh. Indian Journal of Pulses Research, 12 (2) : 197-200.

Yadav, R. L. 1996. Intensive cropping system involving pulses. In Special Group Meeting on Constraints and Opportunities in Pulse Production: Management Options and Researchable Issues (12-13 March 1996), IARI, New Delhi, India.

Yadav, S. K., Singh, S. P. and Bhan, V. M. (1983). Weed control in chickpea. Tropical Pest Management, 29 : 297-298.

❑❑❑

Chapter - 17

Lentil

G.P. Dixit, B.B. Singh & P.K. Katiyar
AICRP on MULLaRP, Indian Institute of Pulses Research, Kanpur 208 024, Uttar Pradesh.

Lentil is one of the most nutritious cool season food legume and ranks next only to chickpea. It is grown throughout the northern and central India for grains, which are used as *dal* (whole or dehulled) and in various other culinary preparations. Lentil contains about 25% protein, 0.7% fat, 2.1% mineral, 0.7% fibre and 59% carbohydrate. It is a rich source of phosphorus and carotene. It is generally grown as rainfed crop during rabi season after rice, maize, pearl millet or *kharif* fallow. It is also grown as intercrop with barley, linseed, mustard and autumn planted sugarcane. In North-eastern parts of the country, lentil is also cultivated as *paira* crop with rice in which seeds of lentil are broadcast in the standing crop of rice 7-10 days before its harvest. It is grown on a wide range of soils from light loamy sand to heavy clay soil in northern parts and moderately deep black soils in central India.

Status of Area, Production & Productivity

In India, lentil is the second most important *rabi* pulse crop next to chickpea. It occupies 1.59 million ha. area with a production of 0.94 million ton. It is mainly cultivated in the states of Uttar Pradesh, Madhya Pradesh, Bihar and West Bengal (Table 1). These states contribute 90% to the total production. It is generally grown as a rainfed crop on marginal lands. In most of the regions, it is cultivated after maize or rice. In West Bengal, it is also grown after jute. In cotton belt of Maharashtra and Madhya Pradesh, it follows cotton. In Bundelkhand region of Uttar Pradesh, it is cultivated under rainfed monocrop situation. In central Madhya Pradesh, eastern plains of Rajasthan and Bundelkhand region, lentil is grown as mixed crop/inter crop with mustard and linseed. In low lands of Uttar Pradesh, Bihar and West Bengal, lentil seeds are broadcasted in standing rice 7-10 days before harvesting (relay cropping) to capitalize on residual moisture and ensure timely sowing. In *tarai* region of Uttar Pradesh it is also intercropped with winter sugarcane.

In the last 3 decades, lentil has registered a progressive increase both in area and production (Fig 1). The area has increased from 0.82 million ha in 1971-75 to 1.59 million ha in 2010-11. Similarly, the production has increased from 0.38 m tonnes to 0.94 m tonnes during same period. The increase in productivity was 42%. The increase in productivity, rather than enhanced area has contributed more towards increased production which has been achieved mainly due to the adoption of new varieties and improved production technology.

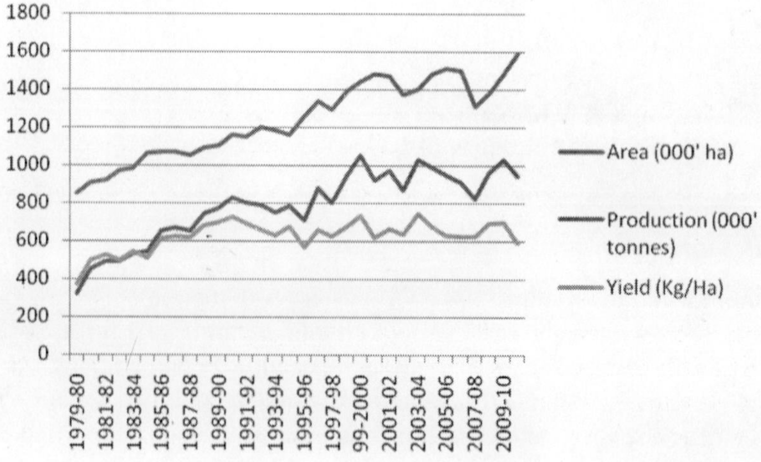

Fig 1 : Trends in Area, Production and Productivity of Lentil in India

Table 1 : State-wise Area, Production and Productivity of Lentil

States	Area (000 ha.)		Production (000 t)		Yield kg/ha.	
	1970-71	2010-11	1970-71	2010-11	1970-71	2010-11
Assam	15.70	23.9	6.30	11.8	401	494
Bihar	140.70	238.6	90.50	214.7	643	900
M.P.	271.50	590.5	108.90	177.9	401	301
Maharastra	12.00	4.0	3.20	2.0	267	500
Punjab	13.30	1.1	5.30	0.7	398	636
Rajasthan	9.10	44.1	7.30	38.4	802	872
U.P.	166.40	586.0	100.50	411.0	604	701
W.B.	94.90	57.4	46.00	53.4	485	929
Uttarakhand	14.20 *	17.0	7.90*	6.0	556 *	653
Jharkhand	21.80*	20.9	14.00 *	17.2	642 *	882
Chattisgarh	18.00 *	13.9	6.20 *	4.2	344 *	302

* Data of 2003-04

Production constraints

Apart from agro-ecological constraints as in other pulses, lentil crop is prone to diseases like rust, wilt and root rot. These biotic stresses have been reported to reduce the productivity of lentil by 20-25%. The other important constraints are terminal moisture and heat stress, non-availability of seeds of improved varieties, non-adoption of improved crop production technologies, inadequate and unbalanced use of fertilizer, delayed sowing and low seed rates, and weed infestation.

A large number of high yielding varieties and improved production technology have been developed but due to poor transfer of technology, the productivity continues to be low.

Poor storage facilities and the vulnerability of all the pulses to stored grain pests results in considerable losses. Thus, lack of on-farm/poor storage facilities is also an important factor for low production.

Another major constraint is the poor processing and marketing facilities. Lack of processing facilities at the village level compels the farmers to sell their produce at relatively low prices. Marketing support is inadequate due to the lack of organized markets and high fluctuating market prices.

Progress in varietal development

Genetic improvement of lentil was initiated systematically around middle of the twentieth century when pure line selections from land races were evaluated by the breeders and based on the multi location testing, high yielding lines were recommended as improved varieties. Some of the popular varieties, which contributed immensely to lentil production in the early phase, are L 9-12 in Punjab, Haryana and North Rajasthan; T 36 and T 8 in Uttar Pradesh and BR 25 in Bihar. L 9-12 having better tolerance to environmental stress spread to the whole northern belt. In central part of India, local varieties dominated the scene till large seeded variety JL 1, a selection from germplasm collected from Sehore district of Madhya Pradesh, was released.

With the inception of All India Coordinated Pulses Improvement Project in sixties, ICAR Institutes/SAUs began to develop varieties with wider adaptability. Breeding lines from different research stations were tested across the country and the promising ones were identified and released for different zones viz., North Hill Zone, North-West Plain Zone, North-East Plan Zone and Central Zone based on the superior performance. Varieties developed during seventies were PL 406 for North-West Plain Zone, PL 234 for Uttar Pradesh, B 77 (Asha) for West Bengal, JLS 1 for Madhya Pradesh and HPL 5 (Vipasha) for Himachal Pradesh. These varieties were developed through selection.

During eighties, varieties were developed through hybridization and mutation. The first variety developed through hybridization, PL 639, was released for northern plains in 1981. Other varieties developed through hybridization were LL 56 and LL

147 for Punjab; large seeded varieties L 4076 for North-West Plain Zone and Central Zone, and LH 84-8 (Sapna) for NWPZ. During this period, some good varieties developed through selection were K 75 (large seeded) for North-East Plain Zone and Central Zone; VL 1 and VL 4 for Uttaranchal. Some of the varieties developed through mutation were PL 77-12 (Arun) for Bihar and B 256 (Ranjan) for West Bengal.

Table 2 : Improved varieties of lentil

Trait	Varieties
Short duration	JL 3, IPL 81, JL1, WBL 77
Large seed size	DPL 62, JL 3, DPL 15, IPL 81, L 4076, VL 507,PL 5, PL 6, PL 7, IPL 406
Small seed size	HUL 57, KLS 218, VL 126, L 4147, PL 4, WBL 77
Rust resistance	IPL 406, VL 126, HUL 57, KLS 218, NDL 1, L 4076, PL 4, LH 84-8, PL 5, PL 6, PL 7,
Wilt tolerant	DPL 62, JL 3, VL 507
Rice – fallow	HUL 57, KLS218, WBL 77, Narendra Masoor 1

During nineties, high yielding and disease resistant small seeded varieties PL 4 and L 4147, and large seeded varieties DPL 15 and DPL 62 for NWPZ; NDL 1 for Uttar Pradesh and WBL 58 for West Bengal were released. These varieties were developed through hybridization. Two wilt tolerant and large seeded varieties, JL 3 and DPL 62 were released for Central Zone.

Very strong collaboration between the Indian Improvement programme and ICARDA, Syria has facilitated the flow of germplasm, elite lines and segregating populations from ICARDA to India. These materials were tested and used extensively in national lentil improvement programme and resulted in development of first extra large seeded lentil variety IPL 406 and multiple disease resistant variety PL 6.

Impact of improved varieties at farmer's field : During 2005 – 10, a total of 954 Frontline Demonstrations were conducted on Lentil under AICRP on MULLaRP in different lentil growing areas. The average yield of improved technology blocks was observed to be 1092 kg/ha whereas the yield of farmer's practice block was 885 kg/ha showing a yield gap of 23% (Fig 2).

Crop production technologies

Soil and seed bed preparation : The seed should be genetically pure, free from diseases and physical damage with 90-95% viability. Seeds should be treated with fungicides like carbendazim, thiram or captan to control seed borne disease. Normally, 2-3 g fungicide/kg seed is recommended.

The seed treatment of legumes with Rhizobium culture is beneficial to utilize the in-built capacity and potentiality to trap atmospheric nitrogen. Seed pelleting with micronutrients is also reported beneficial in areas being deficient in specific nutrient.

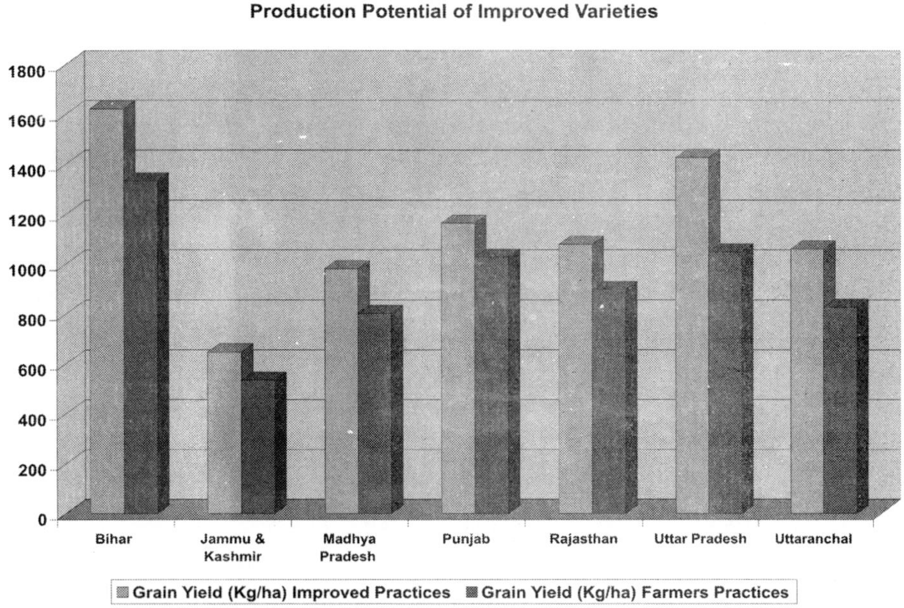

Fig 2 : Production potential of improved varieties

Planting time : Planting time is the most important non-monetary input having significant effect on crop growth and development. Early as well as delayed planting decreases yield. Multilocation studies under AICPIP on planting dates of lentil showed that end of October in the optimal period in North-West Plan Zone whereas in NEPZ can be grown successfully up to second fortnight of November subject to adequacy of moisture at planting. In the Central Zone where moisture is the major constraint, an early planting by mid October is ideal. In *Utera* system, planting time largely depends upon cessation of monsoon rains and maturity of rice crop. In low lands, water from rice field is drained by the end of September or early October and Lentil seeds are broadcast 4-7 days before harvest of rice.

Seed rate and planting method : Adequate plant stand is the basic pre-requisite for higher productivity. To achieve a desired plant density, seed rate is decided on the basis of seed size and germination percentage. The optimum seed rate of small –seeded lentil is 40-45 kg/ha whereas for large-seeded lentil, it is 55-60 kg/ha. Under *Utera* planting where small seeded genotypes are grown, an increased seed rate of 60 kg/ha is recommended.

The recommended spacing is 25 cm x 10 cm for ideal stand. Planting is done in a well prepared seed bed by drilling seeds manually in furrows opened by *desi* plough or seeding through tractor/bullock operated seed drills. Seeding depth influences seed germination and emergence. The effect of seeding depth is more important in dry-land areas where crops are raised on conserved moisture. Usually, 5-7 cm depth of sowing proved adequate for proper germination.

Fertilizer Management : The fertilizer requirement depends upon the initial soil fertility conditions, moisture regime, genotypes, growth and residue recycled. However, a starter dose of 10-15 kg N/ha is essential. Phosphorous is most indispensable mineral nutrient for pulse crops as it helps in better root growth and development. Sulphur is another important nutrient found very effective in increasing the productivity. Application of the following amount of N, P_2O_5, K_2O are recommended:

NWPZ & NEPZ - 20 kg N, 30 –40 kg P_2O_5, 20 kg K_2O, 20 kg S/ha

CZ - 15-20 kg N, 30-40 kg P_2O_5, 20 kg K_2O, 20 kg S/ha

The problem of micronutrient deficiency is likely to be faced in lentil growing areas. The most critical micronutrient is Zinc. Several studies indicated yield improvement with application of 10-15 kg ZnSO4 per ha, or a spray of 0.5% $ZnSO_4$.

Irrigation Management : Irrigation management is important for successful cultivation of lentil as the crop face acute moisture stress during their terminal growth period. In lentil, the most critical growth stage for moisture stress is pod formation followed by the initiation of flowering. In case of failure of winter rains, one to two irrigations are required for enhancing the productivity. The importance of irrigation is increased under late planting due to poor root development. It is advisable to avoid irrigation during active flowering period otherwise flower shedding and reversion to vegetative growth may occur.

Weed management : Lentil being poor competitor to weeds, especially during initial growth stages, suffer considerable yield loss. The important weeds are *Phaleris minor, Avena ludoviciana, Lolium temulentum, Trifolium* sp., Chenopodium album, *Melilotus* spp., *Lathyrus, aphaca, Convolvulus arvensis, Anagallis arvensis, Medicago dentculata, Rumex dentatus, Fumaria parviflora, Cyperus rotundus, Cynodon dactylon*. The degree of loss due to weeds depends upon the nature and magnitude of weed infestation. In lentil, the initial 30-60 days are most critical for weed management. Various methods of weed management such as cultural, mechanical, chemical etc. have been used with different degree of success. Cultural practices are manipulated in such a way that they become more favourable for crop growth and less to weeds e.g. delayed sowing of lentil is reported to reduce the infestation of *Orobancha.* Among mechanical means, hand hoeing and manual weeding are most common practices for weed control. One or two hand weedings at critical crop-weed stage provide satisfactory control of weeds. Among chemicals, use of herbicides such as Alachlor, Pendimethalin and Fluchloralin is highly effective against seasonal weed flora. Pre-emergence application of pendimethalin 1.0 kg/ha or pre-plant incorporation of fluchloration 0.5 kg/ha also proved quite effective in controlling weeds.

Disease management

Many diseases caused by fungi, bacteria and viruses are reported to affect lentil. However, the potentially damaging diseases are wilt, root rots and rust. Wilt and root rots are wide spread causing heavy losses in certain agro ecological niches, whereas rust assumes serious proportion under certain environmental conditions. In India, this disease is a serious threat to lentil cultivation in areas with high humidity and relatively low temperatures (20-22°C) at the flowering/podding stage of the crop.

1. Wilt : Wilt caused by *Fusarium oxysporum* f. sp. *Lentis* is the most important biotic factor responsible for constraining the lentil production. Under natural condition, the disease incidence as high as 50-78% has been reported. Pathogen is both seed and soil borne and can infect the plant at any stage of the growth. Wilt disease results in death of the plant. There is indication of existence of variability in terms of virulence and other cultural, morphological and nutritional requirements of *F. oxysporum* f. sp. *Lentis*. However, existence of races in *Fusarium oxysporum* f. sp. *Lentis* is not confirmed.

For wilt management management following strategy to be followed:

- Adjusting sowing dates in such a way that the crop matures before optimum conditions for disease build up are developed.
- Deep ploughing during summer, field sanitation and crop rotation with non host crops may reduce the inoculum level.
- Seed treatment with thiram (3 g/kg) or carbendazim + Thiram (1+2 g/kg) can help reducing wilt incidence and ensure better plant stand.
- Seed dressing with *T. virens* (*Gliocladium virens*) + vitavax (0.5 g/kg) has been reported helpful in reducing lentil wilt incidence in the field.
- Resistance against wilt has been identified and successfully used for the development of wilt resistant/tolerant varieties of lentil. Few wilt resistant / tolerant varieties in seed chains are DPL-62, JL-3, VL-507 etc.

2. Root rots : Three pathogens are important in causing root rot in lentil. *Rhizoctonia solani* causes wet root rot, *R. bataticola* causes dry root rot and *F. solani* causes black root rot in lentil. These pathogens can infect plant at any stage; however, *R. solani* causes more damage by causing seedling mortality where as *R. bataticola* attacks crop mostly at later stage of growth. *F. solani* becomes serious in soils with high moisture content.

Approaches adopted for management of wilt disease can also be used for management of root rots. However, not much information on sources of resistance against this disease is available.

3. Rust : Rust disease caused by the fungus *Uromyces viciae fabae* is the most important foliar disease of lentil, which under conditions favourable for its development may devastate the whole crop in certain areas. Losses, however, depend on the stage at which the crop is infected and rather more on weather conditions prevailing after the infection.

For management of rust disease, the following strategy is recommended.

- Clean cultivation, destruction of infected plant debris help reducing inoculum. In NEPZ, normal and early sowing reduces intensity or rust disease.
- One to two foliar sprays of mancozeb @ 0.2% at 50 days after sowing are good enough to control the rust disease in NEPZ in susceptible and moderately susceptible varieties.
- Several rust resistant lines have been developed. DPL 15, DP 62, Pant L 4, IPL 81, IPL 406, HUL 57, KLS 218, PL 6, PL 7 etc. are rust resistant.

4. Nematode Management : Out of many nematodes reported associated with lentil, only root knot nematodes (*Meloidogyne* spp.) and reniform nematode (R. *reniformis*) have been found damaging. If the crop is grown on a field infested with nematodes and preceded by a nematode susceptible crop, the losses may be more due to build up of the nematodes populations. The avoidable losses have been estimated up to 16.6% due to root knot nematodes.

Screening of the accessions of lentil against root knot nematodes have been carried out. Out of 226 accessions screened for resistance, 18 accessions have been reported to be resistant to *M. javanica*. The resistant accessions are: DPR-14, DPL-47, LH 90-103, LH 90-110, DPl 53, DPL 54, DPL 55, HUL 44, IPL 63, IPL 398, L 4603, LH 90-103, NDL –93, IPL-10, IPL 406, LH 90-54, KLS 214 and L 4628.

Much work has not been done on the management of the root knot and reniform nematodes in lentil. However, the management methods worked out in other crops can be used to control the nematodes and avoid the losses caused by them in lentil.

Future thrust

Lentil is mostly grown after rice crop in *rabi* season. This leads to agronomic difficulties in establishing lentil crops after rice due to formation of a hard plough-pan, poor drainage and micro-nutrients deficiency in rice fields. Usually the crop is grown as relay crops in order to utilize the soil moisture of rice fields for proper germination. Under such situation, poor plant establishment is the major constraints in addition to excessive weed competition and losses caused by various diseases. The plant growth is stunted due to poor root development on such hardy soil. In the later stage, the crop usually suffers from terminal moisture stress and high temperature resulting into the forced maturity. There is need to identify genotypes tolerant to wider range of environmental variations particularly, temperature and soil moisture.

Breeding for high yield : A narrow diversity limits the scope for new recombinations. The recent focus on hybridization of diverse genotypes from different agro-climatic zones of India, and involvement of exotic lines in breeding programmes has helped in broadening the genetic base and now a wide range of variation can be seen in breeding material generated at various research centers. Wide hybridization involving closely related species is yet to be explored.

A substantial area of lentil is sown under late sown/*utera* condition in rice belt of NEPZ. So far no variety has been released specifically for such situations. Early maturing genotypes with higher biomass are advocated for such situations.

The existing plant types in different agro climatic zones are the most suitable plant types as far as survival of crop is concerned. There is need to understand vegetative and reproductive components to develop the physiologically efficient plant types for higher yield potential. It is suggested that an efficient plant type will be one which has rapid growth, flower is early and is harmonious with environments conducive to seed setting and development. For mechanical harvesting, tall, erect, non lodging habit and non shattering of pods are suitable traits.

Breeding for multiple resistance : Several sources of rust resistance have been identified. However, in case of vascular wilt and root rot, only tolerant lines have been identified because of race complexities. A survey of the wilt reaction of 220 wild lentil accessions revealed resistance in the subspecies *orientalis, nigricans*, and *ervoides*.

Lentil production is limited by lack of moisture and unfavourable temperatures throughout its distribution. Progress has been made in breeding for tolerance to drought through selection for an appropriate phenology and increased water use efficiency Water logging and salinity are also locally important. The early maturing lines are more sensitive to water logging. Genotypic differences for water logging have been reported. Lines ILL 5845, ILL 6451, ILL 6788 and ILL 6793 are relatively tolerant to salt. Available sources of tolerance to abiotic stresses should be utilized in hybridization programmes and breeding lines may be selected for different agroclimatic regions.

Accelerated productivity gains are required for enhancing total production. Apart from productivity, the thrust has to be on increasing the seed replacement rate. The seed production programme has to be streamlined. Specific measures for increasing seeds availability are to be implemented. Effective steps have to be taken for supply of critical inputs such as biofertilizers, phosphatic fertilizers, sulphur and micronutrients like Zinc.

Transfer of available technology can bridge the vast gap which exists between realizable and realized potential in terms of productivity enhancement through accelerated efforts on farm-worthy technology transfer programmes.

References

Ahlawat, I.P.S, Singh, A. and Saraf, C.S. 1981. Effect of winter legumes on the nitrogen economy and productivity of succeeding cereals. Experimental Agriculture 17: 57-62

Ali, M. 2004. Role of pulses in crop diversification in India. In: Role of legumes in crop diversification and poverty reduction in Asia. (Eds. CLL Gowda and S Pande). Proceeding of the Joint CLAN Steering Meeting, 10-12 November 2003, International Crop Research Institute for the Semi-Arid Tropics, India. pp 42-56

Singh, B.B., Katiyar, P.K. and Dixit, G.P. 2007 . Lentil . In: National test guidelines for conduct of tests for distinctness, uniformity and stability (eds. Dinesh kumar and A.B. Mandal), Directorate of Seed Research, Mau pp. 151-156,

Dixit, G.P., and. Katiyar, P.K. 2004. Genetic base of lentil (Lens culinaris) varieties and breeding lines developed in India. Indian Journal of Agricultural Sciences 74(11): 585-589.

Dixit, G.P., Katiyar, P.K..Singh, B.B and Kumar, Shiv. 2009. Lentil varieties in India. AICRP on MULLaRP, IIPR Kanpur

Erskine, W, Chandra S, M, Malik, I.A, Sarkar, A., Sharma, B, Tufail, M and Tyagi, M.C. 1998. A botleneck in lentil: widening the genetuic base in South Asia. Euphytica 101: 207-211

Erskine, W., Ellis, R.H, Summerfield, R.J., Robberts, E.H., and Hussain, A. 1990. Characterization of responses to temperature and photoperiod for time to flowering in a world lentil collection. Theoretical and Applied Genetics 80: 193-199.

Erskine, W, Tufail, M., and Russel, A. 1994. Current and future strategies in breeding lentil for resistance to biotic and abiotic stresses. Euphytica 73: 249-135

Gadhari, P.D. and Laxman, Singh. 1979. Pulse crops of Madhya Pradesh. Jabalpur, India: Directorate of Extension Services, Jawaharlal Nehru Krishi Vishwa Vidyalaya (JNKVV), Jablpur.

Khare, M.N. 1981. Diseases of lentil. In: Lentils (Eds., C Webb and G Hawtin). Commonwealth Agricultural Bureaux England 163-172.

Ladizinsky, G. 1979. The origin of lentil and its wild genepool. Euphytica 28: 279-187

Lal, S. 1984. Constraints and opportunities in pulse production in eastern Uttar Pradesh. In: Pulse Production constraints and opportunities (Eds., HC Srivastava, S Bhaskaran, KKG Menon, S Ramanujam and MV Rao). Oxford & IBH Publishing Co. Pvt. Ltd., New Delhi, India. pp 89-106.

Radha, Y.R, Seetharaman, G, Ram, S., and Subbiah, G. 1989. Efficiency of input use –a comparative study in rice-rice and rice-pulse farming systems. Economic Affairs 34: 219- 224.

Roy Sharma R P and Singh Ambika. 1969. Residual effect of legumes in winter legumes-maize rotataion. Indian Journal of Agricultural Sciences 36: 530-539.

Sarkar, A, Furman, B., and Erskine, W. 2007. Gene mining and use to improve yield and quality in lentil. In: Proceeding of 18th Encarpia Conference on Plant Genetic Resources, Abs. no. 35, Piestany, Solvakia, p. 41.

Singh, B.B., Dixit, G.P., and Katiyar, P.K. 2006. National Test Guidelines for the conduct of tests for Distinctness, Uniformity and Stability in Lentil, Peas and Rajmash. AICRP on MULLaRP, IIPR, Kanpur

Singh, B.B., Dixit, G.P. and Ali, Masood. 2009. Improved varieties and technology led to spectacular increase in lentil production. Agriculture Situation in India. Vol. LXV, No.10, pp. 645-648

Singh, K.M., and Singh, R.K.P.1995. An economic analysis of lentil cultivation in North-East alluvial plains of Bihar. Economic Affairs 40: 157-163.

❑❑❑

Chapter - 18

Fieldpea

G.P. Dixit, B.B. Singh & P.K. Katiyar
Indian Institute of Pulses Research, Kanpur, Uttar Pradesh.

Fieldpea is an important *rabi* pulse crop which is highly productive and is grown for food, feed as a vegetable. There are two types of peas grown in India *viz.*, vegetable type in which grains are very sweet when green and become wrinkled on drying, and grain type where grains are generally white and round. Green seeds of the grain type are not so sweet as in the case of vegetable type. The dry grains are consumed in various forms such as *chat, chhola, dal*, vegetable and flour. Green seeds are used as fresh, frozen or canned vegetables. The cultivation of grain type is confined to northern and central parts of the country.

It is high protein (20%) crop with all the essential aminoacids important for normal activity of living organisms. The fractional composition is also favourable the most easily assimilated water soluble protein from mature pea seeds varies from 36 to 87% and in green matter from 65 to 82%. Pea contains 4 mg pro-vitamin A, 300 mg vitamin C, 3 mg B_1, 1.5 mg B_2 and 1.2 mg pantothenic acid per 1000 g fresh seed weight. Also, it contains 1.1% fat, 2.2% minerals, 4.5% fibre and 56.5% carbohydrate. It is rich in phosphorus. Green pea and the immature pods of pea are typified by a high level of active lipotropic antisclerotic substances-choline and inositol. Choline deficiency may lead to the development and growth of malignant tumors.

Trends in Area, Production & Productivity

In Indian, fieldpea occupies 0.72 million hectare area with an annual production of 0.59 million tonnes (Fig 1). It is mainly cultivated in Uttar Pradesh, Madhya Pradesh, Bihar, Maharashtra and Rajasthan (Table 1). In Uttar Pradesh the area and production have reduced from 724 th. ha and 688 th. tonnes in 1970-71 to 304 th. ha and 353 th tonnes in 2010-11, respectively. But, the productivity has increased for 950 kg/ha to 1761 kg/ha during the same period due to availability of high yielding and powdery mildew resistant varieties. In M.P. the area has increased from 78 th. ha to 288 th. ha and production from 21 th.tones to 146 th. tones during 1970-71 to 2009-10. The productivity has also increased from 279 kg/ha to 508 kg/ha during the period as several high yielding varieties with resistance to powdery mildew have been evolved for the state. In Bihar, the area has reduced for 63 th. ha to 20 th. ha and production from 36 th tones to 21 th. tonnes. But, the productivity has increased from 370 kg/ha to 1051 kg/ha due to availability of powdery mildew resistant varieties with high yield potential.

Table 1 : State-wise Area, Production and Productivity of Fieldpea

States	Area (000 ha)		Production (000 t/ha)		Productivity (kg/ha)	
	1970 – 75	2010 – 11	1970 – 75	2010 – 11	1970 – 75	2010 – 11
Uttar Pradesh	563.78	304.00	439.50	353.00	732	1761
Madhya Pradesh	86.10	226.50	23.44	52.50	272	276
Bihar	25.32	20.00	13.04	21.00	751	1051
Maharashtra	8.34	14.80	2.7	5.00	317	338

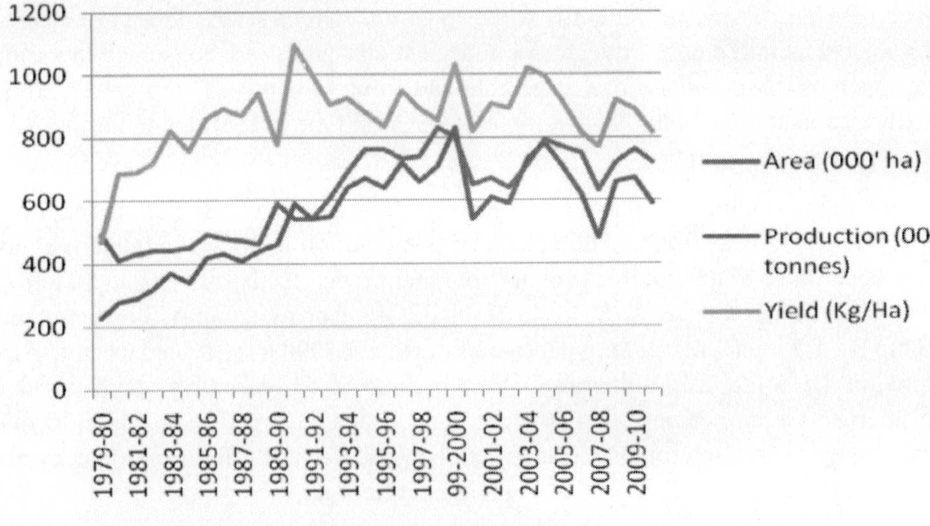

Fig 1 : Trends in Area, Production and Productivity of Fieldpea in India

Yield gaps

During 2005–10, a total of 376 Frontline Demonstrations were conducted on Fieldpea under the ambit of AICRP on MULLaRP in different pea growing areas. The productivity of blocks with improved technology was observed to be 1419 kg/ha whereas the yield of farmer's practice block was 1171 kg/ha only showing a yield gap of 21 %.

Production constraints

Fieldpea is the most demonsticated pulse crop. Unlike other pulse crops, its cultivation is also distributed to the rainfed areas and on marginal and sub-marginal lands generally poor is soil fertility. Apart from agro-ecological constraints, fieldpea is prone to diseases like powdery mildew, rust and root rot. These biotic stresses have been reported to reduce the productivity of pea crop by 15-20%. The other major constraints are terminal stress, susceptibility to water logging, non-availability of seeds of improved and high yielding varieties, non-adoption of improved crop production technologies, insufficient and unbalanced use of fertilizers, untimely sowing and low seed rates as well as the least attention to timely weed control.

Inadequate transfer of technology programme is yet another important constraint. Improved varieties have been developed alongwith matching production technologies for adoption. But due to lack of proper extension programme, the transfer of technology process has been ineffective.

Poor storage facilities and the vulnerability of all the pulses to store grain pest results in considerable losses. Thus, lack of on-farm/poor storage facilities is an important factor for low production.

Another major constraints is poor processing and marketing facilities. Lack of processing facilities at the village level compel the farmers to dispose of their produce at relatively low prices. Marketing support is an adequate due to the lack of organized markets and high by fluctuating market prices.

Varietal development

In early phase of varietal development, local cultivars with wider adaptability were selected and released as varieties. Some of the varieties developed through direct selection in field pea are Type 163, Harbhajan (exotic), BR 2, BR 12, Hans (L 116), B 22, Kinnauri and RPG 3. Type 163, a selection from local genetic stock of Bulandshahar, has semi-spreading and tall type growth habit with wider adaptability and higher average yield of 18-20 q/ha. BR 2 and BR 12 with spreading tall growth habit and average yield of 18-20 q/ha are selections from local germplasm of Bihar. Kinnauri, a selection from Kinnaur local (Himachal Pradesh) has white round seeds with average yield of 15-20 q/ha and is suitable for medium and high hilly regions.

During 60's and early 70's, the area under pea reduced to almost half. This happened primarily due to widespread occurrence of powdery mildew disease and a ruling tall variety T 163 succumbed to this disease. At this stage, need for powdery mildew resistant and input responsive variety of field pea was felt. Hybridization programmes were initiated to incorporate resistant gene in Type 163. The first powdery mildew resistant variety, Rachna with plant type similar to that of Type 163 was released in 1980. This variety was developed from cross T 163 X Type 10. Another variety, Pant P 5 was developed from the same cross. L 116 in combination with 6587 led to the development of DMR 7. Later, some more powdery mildew resistant and high yielding varieties like HUP 2, KFP 103 and JP 885 were released for different zones.

Recently, a number of varieties have been developed which combine resistance to powdery mildew with dwarf stature and leafless trait (Table. 2). Each of these characters has played a significant role in reversing the negative trend of area under the crop. The new plant type with leaflets converting into tendrils and retaining the normal stipules has several advantages such as penetration of sunlight to lower portion of the plant, natural mechanical support to prevent lodging and network of interlocked tendrils above crop canopy to prevent bird damage. Incorporation of the dwarfing gene has also enhanced the productivity through response to inputs like fertilizers and irrigation and high plant population. HFP 4 (Aparna) was the first dwarf, leafless and powdery mildew resistant variety developed from the cross of T 163 with an exotic line EC 109196. Later, HFP 4 in combination with EC 109185 led to the development of variety HFP 8909. It also resulted in the development of variety KPMR 144-1 from hybridization with Rachna.

Table 2 : Varieties resistant to powdery mildew

Dwarf	Tall
Aparna, Uttara , Jayanti, Sapna, Malviya Matar 15, Swati, Pusa prabhat, IPFD 1-10 and IPFD 99-13.	Rachna, Pant P 5, Malviya Matar 2, JP 885, Alankar, Shikha, Ambika, IPF 99-25, VL Matar 42, Pant Pea -42 and HFP 9907 B.

A dwarf variety HUDP 15 is the product of the cross (PG 3 X S 143) X FC 1 and has resistance against powdery mildew and fairly good tolerance to rust. Recently, several short duration varieties like Adarsh, Vikas and Prakash have been developed which escapes terminal stress and have the yield potential of about 2.5 ton per hectare.

Fertilizer management : The fertilizer requirement depends upon initial soil fertility conditions, moisture regimes, genotypes, growth and yield of crop, residue recycling etc. In general, the requirement is 30-50 kg N, 2-7 kg P_2O_5, 12-30 kg K_2O, 3-10 kg Ca, 1-5 kg Mg, 1-3, 1-3 kg S, 200-500 g Mn, 5 g B. 1 g Cu and 0.5 g Mo for producing one tonne of biomass per hectare.

Fieldpea is generally cultivated under irrigated conditions and therefore, responds well to fertilizer application. Basal placement of 20 kg N plus 60 kg P_2O_5 per hectare at the time of sowing is adequate to meet its requirement, however, dwarf types gave more yields at 40 kg N + 60 kg P_2O_5 per ha. Application of 20 kg Sulphur per hactare has also been found effective. Among micronutrients, significant response to molybdenum application has been observed.

Irrigation management : Water management is very important for successful cultivation of fieldpea as the crop face acute moisture stress during its terminal growth phase. In general, a pre-sowing irrigation helps in ensuring optimum plant population and growth when profile moistue at planting is inadequate. On light textured soils, 1-2 irrigations at branching and/or pod development prove highly productive if winter rains fail. Only light irrigation (40-50 mm) should be given as excess of water enhances vegetative growth and depresses the grain yield. Moisture stress at early stage drastically cuts the biological nitrogen fixation which is ultimately reflected into poor yield. Multilocation studies under All India Co-ordinated Pulse Improvement Project (AICPIP) on water management in fieldpea revealed that in North East Plains, one irrigation at flower initiation stage (50 DAS) was found most beneficial, whereas in central India, two irrigations – one each at branching and flowering were found optimum. In NWPZ, branching stage and flowering stage were found most critical. It has been observed that dwarf genotypes respond better to irrigation than tall types.

Disease management : A number of fungal, bacterial and viral pathogens are reported to attack fieldpea. Generally, powdery mildew (PM) and rust are considered as economically most important; however, root rot caused by *Fusarium solani* is emerging as a potentially damaging disease especially under high soil moisture condition.

Powdery mildew : This disease is caused by the fungus *Erysiphe pisi* that attacks aerial parts of the plant. Powdery mildew can cause heavy losses to the pea crop, however losses depend on the stage at which crop is attacked. Early sown (second fortnight of October) crops escape infection by Powdery mildew fungus. Cultivation of early maturing varieties provides another option for avoiding losses due to Powdery mildew. Intercrops like mustard, linseed, wheat, barley are reported to delay build up of powdery mildew. A number of fungicides are reported to provide good protection against PM in fieldpea. One to two spray of carbendazim (0.05%) or two to three sprays of wettable sulphur (0.3%) are, however, most commonly used chemicals employed to avoid losses caused by Powdery mildew.

Rust : Rust disease of fieldpea is caused by the fungus *Uromyces viciae* fabae. The disease causes, considerable damage in areas with high humidity. It is autocious fungus and completes whole life cycle on peas. The primary sources of inoculum come from the infected plant debris in soil. The disease generally occurs in February and its further buildup depends much as prevailing weather conditions. Under favourable conditions for its development this disease can cause heavy losses. This disease is more serious in areas with high humidity with moderate temperatures.

Cultural and chemical management strategies employed against Powdery mildew can also be effectively used to manage rust disease. As far as the sources of resistance against rust are concerned, high level of resistance has eluded the researchers. There are however many lines reported to the tolerant to rust *viz.,* PJ 207508, C 12, Wisconsin, DMR 3, Pant P 5, Pant P 8, Pant 9, HFP 8711 and HUDP 15. None of the released varieties posses high level of resistance against rust. Some rust tolerant varieties of fieldpea released by All India Coordinated Pulses Improvement Project are Pant P 5, Swati (KFPD-24), HUDP 15 (Malviya Matar 15) and IPFD 1-10.

Among other diseases, root rots caused by *Pythium* spp., *Aphanomyces euteiches* (. sp. *pisi, Fusarium solani* (Mart.) Sacc. f. sp. *pisi* (Jones) Synd. and Hans, bacterial blight caused by *Pseudomonas syringae* pv. *pisi,* downy mildew caused by *Peronospora pisi* Syd., black spots caused by *Ascochyta pisi* Lib., *Mycosphaerella pinodes* (Berk and Blox.) and *Phoma medicaginis* var. *pinodella* (Jones) Boerema, and pea seed borne mosaic virus are most important. Their significance in causing yield damage, however, varies with the agro-ecological condition. Seed treatment with carbendazim @ 2-3 g/kg seed reduces seedling mortality due to root rot and help establishment or letter plant stand. Avoiding high soil moisture can also help in limiting this disease

Some excellent information and breeding materials are available for various root rot resistances from the work of Dr. J. M. Kraft at the Washington State University and from the University of Wisconsin. Resistance is also available against downy mildew, powdery mildew, bacterial blight, and pea seed borne mosaic virus. Amongst the organisms causing black spot disease complex, sources of resistance to *Ascochyta pisi* and *Phoma medicoginis* var. *pinodella* are available and have been used in breeding programmes. However, breeding against *Mycosphaerella pinodes,* which also happens to be the most widely distributed and most destructive, has met with little success. Clulow *et al.* (United Kingdom) and Nasir *et al.* (Germany) have recently reported genetic analysis of partial resistance. Critical examination of the above two studies in light of the past and recent work in Australia highlights several persistant problems. The level of resistance in parental lines is generally insufficient. There are doubts about the durability of resistance in view of the large pathogenic variation that have been found to occur. Resistance sources are wild pea collections, some of which are species other than *P. sativum,* and they carry many undesirable agronomic traits. Limited experience in handling segregating populations highlights difficulties, in selection for resistant plants with desired agronomic traits. Resistance to *M. pinodes* therefore poses the greatest challenge in achieving high yields in the Mediterranean and temperate regions.

Future Thrust

Crops like fieldpea with higher productivity may be expended horizontally irrespective of their consumption statistics and usage. The higher yield potential of new varieties and the innovative crop production and protection technologies would be required to further increase the production in the traditional growing areas. Development of tailor made maturity duration varieties, which fit well in the new emerging cropping systems, will further increase the scope for horizontal expansion of areas. There is ample scope of bringing more area under extra-early maturing fieldpea varieties in northern India. The new tools of biotechnology would enable researchers to effectively go for the transfer of genes for insulating varieties from different diseases and pests. The Indian workers are already utilizing the "af", "le", and "er" genes extensively in breeding programme. A number of powdery mildew resistant varieties have been evolved and the production of fieldpea has shown rapid increase with these varieties spreading to newer areas. Neverthless, there are few more areas that need greater attention now. First of all, the structure of pea plant is such that it is destined to lodge at one stage or the other. The stem of pea plant is very thin. Coupled with huge mass of foliage in the upper tiers and accumulation of massive pods, the plant becomes top-heavy, and the thin base cannot hold it in upright position for long. Any genetic system that can make the base of pea plant anatomically strong and thick enough, as well as keep the plant erect till full maturity, will undoubtedly boosts its yield potential.

The pea crop in India has a limited scope become its consumption is highly restricted and the uses of grain is not diversified. Therefore, new uses of fieldpea have to be found if the popularity of this crop is to be increased. Finding export markets would be a definite way to increase pea production in India. Another aspect of pea breeding could be to initiate breeding programmes for dual purpose (feed-fodder). There is no doubt that the demand for cattle/poultry feeds and fodder will increase manifolds in this country. A cheap pulse like fieldpea could come to the rescue in such a situation.

Earliness is yet another trait of economic importance. Early varieties can be planted with higher crop density, and the crop stand can be further increased by selecting dwarf varieties. Perhaps this would be the most ideal situation to maximize yields not only per unit time, but also per unit area. Thus, the future needs of pea breeding will be to develop dwarf, non-lodging and disease free varieties with white and green seed (for human consumption) as well as red seed coat (for feed and fodder purpose).

Accelerated productivity gains are required for enhancing total production. Apart from productivity, the thrust has to be on increasing the seed replacement rate. The seed production programme has to be streamlined. Specific measures for increasing seeds availability are to be implemented. Effective steps have to be taken for supply of critical inputs such as biofertilizers, phosphatic fertilizers, sulphur and micronutrients like Zinc.

Transfer of available technology can bridge the vast gaps which exist between realizable and realized potential in terms of productivity enhancement through accelerated efforts on farm worthy technology transfer programmes.

References

Choudhury, P.R., Hasan Tanveer and G.P.Dixit. 2007. Identification and detection of genetic relatedness among important varieties of pea (Pisum sativum L.) grown in India. Genetica 130(2): 183-191

Dixit, G.P., Hasan Tanveer and Subhash Chandra . 2006. Generation mean analysis for grain yield related traits in fieldpea. Indian Journal of Genetics and Plant Breeding 66(2): 147-148

Dixit, G.P. and P.K. Katiyar .2006. Genetic base of Indian fieldpea varieties and breeding lines developed in India. Indian Journal of Genetics and Plant Breeding 66(4): 316-318.

Hedley, C.L. and Ambrose, M.J. 1981. Designing "leafless" plants for improving yields of dried pea crop. Advances in Agronomy 34: 225 – 227.

Katiyar, P.K., and Dixit, G.P. 2009. Multivariate analysis for genetic divergence in fieldpea germplasm. Indian J. Agricultural Sciences. 79(3): 181-183

Choudhury, P.R., Hasan Tanveer and G.P.Dixit. 2007. Identification and detection of genetic relatedness among important varieties of pea (Pisum sativum L.) grown in India. Genetica 130(2): 183-191

Dixit, G.P., Hasan Tanveer and Subhash Chandra . 2006. Generation mean analysis for grain yield related traits in fieldpea. . Indian Journal of Genetics and Plant Breeding 66(2): 147-148

Dixit, G.P. and P.K. Katiyar .2006. Genetic base of Indian fieldpea varieties and breeding lines developed in India. Indian Journal of Genetics and Plant Breeding 66(4): 316-318

Hedley, C.L. and Ambrose, MJ. 1981. Designing "leafless" plants for improving yields of dried pea crop. Advances in Agronomy 34: 225 – 227.

Katiyar, P.K., and Dixit, G.P. 2009. Multivariate analysis for genetic divergence in fieldpea germplasm. Indian J. Agricultural Sciences. 79(3): 181-183.

Sharma, B. and Khan, T.N. 1996. Creating higher genetic yield potential in Fieldpea (Pisum sativum L.). Indian Journal of Genet. 56(4): 371 – 388.

Snoad, B., Fruscinate, L and Monti, L.M. 1985. The effects of three Genes which modify leaves and stipules in the pea plant. Theoretical and Applied Genetics 70: 322 – 329.

Singh, D.P. 1997. Tailoring the plant type in pulse crop. Plant breeding abstract 67(9): 1213 – 1220

Sinha, S.K. 1977. In: Food Legumes: distribution, adaptability and biology of yield. Food and Agriculture Organization of United Nations. FAO, Rome, Italy

Srivastava, R.P., Kumar, Lalit and Dixit, G.P. 2009. Nutritional composition and fatty acid profile of important genotypes of fieldpea . Journal of Food Legumes 22(2):115-117

Tanveer , Hasan , Choudhury, P. R., Dixit, G.P. and Jha, G.K. 2010. Detection of molecular divergence and development of DNA fingerprints in fieldpea cultivars. Indian Journal of Genetics and Plant Breeding 70(2): 132-139

□□□

Chapter - 19

Lathyrus

G.P. Dixit, B.B. Singh & P.K. Katiyar
Indian Institute of Pulses Research, Kanpur, Uttar Pradesh.

Grasspea (*Lathyrus sativus* L.), also called chickling pea, is an important crop of economic significance in India, Bangladesh, Pakistan, Nepal and Ethiopia. It is also cultivated in China, many countries of Europe, the Middle East and northern Africa. It is a food, feed and fodder crop belonging to the family *Leguminacaae* and the tribe *vicieae* (Smartt, 1990).

Grasspea is very tolerant to drought conditions and has been grown successfully in areas with an average annual precipitation of 380 to 650 mm. Despite its tolerance to drought, grasspea is not affected by excessive rainfall and can be grown on land subject to flooding (Sinha, 1977). It has a very hardy and penetrating root system and, therefore, can be grown on a wide range of soil type, including very poor soil and heavy clays. This hardiness together with its ability to fix atmospheric nitrogen makes the crop ideally to grow under adverse conditions (Campbell *et al.* 1994). In India, Bangladesh, Nepal and Pakistan, it is often broadcast into a standing rice crop one to two weeks before the rice is ready for harvest where it flourishes on the residual soil moisture after the rice has been harvested.

Despite all the advantages compared to other crops, relatively little efforts have been made to improve this very hardy pulse crop. The main reason has been the knowledge that excessive consumption of grasspea can lead to a neurological disorder, called Lathyrism in man and domestic animals (Jackson and Yunus, 1984). It appears when *Lathyrys sativus* forms the main component of the diet contributing to at least 30% of the caloric intake for a period of 3 to 4 months (Campbell *et al.* 1994).

Economic importance in India

Grasspea is the third important cool season pulse crop of India occupying an area of 0.52 million hectares with annual production of 0.35 million tonnes (Fig. 1). Its cultivation is mainly concentrated in the states of Madhya Pradesh, Chhatisgarh, Maharashtra, Bihar and West Bengal (Table. 1). In M.P., area has reduced from 682.1 th. ha. to 43.4 th. ha. and production from 220.3 th. t to 31.7th. t. Similarly, in Bihar area has reduced from 447.8 th. ha. to 87.1 th. ha. and production from 277.4 th. t to 82.2 th. t. In W.B. area has reduced from 134.7th. ha. to 28.1th. ha. and production from 49.5th. t to 19.7 th. t, Approximately, 70% of the acreage is concentrated in Chhatisgarh and Vidarbha region of Maharashtra. This is a rice growing region having canal irrigation for rice only and no water is available for winter crops, thus, grasspea is the only alternative/ bonus crop grown under paira (relay) cropping. Under this situation, it resists extreme harsh conditions of excessive moisture at sowing and moisture stress at later states of crop growth, and yields 3 to 4 quintals of grain per hectare. In fact, grasspea cultivation in this region is prevalent due to its hardy nature and minimum cultivation cost. It is a good cattle feed too. The young plants are used as a fodder for cattle. Often, the animal feed value of the crop is more important than human food in determining the production of this crop.

Production potential

As a result of an aggressive research programme on grasspea in India, recently two varieties viz. Bio L 212 (Ratan) and Prateek, have been released for cultivation. These varieties have high yield potential (1.2 to 1.5 t/ha) and low level of ODAP. These newly bred varieties are being demonstrated on farmers' fields under Front Line Demonstrations programme. These on-farm demonstrations are conducted by Government as well as Non-Government Organizations (NGO's) engaged in the promotion of new technologies. The front-line demonstration is conducted to demonstrate the impact of varieties and other production package for increasing the production. During 2005 – 10, a total of 95 Frontline Demonstrations were conducted on *Lathyrus* under AICRP on MULLaRP in in the states of Madhya Pradesh, Chhatisgarh, Maharashtra, Bihar and West Bengal. The average yield of blocks with improved technology was observed to be 867 kg/ha whereas the yield of farmer's practice block was 693 kg/ha only showing a yield gap of about 10 %.

Table 1 : State-wise Area, Production and Productivity of *Lathyrus*

State	Area		Production		Yield	
	1980 – 81	2009-10	1980 – 81	2009-10	1980 – 81	2009-10
Bihar	447.8	87.1	277.4	82.2	619	944
M.P.	682.1	43.4	220.3	31.7	323	730
Maharashtra	78.8	9.9	14.5	4.3	184	434
W.B.	134.7	28.1	49.5	19.7	367	701
Chattisgarh	250.8*	280.9	108.9*	175.6	434*	625

*Data of 2000-01

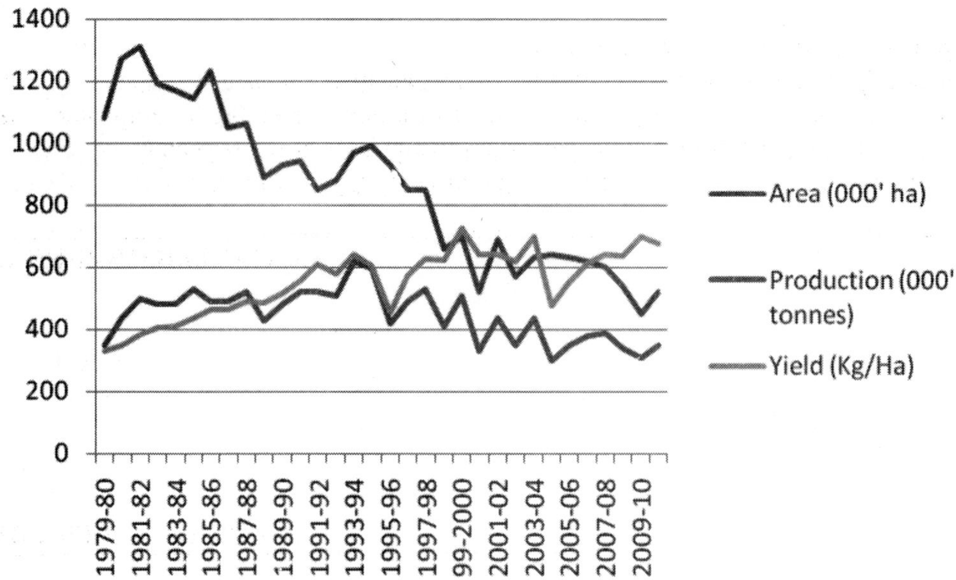

Fig 1 : Trends in Area, Production and Productivity of *Lathyrus* in India

Varietal development : It has been observed that the regions where irrigation facilities are extended to winter crops, the area under grasspea is declining. Also, because of toxin content and poor productivity, it is being regularly substituted by other crops posing damage of genetic erosion. Therefore, it is of great concern to collect and conserve the available variability from different regions.

A systematic collection of grasspea in India was initiated in 1967 and about 600 accessions were collected and evaluated for ODAP content (Mehra *et al.* 1995). The subsequent collections were made from Bihar, Eastern U.P., West Bengal and Gujarat in 1969 and later, in 1975, from tribal areas of Bihar. About 1187 accessions were collected from Madhya Pradesh during 1989-91. These landraces were evaluated and characterized for different economic traits to facilitate breeding programme for reduced neurotoxin content.

At present, 2720 accessions of grasspea have been conserved as base collections in National gene bank of NBPGR. Besides this, active collections are being maintained at IGAU, Raipur; IIPR Kanpur and NBPGR Akola. Most of these accessions have been evaluated and characterized for different economic traits (Table 2 & 3).

The first phase in the crop improvement programme between 1940 and 1960 included the collection of local cultivars and isolation of single plant progeny superior in yields. As a result, a number of strains were isolated and recommended for cultivation by various State Departments of Agriculture, like 'BR 13' 'LC 76' (Bihar); 'T 12' ;(Gujarat) No. 91 and; No. 11 (M.P.); and B 1 & B 19 (West Bengal).

Table 2 : Accessions identified for economic traits

Character	Promising Accessions
Earliness (< 100 days)	IC 120438, IC 120446, IC 120447, IC 120448, IC 120596, IC 120507, IC 120420, EC 200325, EC 209076, EC 208952, RLK 10012, RLK 10013, RLK 10031, RLK 10037, RLK 10050, RLK 10048, RLK 266, RLK 287.
Number of pods (> 50/Plant)	Bio L 212, Bio L 239, Bio R 234, DL 265, RLK 430, IC 12507, IC 120497, IC 120537, IC 120422, NIC 18768, NIC 18849, NIC 18851, NIC 18890.
Grain yield (>10 g/plant)	JRL 47, Pusa 534, RJK 49, RLK 204, RLK 393, RLK 658, RLK 1009, RLK 1081, IC 120479, IC 120512, IC 120530, IC 120531, IC 120535, NIC 18768, NIC 18890, S 270, P 72, P 176
Seed size (> 12 g/100 seed)	EC 209017, EC 209026, EC 209044, EC 200322, RLK 143, RLK 148, RLK 158, Bio L 208, Bio R 227, LS 8246, Sel. 505
ODAP content (< 0.2%)	Bio L 212, Bio R 202, Bio R 231, Bio I 222, Bio L 208, LS 157-12

Table 3 : Variation for agronomic characters in grasspea germplasm collected in Madhya Pradesh, India

Character	Mean	Range		C.V (%)
		Min.	Max.	
Days to 50% Flowering	62.20	47	94	12.108
Days to maturity	107.71	86	127	4.753
Plant height (cm)	33.93	15.4	68.4	23.172
Branches/Plant	9.26	1.8	28.4	33.657
Pods/Plant	19.38	2.47	59	45.775
Pod length (cm)	2.97	1.88	5.18	11.587
Pod width (ch)	0.88	0.26	1.3	11.013
Seeds/pod	3.27	1.6	4.6	12.756
Seeds/plant	54.70	6.2	200	50.159
Biological yield (g)	8.31	0.4	51	55.933
Grain yield/plant (g)	3.78	0.62	19.81	59.832
ODAP content (%)	0.438	0.128	0.872	38.527

The second phase (1965-1990) witnessed the development of improved varieties which were relatively free of the neurotoxim. The obvious, seemingly simple and straight forward solution at the initial stages was to ban the cultivation of crop altogether. But this suggestion was considered impractical by both the State and Central Governments. Following this, the research programme on identifying the toxin in the pulse and the way of eliminating or reducing it was initiated. Two approaches were suggested:-

1) The development of varieties of lathyrus which are either low or completely free of ODAP through breeding.

2) The detoxification of *dal* through leaching as ODAP in a water soluble compound.

Developing varieties for low levels of toxin : The extensive testing and evaluation resulted in the recommendation by the Pulses Research Workshop in 1974 for the release of variety Pusa 24 with very small amounts of ODAP (0.2%) and suitable for upland cultivation. Further research work helped in identifying 3 varieties 'LSD 1', 'LSD 3' and 'LSD 6' with low neurotoxin, ranging from 0.15 to 0.2%. While 'LSD 6' was recommended for upland cultivation LSD 1 and LSD 3 were suitable for *Utera* cultivation in rice fallows. Recently, three varieties *viz,* Ratan (BioL 212), Prateek (LS 157-14) and Mahateora (RLS 4595) have been developed with yield level of 1.2-1.5 t/ha and very low toxin level (< 0.1 %).

About 90% of neurotoxin is known to seep out into boiling water, but has limited practical scope. This is because there are some intricate problems associated with it like marketing and distribution of detoxified and non-toxified stocks, keeping quality etc. That is the reason why the Govt. agencies have not accepted the responsibility of detoxification.

Screening against diseases & Pests : Thrips (*Caliothrips indicus*) is a major pest of grass pea in India. Approximately, 1200 accessions have been screened against thrips at Raipur. None of these accessions found resistant to this pests, however, RLK 273-1 and RLK 273-3 were moderately resistance. Powdery mildew and downy mildew are the two major diseases of this crop. On the basis of field screening at Raipur, IPLy 2-10, RPL 26 and RL 41 have been found tolerant to powdery mildew. In addition, a number of collections from surrounding Raipur area have shown resistance to downy mildew.

Agronomy

Grasspea is a winter season crop adapted to areas with arid or semi-arid conditions. In Indian conditions, the crop is generally sown in October or November and harvested in February/March. It is usually grown after a crop of rice and is broadcasted into the standing rice crop about two weeks before harvest. (*Utera* crop). When the water is drained to allow for harvesting of the rice, the seeds germinate and the crop utilizes the remaining moisture for growth. The crop, thus, germinate and grows on the residual moisture. The crop has ability to with stand high moistue during early stage of growth and prolonged drought at grain filling/maturity stage. Normally the smaller seeded type genotypes are preferred for *Utera* crop as the farmers believe that they remain dormant under flood conditions for a longer period and give better germination when the water is drained. The cultivation of grasspea is confined to these areas where the cultivation of other crops is not possible and therefore, it acts as a "bonus crop". The crop is considered to require low or zero inputs and thus utilizes the residual soil nutrients. When the crop is sown on rainfed lands, the soil is usually prepared by two ploughings.

Its sowing time is largely determined by the time of recession of monsoon rains, type of soil and the status of soil moisture in the month of October. In rainfed areas of M.P., grasspea in grown as mixed crop with wheat, chickpea, barley and linseed.

The productivity of the crop is dependent on its sowing time, which in turn is determined by the soil moisture status when the crop is sown and the extent of winter rains during the crop season. Though the crop is considered to be resistant/tolerant to drought conditions, the moisture regime still plays a significant and predominant role in enhancing or decreasing the productivity of the crop.

Grasspea has a substantial area in the Indi-Gangetic Plains of India. Among the various nutrients, phosphorus and molybdenum have been found to influence the productivity of pulses. A study was conducted at West Bental for enhancing the productivity of grasspea under different levels of phosphorus and foliar spray of molybdenum. Application of Phosphorus increased the yield attributes and showed a marked increase in yield of grasspea. The improvement in yield attributes and yield may be attributed to profuse nodulation, leading to increased N fixation which in turn had position effect on photosynthetic organs and rate. Foliar spray of 0.05% Mo resulted in marked increase in seed yield and overall improvement in yield attributes. The increase in yield could be attributed to greater nitrogenase activity which led to increase in yield. Higher yield might also possibly be due to greater assimilation of carbohydrate and protein synthesis as Mo acts in N assimilation and increases intensity of photosynthesis.

In a study conducted at West Bengal it was observed that in the multiple cropping programmes under rainfed conditions, grasspea gave good yields of green fodder ranging from 15000 to 25000 kg/ha when harvested at flowering. It the preceding rice-straw height was greater (about 30 cm), the green fodder was also increased. Conversely, if the preceding rice-straw was cut at soil level, the seed yield of grasspea was higher. Grasspea, as a paira crop under rainfed multiple cropping systems, had some effect on improving soil fertility, particularly in terms of organic matter, total nitrogen and available phosphorus. Higher yields of green gram after grasspea reflected higher soil fertility in the rainfed crop systems.

Utilization

In India, Grasspea is commonly grown for its grain but is also used for fodder. The crop has a great potential as a fodder crop. In surveys of growers in Indian it is often found that animal feed value of the crop is relatively more important than human food. Often the feed is comprised of ground or split grain or flour and is used as feed for lactating cattle or for bullocks at the time of heavy field use such as during land preparation.

For human consumption, different seed coat colours are preferred in different regions according to tradition and use of the crop. The grains are sometimes boiled whole, but are most often processed through a *dal* mill to obtain split *dal*. The flour is made from grinding either the whole or split seed. The green pods and seeds are eaten directly or the whole pods are cooked and eaten as a vegetable. The remaining plant is then allowed to grow and is harvested for seed. Sometimes, the young pods are boiled, salted and sold or consumed as a snack.

Post harvest handling and processing

Grasspea is an important nutritional source because of its high protein content. However, daily consumption of foods made from the seeds for some period, is known to cause neurolathyrism. The neurotoxic agent found in the seeds is ODAP. Processing studies conducted at IGAU, Raipur revealed that embryo contains higher concentration of ODAP but its share by weight is only 1.6 per cent and that is why, nearly 2.91 per cent of seed ODAP goes with embryo. The major portion (71.02%) of toxin is concentrated in the cotyledons used as *dal*. *Dal* recovery and ODAP distribution under two different methods are presented in Table 4. Seeds were pre-soaked in plain water for 12 hours, before processing. Since ODAP is water soluble, nearly 25% of the seed ODAP leached out with the water.

Table 4 : *Dal* recovery and ODAP distribution in Grasspea seed

Component	Commercial Milling			Laboratory Processing		
	Recovery %	ODAP estimates (%)	ODAP % by weight	Recovery %	ODAP estimate (%)	ODAP (%) by weight
Whole grain	-	0.268	100.00	-	0.412	100.0
Cotyledons	75	0.249	69.4	82.9	0.352	71.02
Embryo + Broken grains	10	0.273	10.0	1.60	0.751	2.91
Husk	10	0.262	9.7	15.40	0.037	1.38
Losses	5	-	-	0.10	-	24.69

Future thrust

- Collection, conservation and evaluation of germplasm from unexplored regions.
- Breeding for low or near-zero level ODAP genotypes
- Breeding for high yielding ability through component breeding approach
- Breeding for increased forage/fodder production
- Physiological studies to determine drought tolerance mechanism in grasspea
- Studies on nitrogen-fixing ability
- Studies on role of Zinc in ODAP metabolism
- Identification of resistant sources against thrips

- Identification of resistant sources against powdery mildew and downy mildew
- Identification and elimination of enzymes responsible for ODAP production
- Work on chemical detoxification needs to be strengthened
- Development and refinement of production technology for *paira* cultivation.

References

Campbell, C.G., R.B. Mehra, S.K. Agrawal, Y.Z. Chen, A.M.A. El-Ali, H.I.T. Khawaja, C.R. Yadav, J. Toy and W.A. Araya. 1994. Current Status and future strategy in breeding grasspea (Lathers sativus). Euphytica 73: 167-175.

Erskine, W., Joseph, Smartt and Muehlbauer, F.J. 1994. Mimicry of Lentil and the domestication of Common Vetch and Grasspea. Economic Botany 48(3): 326-332.

Jackson, M.T and A.G. Yunus 1984. Variation in the grasspea (L. sativus L.) and wild species. Euphytica 33:549-559

Mehra, R.B., D.B. Raju and K. Himabindu. 1995. Evaluation and utilization of Lathyrus Genetic Resources in Asia. Proceedings of a Regional Workshop, 27-29 December 1995, IGAU, Raipur, India (R.K. Arora, P.N. Mathur, K.W. Riley and Y. Adham, eds.) IPGRI Office for South Asia, New Delhi, India.

Pandey, R.L. and O.P. Kashyap. 1993. Studies on Socio-economic strata and Lathyrus consumption in Rural Madhya Pradesh. In Lathyrus sativus and Human Lathyrism: Progress and Prospects (Yusuf H.K.M.,

Lambein, F., eds.). Proceedings of second Int. Colloq. Lathers/Lathyrism. Dhaka, 10-12 Dec. 1993, University of Dhaka, pp 47-50.

Smartt, J. 1990. Grain Legumes: Evolution and Genetic Resources. Cambridge University Press, Cambridge.

❑❑❑

Chapter - 20

Guar

D. Kumar and K.S. Solanki
Central Arid Zone Research Institute, Jodhpur 342 003, Rajasthan.

Guar [*Cyamopsis tetragonoloba* (L.) Taub.], internationally known as guar, is extremely drought hardy, deep rooted, summer annual legume of great economic and adaptive significance. Virtues of guar effectively match with low and erratic rainfall pattern and high ambient temperature habitats of arid regions. The crop may thrive very well in rainfall range of 250-450 mm with 3-4 spells, temperature range of 25-40°C, RH values of 50-65%, longer days with 8-9 hr sunshine, particularly at maturity. Guar is ideally suited to light to medium textured soils, with no water logging conditions. The crop tolerates less inputs and restricted after care, matching with the arid farmers' livelihood conditions. It is grown for grain, vegetable, fodder, cattle feed and green manure purposes, mainly in rainfed habitats of Rajasthan, Haryana, Gujarat and to some extent in Punjab and Madhya Pradesh. This crop has recently been introduced in non-traditional regions of Annantpur, Kadappa and Chittoor dry region of Andhra Pradesh. Only during current summer season of 2012, guar has been successfully cultivated with 4-5 irrigations in parts of Chattirgarh and Vidarbha and in non traditional summer seasons of Rajasthan. The cultivation in non-traditional regions and seasons have increased possibility of two crops in rainy and summer seasons. Even though, guar is cultivated on marginal lands, with marginal inputs in these regions, yet it has occupied the status of international, export oriented cash crop. The same has been due to presence of natural polysaccharide galactomannan gum content in seed endosperm. The gum content has wide industrial uses. Almost 80-85% guar gum produced in India is exported and rest is used for inland commercial demands. Of late, increasing demands of guar gum for USA alone to the tune of 5.0 lakh tonnes annually has led to sky rocketing prices resulting in economic benefit of Rs. 1.0 lakh to 1.5 lakh/ha in 90-100 days only. Hence, guar is rated as a source for providing livelihood security to arid farmers in practical reference. Relevant information on guar have been provided in this chapter.

Present Scenario

i) Area, production, productivity

Total global production of guar is realized to the tune of 12-15 lakh tonnes annually. However, India is the most important guar producer at the global level and contributes to almost 75-82% of total guar grain production, followed by Pakistan (15%), USA, South Africa, Sudan and Australia. In Pakistan, guar is mainly grown in Punjab and Sindh provinces, which contribute almost 80% of total guar production in Pakistan under irrigated conditions.

The trends in are a, production and productivity have greatly fluctuated over the years in India and Rajasthan as well (Fig. 1). Being a mega state, Rajasthan decides trends at the country level. The year 2003-04 witnessed maximum productivity in India and Rajasthan. It was followed by 2010-11. Average productivity of guar greatly fluctuates due to rainfall pattern in Western districts of Rajasthan. For instance, during severe drought year of 2002-03, the production sharply dropped to 2.0 lakh tonnes, whereas the same dramatically climbed to 15.0 lakh tonnes during 2003-04, probably due to high price rise during the previous year and increased area during 2003-04 (Fig. 1).

Rajasthan alone grows country's 70-75% guar and produces country's 60-62% guar grains. Haryana contributes 22%, Gujarat contributes 12% and Punjab contributes 2.0% towards total guar area in India (Fig. 2). In Rajasthan, six districts (Bikaner, Churu, Barmer, Jaisalmer, Hanumangarh and Jodhpur) occupy more than 60% state's guar area. Punjab and Uttar Pradesh contribute marginally. As per 2007-08 figures, there was mismatch between area and production of guar at the districts level in Rajasthan. For instance maximum area was occupied by Bikaner (23.0%) but maximum production (28.0%) was realized from Hanumangarh district (Fig 2).

Critically reviewing the growth and instability in area, production and productivity of guar in India and Rajasthan, it becomes evident that cv. values for India area 26.61, 43.73, 32.62% and for Rajasthan 29.18, 61.13 and 51.13%, respectively, for the period of 1970-71 to 2007-2008, indicating somewhat stability in area. Thus, production could be increased by imparting stability in production via productivity.

In Gujarat state, drought prone districts, Kachch and Banaskantha, are quite important for guar. These districts occupy almost 41.7% guar area of the state. More important is that productivity of guar in Kachh (610 kg/ha) and Banaskantha (604 kg/ha) districts is higher than the productivity of the state and the country as well.

ii) Guar gum

Guar gum is a cold water soluble polysaccharide. This ability to hydrate without heating makes it very useful in many industrial uses. Dissolved guar gum forms a line of high viscosity and viscosity is a function of temperature, time and concentration.

Solutions with different gum concentrations can be used as emulsifiers and stabilizers because they prevent oil droplets from coalescing. Guar gum is also used as suspension stabilizer and is an economical thickener and stabilizer. Being non-ionic, it is not affected by ionic strength or pH but will degrade at low pH and high temperature (pH 3 at 50°C). Guar gum retards ice crystal growth, non-specifically by slowing mass transfer across solid/liquid interface. The various derivatives or industrial grades of guar gum are manufactured by reaction of these hydroxyl groups with chemicals.There are two types of guar powders:

(i) **Food grade guar gum powder :** Used in industries like food, cosmetics, pharma etc. The particle sizes of food grade gum powders are: 200 mesh having 2000-5000 viscosity (cP) and with 300 mesh having 3500-5000 cP values.

(ii) **Industrial grade gum powder :** Used mainly in industries like paper, mining, explosive, fire fighting, oil drilling etc with particle size of 100 mesh with viscosity values of 3000-7000 cP.

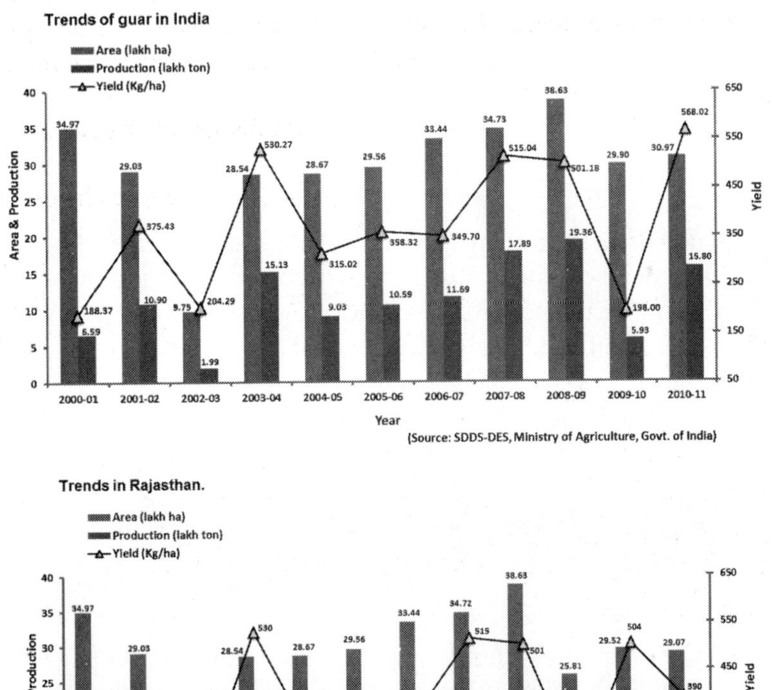

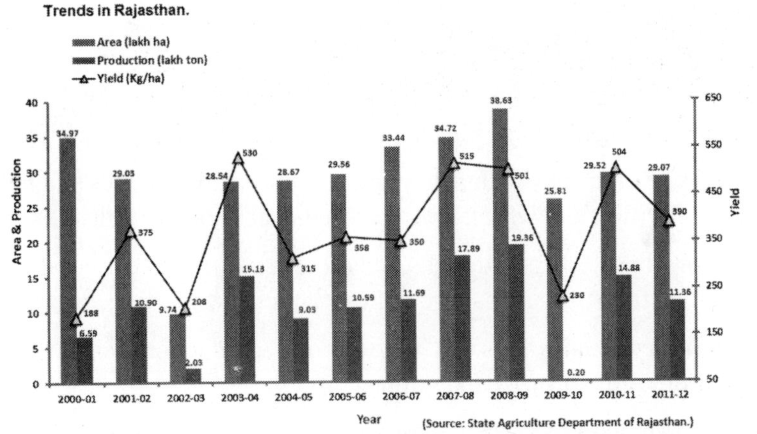

Fig 1 : Trends in area, production and productivity of guar in India and Rajasthan.

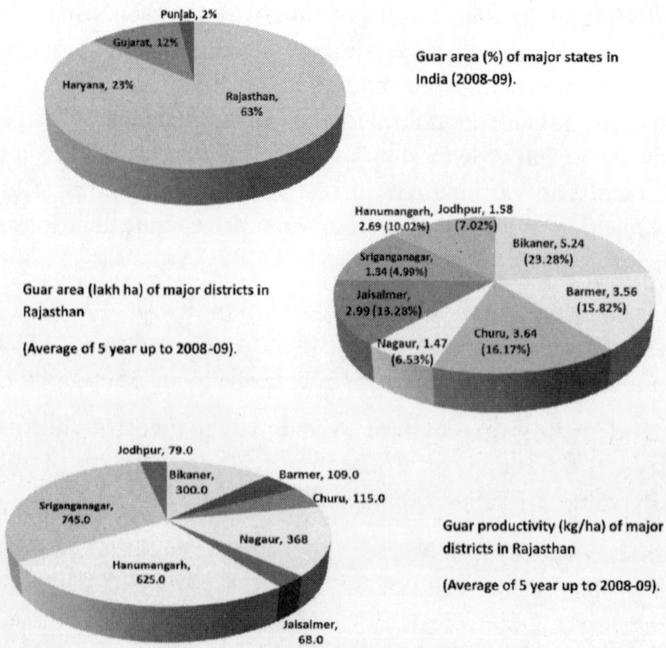

Fig 2 : Area under guar in major states of India and area productivity of guar in the major districts of Rajasthan.

Uses of guar gum : Guar gum is used in almost all systems where water is an important factor. A number of food stuffs and products compositions require the use of colloid to modify the behaviour of water with reference to other components. Guar gum is known to act as a stabilizer, binder, thickener and as a material for water proofing in various food products. The share of guar gum consumption is almost 30-40% is by food industries by petroleum and mining industries 20-25% and almost 18-20% textile industries share of the total. Use of hydrocolloids in food products is given below as per Halis and Feramuz (2007)

Dairy industry : Ice-cream stabilizer, milk shake, Ice milk (Soft ice-cream), ice pops and water ice chocolate, milk drink, flavoured milk drinks, instant dessert puddings, cooked dessert puddings, cottage cheese, cheese spread, whipped cream yoghurt

Beverages : Soft drinks with fruit pulp, soft drinks, fruit juices and nectars, foam stabilizer, beer clarification, wines, juices and vinegar, aging of spirits

Bakery products : Bread doughs, cake batters, fruit cakes, yeast raised,dough nuts, pipe fillings, fruit fillings, bakery jellies, flat icings, cookies, frozen pies fillings

Confectionery : Candy jells and jellies, Frozen confection, candy glaze, chewing gums, cough drops, gum drops, candy mints

Meat and fish products : Fish preservation, canned fish, meat and poultry, Coated jellied meat, preservative coating for meat and poultry, synthetic meat fibers and products

iii) Export of guar gum

India exports almost 75-80% of its guar gum annually including other derivatives. The major derivatives of guar seed (refined splits, gum powder and guar meal) are exported from India. India exported 2.58 lakh ton of guar gum worth Rs. 13.8 crores during 2008-09. The same has dramatically increased to almost 3.5 lakh ton worth Rs. 2800 crore during 2010-11 (Fig. 3). It is due to the demand of guar gum from USA increasing from 75,000 tonnes to 4.5 lakh tonne recently as a consequence of enhanced demand for drilling of petroleum fields in middle east countries. During 2006-07, top importing countries of guar products were; USA (75 th ton), followed by China (27th ton), Germany (17.7 th ton), Italy and Netherlands (4.5 th ton each). Global consumption of guar derivatives has in general been of food grade (50-55%), industrial grade (25-30%), textile, paper, tobacoo, photography (10-15%) and 5-10% in pharmacy grades (Kuamr, 2010).

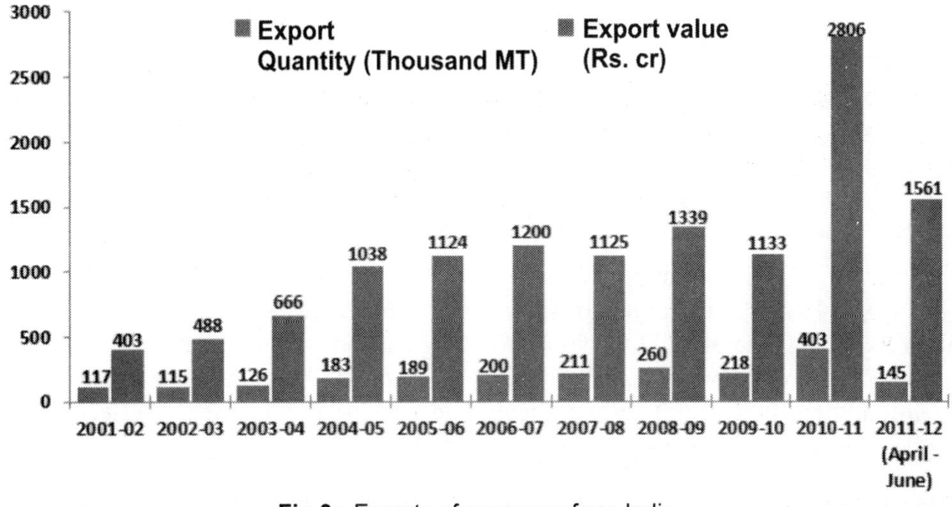

Fig 3 : Exports of guar gum from India

iv) Guar meal

The germ and outer seed-coat of guar seed together constitute guar meal. Removal of gum from guar seeds increases the protein content of the residual byproduct i.e. guar meal. It is light, grayish material with beany flavour. The guar seeds result in 62-68% of guar meal having a rich source of crude protein content of about 35-46%. It contains about 1.5 times more protein than guar seed and compares well with other vegetable protein sources like oilseed cake used in poultry diets. It was observed that the total protein of defatted guar meal consisted of 4 fractions, 2 of high and other 2 of low molecular weight proteins and detoxification of guar meal did not cause any significant change in the protein while autoclaving resulted in drastic changes. Guar seed contains 4-5% oil content. It was observed that the guar oil contained 55.1% linoleic acid compared with 51.8% in sunflower oil. The total unsaturated fatty acids were 78.7 and 92.0% in gaur and sunflower oil, respectively. Guar oil contained 3.36% linolenic acid. The iodine value and refractive index of the guar oil are well comparable with that of sunflower oil (Rodge, 2009).

Yield gaps

Yield gaps are generally a cause of great worry in rainfed crops, the same is true for guar also. In general gaps in production when compared between states and states vs country, simply appears due to variation in area and productivity (Table 1). For instance, in Haryana 4.5 lakh ton guar is produced from 4.0 lakh ha area, whereas from the mega state of Rajasthan 6-7 lakh ton guar is produced from large area (28-30 lakh ha). Gujarat is also producing 1.29 lakh ton guar from 1.95 lakh ha area. The obvious reason is the productivity of Haryana is very high i.e. 1200 ka/ha which has increased 206% in past 5 years only. The productivity of Punjab is also high (669 kg/ha). On the contractory productivity of Rajasthan is meager (370 kg/ha) even lesser than national productivity. Critical analysis of productivity of major district of Rajasthan reveals that productivity of Hanumangarh and Sriganganagar is very high (870 and 807 kg/ha, respectively). Productivity is quite low in Jodhpur, Churu, Barmer (144, 193 and 207 kg/ha, respectively). Thus, Barmer and Churu districts having large areas (3.54 and 3.47 lakh ha, respectively, need to emphasize on increasing productivity to be at least at par with dry district of Bikaner (415 kg/ha). Attempts in Jodhpur district may be on increasing area as well as productivity (Table 2).

Out of 333 FLDs conducted during 2002-2006 in 6 states of Rajasthan, Gujarat, Haryana, U.P., A. P. and Punjab, on over all mean basis, yield of promising variety was observed to the tune of 1243.5 against 978.5 kg/ha, showing, increase by 27.08% over the local variety (Anomunous, 2007). However, potential yield of guar is estimated to be around of 2000 to 2200 kg/ha indicating large gap in yield. Hence reflecting large gaps in realized and realizable grain yields of guar.

Table 1: Area, production and productivity of guar in three main states (2008-09)

India/state	Area (lakh ha)	Production (lakh ton)	Productivity (kg/ha)
India	33.00	12.50	490
Rajasthan	28.08	6.58	370
Haryana	4.00	4.50	1220*
Gujarat	1.95	1.29	669

* increased 206% during past 4 year

Table 2: Area, production and productivity of guar in major districts in Rajasthan during 2008-09.

India/state	Area (lakh ha)	Production (lakh ton)	Productivity (kg/ha)
Bikaner	6.05	2.510	415
Barmer	3.54	0.734	207
Churu	3.47	0.670	193
Hanumangarh	3.19	2.270	870
Sriganganagar	1.80	1.450	807
Jodhpur	1.49	0.215	144
Nagaur	1.29	0.618	541

Cropping System and traditional wisdom

Guar, basically a crop of arid habitats as mentioned earlier, is largely grown in western dry tracts of India, mainly in Rajasthan, Haryana and Gujarat. It requires well drained, light textured sandy to loamy plain lands having even slopes of almost 5-10%.

Guar, basically raised as rainfed crop, is planted after first showers of 30-40 mm in the first to second week of July. Early planting may cause profuse growth resulting in poor harvest but late planting may not attain usual growth and result in severe yield reductions (Virender Kumar, 2009). However, results reported by Yadav *et al.* (2002, 2003) on early maturing guar varieties (HG 365, HG 563 and RGC 936) revealed that planting after 25 June resulted in higher grain yield. Grain yield drastically reduced and disease incidence was more when planting was done during last week of July.

Generally arid farmers grow guar crop by traditional broadcasting method which is unscientific; allows more weeds, plant competition for nutrients and moisture; resulting in poor yield and should therefore, be discouraged. Planting in rows as paired rows (Singh *et al.*, 2003) or solid rows resulted in higher yields than broadcasting method of sowing. For obtaining high yields, optimum plant stand is desired which depends on inter and intra-row spacings which are different for branched and unbranched type of varieties. Under late plantings, close spacing may be recommended. Kumar *et al.* (2003a) reported that for early maturing varieties (HG-563 and RGC-936), 60 cm inter row spacing gave higher yields over 40 cm intera row spacing. Generally, 15 and 10 cm inter row spacing were optimum for branched and unbranced guar varieties.

Cropping Sequence involving Guar

Cultivation of guar improves N status of the soil thereby, reduces the nitrogen requirement of the succeeding crops (Faroda and Singh, 2003). Guar fixes almost 30-70 kg N/ha and leaves a residual effect equivalent to about 15-20 kg N/ha (Rao, 1995). In the areas where irrigation facilities are available and land is sown during *rabi*, guar is most preferred crop in rotation with wheat or mustard, particularly in northern India. In rainfed conditions, most prominent rotations are: pearl millet + guar – fallow, guar –

fallow/mustard, sesame + guar – taramira, guar-taramira, guar-mustard, guar + grass – fallow etc. Five year consecutive study (2000-01 to 2004-05) on guar-based double cropping system under irrigated condition at Hisar revealed that guar-mustard system was as more profitable (Rs. 21, 908/ha) than guar-wheat which gave Rs 19296/ha. A long term study revealed that pearl millet – guar cropping system gave 11% higher seed yield over continued cropping of pearl millet only. Contents of organic carbon and available soil phosphorus were also improved by 12 and 25%, respectively (Singh *et al.*, 1985). In a study conducted by Kathju *et al.* (1987), it was reported that yield level of pearl millet was much higher (13.20 q/ha) when guar was included in the rotation for three years compared to pearl millet cropped continuously (8-10 q/ha). Grain yield of pearl millet increased by almost 35% in arid situation in guar-pearl millet sequence compared to continued pearl millet (Saxena *et al.*, 1997). On the basis of a eight year consecutive study with different pearl millet based cropping systems, it was concluded that the amount of net C balance in guar based cropping system was higher than manure based cropping system (Narain and Praveen Kumar, 2003).

Guar intercropping

(a) Guar with cereals

In general, intercropping of guar with pearl millet was found profitable in good and sub-normal rainy years, but unprofitable in drought years. Studies carried out at Hisar and Gwalior, on intercropping of pearl millet with guar revealed that higher returns were obtained with the sole crop of guar, implying that intercropping of pearl millet in any proportion had no economic significance. However, amongst different intercropping systems, 3:1 ratio of guar: pearl millet gave highest guar equivalent and economic return, closely followed by 2 : 1. At Bawal, two rows of guar sown in inter-spaces of pearl millet planting system of rows 100 cm apart, gave higher total productivity and gross returns compared to sole crop of pearl millet. Further studies at Bawal indicated that intercropping one row of pearl millet in inter-spaces of paired rows of guar increased the guar yield by 27.2% compared to sole stand in paired row, but produced 8.7 q/ha additional yield of pearl millet (Singh *et al.*, 2003). Water use efficiency of intercropping system under rainfed condition same with protective irrigation was 58.8-89.5 and 63.9-1.8 kg grain/cm water, respectively compared with 51.3-84.6 and 53.3-86.2 kg, respectively, in pearl millet pure stands. At Bikaner, strip-cropping of pearl millet : guar in 4 : 4 row ratio gave highest pearl millet yield than 4 : 6 or 4 : 8 ratio (Yadav, 1992), whereas Singh and Joshi (1994a) observed the superiority of alternate row planting (1 : 1) of pearl millet : guar compared to other planting systems. Although guar yield was highest in 4 : 4 row strip-cropping of pearl millet : guar, but the net returns were highest in alternate-row intercropping system (Singh and Joshi, 1994a). At Bikaner, growing of guar and pearl millet in 2 : 2 planting system showed the highest yield advantage (Yadav and Beniwal, 2003). At Jodhpur, higher total productivity and water use efficiency were obtained with 1:1 row intercropping than 4:4 strip cropping or sole pearl millet

(Singh and Joshi, 1997). Evaluation of different pearl millet based cropping systems (sole cropping, mixed cropping, row intercropping and intercropping) in drought amelioration, revealed that strip-cropping in companionship of guar gave better yields and covered more risk compared to other systems (Singh and Joshi, 1994b).

(b) Guar with oilseeds

Intercropping of guar with castor yielded more per unit area without affecting the yield of main crop. Parallel cropping of castor + guar in 1:2 system gave higher yield of both the crops. Guar parallel cropped with castor produced highest dry matter and seed yield than other legumes under rainfed conditions. At Bawal, intercropping two rows of guar in castor grown in paired row (30/120 cm), increased the castor equivalent yield, land-equivalent ratio and net returns compared to pure castor crop. Studies conducted at CRIDA, Hyderabad revealed that the average yield of castor in intercropping system was 80% of that in sole castor and further showed that uniform planting of castor + guar in 1:1 ratio and application of 10 kg N and 30 kg P_2O_5/ha to guar component significantly increased the yield advantage of the intercrop system and additional gross income (Rs. 1564/ha) over the sole crop of guar.

(c) Guar with pulses

At Morena higher pigeonpea equivalent yield with guar intercropping (25.9 q/ha) compared to sole pigeonpea (20.8 q/ha) was observed, but Yadav *et al.* (1997b) found that amongst all intercrops, guar yielded lowest and caused maximum yield reduction in base crop pigeonpea, thus proved most uneconomical. Intercropping of green gram in guar proved more compatible as witnessed by higher value of guar equivalent yield (17.69 q/ha), net returns (Rs. 6846/ha) and B : C ratio of 1.71 and also possessed better LAI, LER, Area Time Equivalent Ratio (ATER) and crowding coefficient (Singh *et al.*, 2003).

(d) Guar with trees species

Based on average of three years' (1995-98) research at S. K. Nagar, higher yield of guar was obtained when raised in association with *khejri* (*Prosopis cineraria* L.) than under sole cultivation, and higher return from *khejri* based agri-silvi system of agro-forestry (Rs. 10078/ha) compared to sole cropping of guar (Rs. 8648/ha). Beneficial effects of *khejri* on soil chemical composition was noticed with higher content of organic carbon and increase in available N, P and K compared to soils with no *khejri* cover. Under agri-silviculture system, guar yield was adversely affected by subabul [*Leucaena leucocephala* (Lamk) de Wit] with advancement in its age and canopy development. However, the reduction in seed yield was the least, when *subabul* was topped at 1.5 m over central leder system, open system and no-topping system. The content and uptake of N in seed and stalk of guar was maximum in the 1.5 m topping system and comparatively more in the first year, because of less competition offered by sababul for light, moisture and minerals. Guar yielded 3.0 q/ha seed when grown in association with established (5 years old) ber orchard after light pruning under rainfed conditions at Bawal.

Main constraints

The main concern in guar is too much fluctuation in area and productivity ultimately showing inconsistency in production. The statistics on guar during the drought year of 2002 and following year 2003 justify the same. The area, production and productivity of guar during 2002 declined to 5.56 lakh ha, 28 th ton and 50.36 kg/ha, whereas the same dramatically jumped to 22.78 lakh ha, 11.63 lakh ton and 511 kg/ha during 2003, respectively. In guar realizable yield is 2000 kg/ha but hardly 50% of the same is being achieved. Hence, major constraints are large which are summarized below:

(a) Practical constraints

Growing conditions of which crop in question are fragile and uncontrollable coupled with farmers mind set is difficult to be changed. Farmers are not interested in making required investments in cultivation of guar. Inputs are becoming very costly. Therefore, the constraints are general in nature but have been specifically and judiciously listed below:

- **Pure seed :** The pure seed of desired variety is still a major constraint. Public seed sectors are not providing desired quantity of seeds and private sectors are interested in production of hybrid seeds only. Farmers have to sow their old seed stock which is a mixture of different kinds. There is wide publicity of improved seeds, many seed companies are available also, but for such local crops, big companies and sectors are not coming foreward. Guar growers have high faith in seed and they do not for any other input. If seed is made available in desired quality and quantity, it can definitely increase grain yield by about 25-30% keeping other factors constant. Nearly 36,000 ton certified seed is required to sow almost 30 lakh ha land each year, in case no increment in sown area is expected.
- **Planting device :** Tractor mounted seed planters are essentially required immediately after 2-3 days of rainfall (30-40 mm) to cover large acreage of (25-30 lakh ha) of guar. Delayed sowing may lead to poor plant stand. Farmers capacity is not adequate to purchase tractor mounted seed drill. Hence, same may be made available in advance on rental basis at village Panchayat level. Large areas in Rajasthan (25-28 lakh ha) are difficult to be sown by traditional device in 2-3 days only.
- **Line sowing :** Farmers largely plant their crops by broadcast method. Improved sowing method at desired spacings (40-50 cm between rows) may increase grain yield of guar by about 30-35%. It is for minimizing competition between

plants for soil moisture, nutrient supply, distribution of radiation and making interculture operations easy. The line sowing method, particularly, in rainfed situations is very crucial.

- **Weeding:** Arid farmers cultivating guar do not believe in interculture operations which conserves soil moisture, makes it available to the roots and clears shelters of insects and pests. It is a bad practice as it reduces grain yield by 35-40%.
- **Patchy plant stand:** Ultimate agronomic grain yields are not realized, if optimum plant stand is not achieved with reference to plant type or the variety, rainfall intensity and soil type. It is influenced by sowing methods, soil moisture, seed treatment etc.
- **Supplementary irrigation:** In case of absence of rainfall for about 40 days, one supplementary irrigation at podding stage, from the water collected in the pond near fields only, may revive plant life and may give 40-45% more grain yield.

(b) Technical constraints

1. Growth period: No variety is maturing in less than 90 dayts. *S. serrata* wild type is a source of earlyness (45-50 days) but absence of cross ability between *S. tetra* and *S. tetragonoloba*, thus overy reseue or protoplasmic fusion area required by non-conventional measures.

2. Need based improvement in plant type: Emphasis is needed for developing varieties for both rainfed and assured water availability conditions.

- **For rainfed situation**: The characteristics required are initial fast growth, lesser number of secondary branches, enhancing length of main shoot with dense podding behaviour and better source-sink relationship, less biomass and more reproductive area, more synchronization in flowering, podding and maturity behaviour, less loose canopy and semi-spreading to erect canopy. Early maturity may range from 70 to 75 days.
- **For assured situations:** Medium maturing growth period of 85-95 days, better inputs utilization potential, more synchronized maturity, erect upright growth behaviour, adequate number of secondary branches and longer main shoot with dense podding behaviour are required. Growth behaviour with unbranched types may suit mechanical harvesting; therefore, podding from 2nd to 3rd nodes may be preferred.

3. Disease susceptibility: The major concerns are root rot and bacterial leaf blight.

- **Root rot (RR):** It is most prevalent in rainfed situations (high ambient temperature and low soil moisture), being soil born in nature, it is difficult to predict and control. Seed treatment, foliar sprays are effective and being used but no resistant variety is available in guar. There is need for a stable and effective resistant source against RR pathogens.

- **Bacterial Leaf Blight (BLB):** It is most prevalent under high humid and warm temperature (25-32°C). Currently, no variety is showing stable resistance. HG-75 and *S. senegelensis* wild sp. are source of resistance. The latter is also a source of resistance to BLB with more number of pods.

4. Improved gum qualities: High gum (> 32%) content with varied viscosity levels in guar seeds are required for use in manufacturing of different products of food and industry.

Technologies for doubling the production

For the production of guar in fragile agro-eco-system, the following regional and environmental strategies need to be adopted to take care of the weakness so that the production can be doubled.

a. Policy options

It requires administrative, governance and social efforts as discussed below.

(i) Increasing productivity in major guar regions: Rajasthan being the mega state of guar, covers on an average basis, 25-28 lakh ha area annually. Some regions like those of Sikar, Bikaner and Jodhpur having 1.44, 11.88 and 6.62 lakh ha area respectively, with corresponding productivity of 351, 180 and 139 kg/ha, can be priotised for enhancing productivity to the tune of 500 kg/ha, which is feasible (Table 3).

(ii) Increasing area in high productivity zones: These are some regions in eastern-southern zone with more than 500 mm rain and having very high productivity of this crop but area cultivated is less. For instance, regions like Jaipur, Bhilwara, Udaipur and Bharatpur having 699, 554, 655 and 1175 kg/ha productivity and occupy only 73.5, 31.4, 15.4 and 37.0 thousand ha respectively. Every efforts may, thereafter, be made for doubling the area. (Table 4)

Table 3 : The regions showing area, productivity and proposed enhanced productivity of guar in Rajasthan

Regions	Area (lakh ha)	Productivity (kg/ha)	Proposed productivity (kg/ha)
Sikar (Sikar & Jhunjhunu distt.)	1.44	351	600
Bikaner (Bikaner, Churu & Jaisalmer distt.)	11.88	180	500
Jodhpur (Jodhpur, Barmer & Nagaur distt.)	6.62	159	400

Table 4 : Regions showing productivity, area and proposed enhanced area of guar in Rajasthan

Regions	Productivity (kg/ha)	Area (th. ha)	Proposed enhanced area (th. ha)
Jaipur (Jaipur, Ajmer & Dausa distt.)	699	73.5	147.0
Bhilwara (Bhilwara, Chittore, Rajsamand & Partapgarh distt.)	554	31.4	62.8
Udaipur (Udaipur, Banswara, & Dungarpur distt.)	655	15.4	30.8
Bharatpur (Bharatpur, Alwar, Dholpur, S. Madhopur & Karoli distt.)	1175	37.0	74.0

(iii) Introduction of guar in non traditional areas: In drought prone districts of Annantpur, Kadappa, Chittoor of AP, guar has been tried successfully. This crop can supplement groundnut which very often fails them imposing heavy financial burden on AP government as compensation @ Rs. 5000/- acrea. Grain yield of guar have ranged from 350 to 1500 kg/ha on rainfed red soils with little inputs (Anonymous, 2007-08).

(iv) Propagation of summer cultivation of guar: During current summer season of 2012, more than 5000 ha area in Rajasthan, Gujarat, parts of Madhya Pradesh, Chattisgarh, Uttar pradesh. and Vidarbha have been put under guar with 4-5 irrigations. The grain yields in Vidarbha and Rajasthan have been estimated to 10-12 and 9-15 q/ha. This has generated hope for guar cultivation in summer season and taking two crops in one year.

(v) Crop substitution: In Western districts of Rajasthan, groundnut is cultivated with 10-12 irrigations and matures in 130-140 days. In case this crop is replaced by guar even in 1.5 lakh ha area and by applying 1-2 irrigation at critical stages of 40-45 and/or 60-70 days of planting, grain yield of guar can be realized to the tune of 1200-1500 kg/ha. There is every likelyhood of substituting cotton and pearl millet in Gujarat, Rajasthan and Haryana as is these crops proving uneconomic, leading to farming loads.

b. Technological options

i) Appropriate genotypes: Rainfall veries (150-550 mm) and so do soils. Farmers' interests also vary, hence, varieties are needed that can be fitted as per situations. The varietal maturity must coincide with rainfall of the region, so as to have desired production depending on the rainfall pattern (Table 5). The early maturing group of varieties (RGC-936, RGM-112, HG-365) will sustain production even in extreme drought situations. However, in congegemid conditions, late maturing varieties (RGC-986, HG-2-20) will use high soil moisture to give grain yield upto 12-15 q/ha. Single stem varieties (RGC-1066 and RGC-1031) with close planting 25 × 5 cm and assured emergence can be tackled by combine harvesting and threshing, particularly in Ganganagar and Hanumangar areas of Rajasthan, where farmers are ready to put in more inputs.

Table 5: Varietal options for guar under rainfed situations

Varieties	Maturity (days)	Raifnall (mm)	Regions
RGM-112, HG-365 HG-563, RGC-936	85-90	300-400	Gangangar, Hanumangarh, Barmer, Hisar, Rewari, Bawal, Sarsa, Jodhpur Bachau, Radhanpur
RGC-1003, RGC-1002, RGC-884, HG-884, HG-563	90-100	400-500	Nagar, S. K. Nagar, Kothara, Jalore, Sirohi, Jaipur, Hisar, Anantpur, Chittoor, Kadappa
RGC-1038, RGC-1031	101-105	500-550	Nagaur, Alwar, Sirohi, Sawai Madhopur, Pali, Jalore
RGC-986, HG-2-20	110-120	555-700	Jhunujhunu, Bhopalgarh, Canal command areas, Vidarbha region assured inputs
RGC-1066	100-105	500-550	Ganganagar & Hanumangarh
RGC-1031	105-110	500-550	Ganganagar & Hanumangarh

ii) Optimum crop geometry: Early maturing varieties (HG-365 and RGC-936) performed better at 30 than 45 cm row spacing but gum content were more at 45 cm (Kumar *et al.*, 2003a, Yadav *et al.*, 2002) in Hisar and Bawal conditions. However, at Jhansi, 60 cm row spacing (1.1 lakh plants/ha) was optimum for Durgapura Safed; 45 cm row spacing (1.5 lakh plants/ha) for B-19-1-55; 30 cm (2.22 lakh plants/ha) for FS-277 (Ali, 1982).

Fig.4 Use of combine harvester for unbranched variety of guar.

Generally for extreme, drought situation (200-300 mm rainfall) a wider spacing 60 cm with 1.10 lakh plants/ha is recommeded while for moderate rainfall (350-400 mm rainfall) 30-45 cm spacing (1.5 lakh plants/ha) may be used for effective WUE and high production. Under delayed planting, closer spacing of 30 cm may be used compared to 45 cm inter-row spacing (Rana *et al.*, 1991). An intra-row spacing of 15 cm was appropriate for branched and 10 cm for unbranched varieties. For obtaing maximum grain yield, (16-17 q/ha), gum yield (5-5.50 q/ha), gum content (31-31.5%) and grain protein (28.86%) a seed rate of 15 kg/ha may be used for early maturing varieties (Yadav *et al.,* 2003).

iii) Fertilizer Management : Generally application of 20 kg N and 40 kg P_2O_5/ha may significantly increase the seed yield of guar by 25.18%; and water use efficiency by 21.7% and subsequently nutrient status of the soil may also marginally increase (Singh and Khan, 2003). Work at Bikaner revealed that 40 kg P_2O_5/ha increased the yield of guar, by 38.4 and 22.7% compared to control and 20 kg P_2O_5/ha, respectively. Similarly, three years (1992-94) study at S. K. Nagar revealed that guar fertilized with 40 kg P_2O_5/ha recorded highest seed yield which was more than 28% over control. At Agra, there was a linear response of applied P on sandy loam soil, where application of 60 kg P_2O_5/ha increased the seed yield of guar significantly to the level of 38.1 and 13.6% over control and 30 kg P_2O_5/ha (Table 6).

Table 6: Effect of phosphorus on seed yield of guar

P levels (kg P_2O_5/ha)	Seed yield (q/ha)								
	Bawal	Hisar	Durgapura	Gwalior	Bikaner	Jhansi	Jobner	SK Nagar	Jodhpur
0	9.98	14.6	7.06	15.25	12.13	7.52		5.10	6.65
20	11.69	17.8	8.03	17.05	13.51	8.78	12.5	5.75	7.50
40	12.54	17.6	8.34	18.27	16.83	10.38	15.2	6.54	8.10
60	12.26	17.5	8.24	22.05		11.81		6.99	8.50
CD (5%)	1.14	2.3	0.67	1.19	0.81		0.93	0.47	0.59

Source: Virender Kumar, 2009

Gum guar content increased by 29.41% with the application of phosphorus @ 20 kg P_2O_5/ha over control, but higher doses of phosphorus resulted in decreased gum content, though it was higher than the control. Hence, P may play a definite role in doubling the grain yield of guar.

Sulphur: Guar grown on S-deficient soil of Gwalior, responded upto a level of 40 kg S/ha in terms of growth, yield attributes and seed yield. The yield increase at 40 kg S/ha was 10.27 and 28.34% over 20 kg S and control, respectively. Sulphur also markedly enhanced P, S, crude protein and gum content in seed. Based on three years' results (1992-1994) at Bawal, the application of 20 kg S/ha in guar crop resulted in significantly higher seed yield (12.12 q/ha) over control seed yield of 10.82q/ha as well as gum content (29.98%) over control (29.10%). At Durgapura, guar responded up to 20 kg S/ha, whereas no response of S was observed at S. K. Nagar during 1992-94 and at CAZRI, Jodhpur during 1994-96.

Micro-nutrients: Studies carried out at Durgapura showed that foliar spray of 1 kg $ZnSO_4$/ha (0.2%) on guar at 45 days of sowing was better than soil application of 5 or 10 kg $ZnSO_4$/ha. Similarly, studies conducted at Bawal showed that foliar spray of 0.5% $ZnSO_4$ either at 25 or 45 DAS or both stages significantly increased the seed yield of guar over control and soil applied 25 kg $ZnSO_4$/ha at sowing time. Micro-nutrient studies carried out at Hisar, Bawal, Gwaliar, Durgapura and S. K. Nagar revealed that one spray of 0.5% $ZnSO_4$ at 25 or 45 days of sowing gave significantly higher grain yield over control but at par to 25 kg $ZnSO_4$/ha as soil application. Thus, $ZnSO_4$ as spray @ 0.5% may be used for enhancement of grain yield of guar.

iv) Irrigation Management: Recent work on irrigation management has indicated that significantly higher yield of guar could be obtained when the crop was given post sowing irrigation over control (Table 5). Maximum seed yield at Gwalior was obtained when the crop received 3 irrigations at vegetative, flowering and grain development stage. However, 2 irrigations at flowering and grain development stages were optimum at Durgapura, whereas one irrigation at grain development at S. K. Nagar, and at flowering stage at Bawal was sufficient to obtain higher seed yield (Table 7).

Table 7: Effect of irrigation scheduling on seed yield of guar

Treatment	Seed yield (q/hq)				
	2007		2008		
	Hisar	Bawal	Gwalior	Durgapura	SK Nagar
T1- Control	13.45	9.08	14.09	10.83	6.42
T2- One irrigation at vegetative stage (25-30 DAS)	13.57	9.40	15.99	11.35	7.75
T3- One irrigation at 50% flowering (40-45 DAS)	14.53	11.25	17.86	11.56	8.21
T4- One irrigation at grain development (55-60 DAS)	13.30	9.31	14.55	11.35	8.70
T5-T+T3	14.82	11.18	16.23	11.66	7.74
T6- T2+T4	14.00	9.56	17.12	11.77	7.70
T7-T3+T4	14.68	11.45	17.43	13.33	7.36
T8- T2+T3+T4	15.13	11.50	20.13	13.33	8.54
CD at 5%	NS	1.56	1.87	0.95	1.14

Source: Anonymous (2008-2009)

v) Weed Management: Season-long competition by weeds in guar may cause reduction in yield by 30 to 50% and the losses may go up by 70-90% depending on the infestation. Weed control alone was responsible in increasing the seed yield by 68% at Bawal and 61% at Hisar. Crop-weed competition is critical between 15 to 30 days of sowing on loamy sand and sandy loam soils. Removing weeds at 20 or 30 days increased the number of pods/plant, water use efficiency and seed yield (Yadav, 1998). Studies conducted in different agro-climatic zones and on various soil types have established the superiority of mechanical/cultural means over chemicals (Kumar *et al.*, 1996). These methods, besides being environmental friendly, are comparatively more economical than chemicals. Smith *et al.* (1973) summarized 11 years' research on the effectiveness of certain pre-plant, pre-emergence and post emergence herbicides on guar in Texas and Oklahoma and found that trifluralin, nitraline, dinitramine applied prior to planting seeds, were effective against weeds on warm soils in Oklahoma. However, in cold soil, application of trifluralin caused injury to the crop. The results further revealed that herbicides tested did not affect gum and protein content in guar seed, regardless of whether plants were injured. Yadav *et al.* (1998) found significant effect of fluchloralin at 1.0 kg/ha in guar under rainfed condition at Bawal; while Kuamr *et al.* (1996) also working at Bawal found that pre-plant incorporation of fluchloralin at 1.0 kg/ha gave seed yield at par with one hand weeding at 30 DAS. One hand weeding at 30 DAS, was at par with fluchloralin used at 1.0 kg/ha at Jodhpur. Increase in seed yield due to one hand weeding and fluchloralin was 49 and 45%, respectively, over weedy check. However, Yadav *et al.* (1997a) working at Hisar, observed that two hoeings and application of fluchloralin at 1.5 kg/ha were at par in seed yield (18.9 and 18.4 q/ha respectively) and net income (Rs. 2900 and 2845/ha respectively). The weed control efficiency of the two hoeings (80.7%) and fluchloralin at 1.5 kg/ha (79.9%) were better than one hoeing (64.2%). Fluchloralin applied at 2.0 kg/ha gave higher weed control efficiency (83.7), but had toxic effect on the crop and was found to reduce the yield significantly. Pre-plant incorporation of fluchloralin at 1.0 kg/ha reduced 84% weed population but seed yield was at par with one or two hand weedings, whereas pendimethalin caused phytotoxic effect on guar.

vi) Plant protection

- *Bacterial leaf blight* [*Xanthomonas axonopodis* pv. *vignicola*; *X. campestris* pv. *cyamopsidis* (Patel *et al.*,) Dye and X. phaseoli]: The temperature range of 25-30% and relative humidity range of 70-90% favour this disease. The bacterial pathogen is seed born. High amount of nitrogen in the plant may be associated with susceptibility of this disease. The disease appears both as leaf spot and blight simultaneously. Disease spots appear intravenial, are round, water soaked or oily in appearance and well defined on the dorsal surface of the leaves. The blight phase develops in the form of necrotic areas mostly from leaf margin. Black streaks which develop later on, may lead to cracking of stem. These black streaks may also develop on pods. Seed treatment with ceresin wet (0.2%), streptocycline (0.1%) or thiram (0.2%) was found effective in controlling this disease. Gandhi *et al.* (1991) reported that seed born infection due to bacterial blight can also be controlled by dipping the seed in streptocycline solution (200 ppm) for almost 30 min, followed by air drying of seeds. However for secondary spread of disease, spray of streptocycline was essential. A 3-yr consecutive study (2001 to 2003) carried out by Gupta *et al.* (2007) revealed that two spray of $ZnSO_4$ (0.5%) at 25 and 45 DAS were found most effective in reducing the incidence of BLB and AB diseases in guar at Hisar. It may be summed up that for managing guar diseases and having higher harvest, seed treatment with thiram/bavistin @ 2-3 g/kg seed may be treated mandatory for preventing seed born infections. For secondary infection need based foliar spray may be taken-up. Early & late planting may be discouraged from disease managements view point.

- *Root rot complex* [*Macrophomina phaseolina* (Tass) Goid, *R. solani, Fusarium caeruleum, Sclerotium rolfsii, Neocosmospora vasinfecta*]: Among root diseases root rot and wilt are the major diseases. In case of the former, root tissues are decomposed, preventing normal growth of host plants. In case of wilt, leaves and green parts lose their turgidity, become flaccid and ultimately fall down. At a later stage, the whole plant may suddenly die or dry up. *R. solani* is highly pathogenic in guar and may kill 90% seedlings within three weeks. Dry root rot is most serious under dry and warm conditions causing considerable yield losses in guar. The maximum population of *M. phaseolina* (233 sclerotia/g soil) was found after harvest of guar in 0-5 cm soil depth. The same was attributed to high soil temperature (35-40°C) and low soil moisture content (0.5-1%). The pathogen is difficult to managed due to wide host ranges. Natural heating of the soil during summer months led to reduction in *M. phaseolina* population by about 42%. Fungicide like brassicol (0.2%), ditham M-45 (0.1%), foltaf (0.2%) and thiride (0.2%) completely inhibited the fungal growth of *S. rolfsii* causing basal stem rot in guar (Ganeshan, 1997). Seed treatment with vitavax (0.2%), followed by bavistin (0.2%) appeared most effective in reducing the pre-emergence mortality. Mulching the soil with pearlmillet stover (3.5 t/ha), low plant density (1.6 million/ha) and FYM (10 t/ha) alone or in combination effectively conserved the soil moisture and reduced population of *M. phaseolina* and dry root rot incidence (Lodha, 1996).
- *Alternaria* leaf spot [*Alternaria cyamopsidis Rangaswami* & Rao. A. cyamopsidis (Ell. & Ev.) Elliot]. This disease may cause heavy yield losses to the tune of 50-55%. It appears every year in mild to severe form and the pathogen is seed born in nature. Disease infection is observed more in late sown conditions, high humidity (70-90%) moderate temperature range of 25-31 C. Early sown crops particularly, during last June may escape from this disease. Seed treatment with thiram (0.3%), dithane Z-78 (0.3%), tillex (0.3%) were effective against this disease. Foliar application of iprodione (0.2%) also proved effective in controlling this disease (Gupta *et al.,* 1999). Soil application and foliar spray of boron and zinc individually or in combination provided good control of this disease (Gupta *et al.,* 1999).
- Powdery mildew [*Erysiphe polygoni* DC, *Sphaerotheca fuliginea* (Schlect) Pollacci, *Leveillula taurica* (Lev) Arnaud]: Warm temperature (> 32-33 C), low humidity (< 70-75%) with bright sunny days may prove congenial for severe infection of this disease. Appearance of dirty-white, circular, floury

patches at the time of pod formation and maturity are common symptoms. Patch size may increase with temperature. Dusted white chalk powder appears on whole plant parts. Severely infected plants appear defoliated, weekened, showing pre-mature drying and death of infected leaves (Gupta and Rohilla, 2008). Application of 0.1% calixin and 0.5% kerthane as spray immediately after appearance of disease twice at 15-days interval may reduce intensity and the spread of disease as well. Among the biological control measures, three sprays of *T. viride* were most effective in control of this disease (upto 90.6%).

Strategies to achieve projected targets

Guar is an industrial crop due to presence of gum in its seed. Guar gum has many industrial uses in terms of food grade and industrial grades. It requires high gum content with viscosity profile suited to concerned industrial uses. Besides, the crop suffers from important diseases and terminal stress due to long maturity. The plant type is not of productive but of survival value. It is generally known as poor responder towards better management and inputs. There are large gaps in experimental and farm yields. In view of this historical back ground, the targets and strategies can be set as below:

1. World demand of guar gum at present is 6.0 to 6.5 lakh tonnes per annum. India is fulfilling almost 80% the world demands, hence, including domestic demand, India must produce 5 to 5.50 lakh tonnes guar gum annually. Hence, our target is to produce 5.50 lakh tonnes guar gum from almost 18-20 lakh tonnes guar seed annually. On the other have, production of guar seed is howering around 10-12 lakh ton annually showing a short fall of 6-8 lakh tonnes annually.

2. The same can be achieved if area of guar crop is increased to 40 lakh ha against 30 lakh ha at present and productivity is increased to 500 kg/ha at the country level against 400 kg/ha. The target can also be achieved by increasing area in non-traditional zones of Rayalseema region of Andhra pradesh, Vidarbha region of Maharashtra, area around Raipur in Chattisgarh and central eastern parts of Rajasthan. Area can also be increased by extending its cultivation in summer season with 4-5 irrigations in aforesaid and traditional zones.

3. Productivity of guar can also be increased by using production technologies available and making good quality seed available with reference to rainfall and soil conditions of the zone. One to two supplimentry irrigations and proper crop geometry may increase grain yield by about 20 and 32%, respectively.

Future Thrusts

- Production Technologists for corporate world and subsistence farming communities may be developed separately.

- District level technology with reference to soil type, rainfall pattern, cropped area, productivity, farmers interest, need be developed.
- Horizontal expansion of crops to newer non traditional areas may be taken-up.
- Development of hybrids in guar by finding out suitable sterile lines, and maintainance of their lines.
- Complete mechanization by developing unbranched varieties and commercial production of lines and small combines.
- Extraction and commercialization of edible oil of guar seed having almost 5-6%. The same by many means is as good as sunflower oil.
- Intervention of biotechnological tools for incorporation of golden and rare traits (high gum content and high viscosity).
- Hastening initial growth rate of guar to lessen the initial weed competition.
- Separation of food and industry grade guar gum.
- Application of guar gum in newer and novel food products to enhance shelf life and other qualities as an additive.
- Developing non-traditional breeding strategies for transferring desired genes from wild sp. *C. serreta* into cultivated guar sp. *C. tetragonoloba* for early maturity (53-60 days), more pods and resistance against BLB disease.

Acknowledgements: The financial assistance provided through Emeritus Scientist Scheme by the Education Division of the ICAR, New Delhi and required facilities extended by the Director, CAZRI, Jodhpur for implementation of ES Scheme are duly acknowledged.

References

Anonymous (2007-2008). Annual Progress Report of Network Project on Arid Legume, PC-Unit, CAZRI, Jodhpur, India.

Anonymous (2009-2010). Annual Progress Report of Network Project on Arid Legumes, PC Unit, CAZRI, Jodhpur, India.

Faroda, A.S. and Singh, Raj (2003). Crop production strategies for arid legumes with remunerative efforts and cost. In: Advances in Arid Legumes Research (Eds. Henry A., Kumar, D. and Singh, N. B.) Scientific Publisher, Jodhpur (India). pp 161-171.

Ganeshan, Girija (1997). Fungicidal control of sclerotium basal root rot of clusterbean cv. Pusa Navbahar. Indian Phytopath. 50: 508-512.

Gupta, P. P., Kumar, Rajesh and Jalali, B. L. (1999). Response of different vesicular arbuscular mycorrhizal fungi on cowpea. Plant Disease Research 14(1): 25-30.

Gupta, P.P., Yadav, B. D. and Joshi, U. N. (2007). Effect of micronutrient on major diseases of clusterbean [Cyamopsis tetragonoloba (L.) Taub.]. Journal of Arid Legumes 4(1): 47-51.

Halis Sahin and Feramuz Ozdemir (2007). Effect of some hydrocolloids on the rheological properties of different formulated ketchups. Food Hydrocolloids 18: 1015-22.

Kathju, S., Agarwal, R. K. and Lahiri, A. N. (1987). Evaluation of diverse effect of phosphate application on legumes of arid areas. Tropical Agriculture 64: 91-96.

Kumar, D. (2010). Guar – a golden crop for arid regions. Indian Farming. 60(1): 4-9.

Kumar, V., Yadav, B.D. and Agarwal, S.K. (1996). Performance of fluchloralin and pendimethalin to control weeds in clusterbean under rainfed conditions. Haryana Journal of Agronomy 12: 43-46.

Kumar, V., Yadav, B.D. and Yadav, J.S. (2003a). Spatio-temporal management for grain production of clusterbean genotypes HG 563 in light textured soils of Southern Haryana. In: National Symposium on Sustainability, Advancement and Future Thrust Areas of Research on Forage, held at CCSHAU, Hisar from March 5-6, 2003. pp. 77.

Kumawat, P. D. and Khangarot, S.S. (2001). Response of sulphur, phosphate and Rhizobium inoculation on growth and seed yield of clusterbean [Cyamopsis tetragonoloba (L.) Taub.]. Annals of Biology 17: 189-191.

Lodha, S. (1996). Influence of moisture conservation techniques on Macrophomina phaseolina, dry root rot and yield of clusterbean. Indian Phytopathology 49: 342-9.

Narain, P. and Kumar, Praveen (2003). Integrated nutrient management through arid legumes in Rajasthan. In: Advances in Arid Legumes Research (Eds. enry, A., Kumar, D. and Singh, N. B.), Scientific Publishers, Jodhpur (India). pp 172-178.

Rodge, A. B. (2009). Status and direction of quality oriented research. In: Perspective Research Activities of Arid Legumes in India (Eds. D. Kumar and A. Henry), Indian Arid Legumes Society, CAZRI, Jodhpur. pp 71-95.

Saxena, A., Singh, D. V. and Johsi, N. L. (1997). Effect of tillage and cropping systems on soil moisture balance and pearl millet yield. Journal of Agronomy and Crop Science 178: 251-257.

Singh, J., Kumar, Anil and Singh, Bikram (2003). Production potential and economics of clusterbean under various geometries and intercropping systems. In: Advances in Arid Legumes Research, (Eds. Henry, A., Kumar, D. and Singh, N. B.), Scientific Publishers, Jodhpur (India). pp. 207-211.

Singh, M. and Joshi, N. L. (1994a). Effect of mixed and intercropping systems on dry matter and grain yields of companion crops in arid environment. Annals of Arid Zone 33: 125-128.

Singh, M. and Joshi, N. L. (1994b). Performance of pearl millet based intercropping system under drought conditions. Arid Soil Research and Rehabilitation. 8: 313-319.

Singh, M. and Joshi, N. L. (1997). Pearl millet based cropping system in relation to plant population and moisture availability. Annals of Arid Zone 36: 37-42.

Singh, Raj and Khan, M.A. (2003). Response of clusterbean varieties to fertility levels and cropping systems under arid conditions. In: Advances in Arid Legumes Research (eds. Henry, A., Kumar, D. and Singh, N. B.) Scientific Publishers, Jodhpur (India). pp 225-228.

Singh, S. D., Bhandari, R.C. and Agarwal, R. K. (1985). Long term effect of phasing of phosphate fertilizer on soil fertility and yield of pearl millet grown in rotation with grain legumes. Indian Journal of Agricultural Sciences. 55: 274-278.

Smith, D.T., Wiese, A. F. and Santalmann, P.W. (1973). Weed control research in guar in Texas and Oklahoma 1961 to 72. Texas Agriculture Experimental Station Bulletin 13: 1138.

Virender Kumar (2009). Perspective production technology of arid legumes. In: Perspective Research Activities of Arid Leguems (Eds. D. Kumar and A. Henry) Indian Arid Legumes Society, CAZRI, Jodhpur. pp 119-155.

Yadav, A., Budhiraja, M., Rathee, S. S., Yadav, I. S. and Malik, R. K. (1998). Selectivity of clusterbean varieties to dinitroanilines in different soils. Indian Journal Weed Science 30: 108-112.

Yadav, B. D., Joon, R. K. and Kumar, V. (2003). Response of early maturing guar variety to date of sowing and seed rates. In: Advances in Arid Legumes Research (Eds. Henry A., Kumar D. and Singh N. B.) pp 199-202. Scientific Publishers, Jodhpur (India).

Yadav, B. D., Joon, R. K. and Lodhi, G. P. (1997). Chemical weed control in clusterbean [Cyamopsis tetragonoloba (L.) Taub.]. Forage Research 23: 189-192.

Yadav, B. D., Kumar, Virendra and Saxena, A. (2002). Agronomy. In: Guar in India (Eds. Kumar, D. and Singh, N.B.) Scientific Publishers, Jodhpur (India). pp. 35-62.

Yadav, N. D. and Beniwal, R. K. (2003). Competition relationship of arid legumes and pearl millet intercropping system under rainfed condition of western Rajasthan. In: Advances in Arid Legumes Research (eds. Henry, A., Kumar, D. and Singh, N. B.), Scientific Publishers, Jodhpur (India). pp. 272-274.

Yadav, N.D. (1992). Studies on strip cropping of pearl millet + arid legumes in rainfed arid western Rajasthan. Current Agriculture 16: 89-92.

Yadav, R.P., Sharma, R. K. and Shrivastava, U. K. (1997b). Fertility management in pigeonpea (*Cajanus cajan*)-basal intercropping system under rainfed condition. Indian Journal of Agronomy 42: 46-49.

Yadav, R.S. (1998). Effects of weed removal in clusterbean (*Cyamopsis tetragonoloba*) under different rainfall situations in an arid region. Journal of Agronomy and Crop Science 181: 209-214

❑❑❑

Chapter - 21

French Bean

Dileep Kachroo, Manpreet Kour and Jyoti Kachroo
Sher-e-Kashmir University of Agricultural Sciences & Technology of Jammu, Main Campus Chatha, Jammu - 180 009

Introduction

Among the pulses, rajmash is one of the high potential pulse crops with a yielding potential of 18 to 20 q per ha. *Rajmash* or French bean (*Phaseolus vulgaris* L.) is a short duration non–traditional grain legume and one of the precious and highly relished pulse crops of North India. *Rajmash* is known by various names viz., French bean, *rajma*, haricot bean, field bean, kidney bean, snap bean, pole bean etc. It is an important winter vegetable grown both for tender pods and dry seeds, which form a rich source of crude protein (21.25%), fat (1.7%) and carbohydrates (70%). Besides, it contains 0.16 mg iron, 1.76 mg calcium and 3.43 mg zinc per 100 g of edible part (Jasvinder Kaur and Mehta, 1994). The fresh pods and green leaves are used as vegetable. The antimetabolites of dry beans needs removal by cooking and soaking in water.

Origin

French beans have evolved in the highlands of middle America and Ander from a wild vine over a period of 7000-8000 years. Vavilov (1951) reported Mexico and Central America as the primary and Peruvian-Ecquadrion-Bolvian region of South America as the secondary center of French bean.

Rajmash is being widely cultivated in tropical, sub tropical and temperate regions. It was a traditional temperate region crop but now grown in subtropical regions also throughout the world as well as in the cool, dry season in tropical areas (Barelliet al., 2000). Globally French bean is cultivated on about 28 m ha with a production of 19 million tonnes. Brazil is the leading producer of French beans, producing more than 2.2 million tons, which represents 17.3% of the world's production (Goncalves et al., 2008). Columbia, USA, Canada, Ethiopia, China and Turkey are other leading countries producing French beans.

Present scenario

In India, it is grown on an area of about 1 lakh ha mainly in the states of Maharashtra (60,000 ha), Jammu and Kashmir (10,000 ha), Himachal Pradesh and Uttar Pradesh Hills, Nilgiris (Tamil Nadu) and Palni (Kerala) hills, Chickmagalur (Karnataka) and Darjeeling hills (West Bengal) as reported by (Ahlawat,1996) as they are the potential areas. It may be possible to cover about 0.1 million hectares area in four to five years time. However, these crops have a limited preference from consumers point of view.

Its cultivation in India is mainly done in *kharif* season but it is also cultivated during *rabi* season, where winters are mild and frost free. The genus *Phaseolus* has over 50 species and Rajma (*Phaseolus vulgaris* L) is one of them accounting for 90% of cultivated species worldwide. In India, Both bushy and trailing types *rajmash* are found.

Yield Gap: (Farmers' Production Strategies and gaps in technology)

The seed yield of Rajmash (sole crop) realised at Research farm varies from 10-13 q/ha, whereas in FLD's the yield varies from 6- 8 qt /ha undewr improved technology. On farmer's fields, the yield realised ranges between 2-3 q/ha. The reasons for about 30 percent yield gap between research farm and farmers field is mainly due to sub-optimal doses of fertilisers, pesticides and number of irrigations after meeting the requirements of wheat, paddy and vegetable crops. Only occasionally some farmers applied farm yard manure at the rate of 1- 1.80 t/ha.

The improved variety of Rajmash has recorded a yield increase of 44.4% over local varieties, overall, with the adoption of improved varieties, gross returns increased more than the increase in costs, thereby increasing the cost-benefit ratio. Non-availability of improved varieties and rhizobium culture are major reasons for their non-adoption. Farmer-participatory testing will help refine the technologies, pinpoint and eliminate adoption constraints.

Efficient cropping systems

In north India, it is grown in spring season after potato or mustard. In the hills, it is intercropped between maize and soybean.

In *rabi*, intercropping of potato + rajma (3:2 ratio) is being practiced in central and eastern Uttar Pradesh and northern Bihar. Rajma + linseed (2:1) is also found to be an efficient cropping system.

In subtropical climatic conditions of Jammu division of J&K state, the crop sown in the sequence of Rice (var. PC-19) - Marigold (Var. Pusa Narang) - French bean (var. contendor) proved to be the most economical and earned a net profit of Rs. 1,67,990, system profitability of Rs. 460/ ha/day and crop profitability of Rs 547/- per day, with a benefit- cost ratio of 2.90 and land use efficiency of 84% . This sequence generated on farm employment of 307 days in a year and is having the potential to replace the existing rice-wheat cropping systems

The apparent nutrient use efficiency has been worked out as 41.52 kg/ha with an NPK uptake of 62.04, 14.43 and 85.63 kg/ha, respectively. It has also been realized that the sequence is most sustainable as no negative trends were observed in terms of soil health and there was an improvement of soil organic carbon content from the initial value of 0.55 % to 0.58% was observed over a period of five years.

The agronomic management practices rice-marigold-frenchbean cropping systems are given in Table 1.

Table 1: Agronomic Management Practices

ManagementPractices	**Kharif**	**Rabi**	**Summer**
	Rice(PC-19)	**Marigold (Pusa Narangi)**	**Frenchbean (Contender)**
Date of sowing	10thJuly to 20th July	10thOctober to 20th October	25 Feb. to Ist March
Method of sowing	Transplanting	Transplanting	Line sowing
Seed rate	40 kg/ha	1 kg/ha	85 kg/ha
Spacing	20 cmx10 cm	40 cmx40 cm	60 cmx10 cm
FYM	-	8 t/ha	8-10 t/ha
Fertilizer($N:P_2O_5:K_2O$)	120:60:25	120:60:25	50:100:50
Irrigation	9(no.)	4	4
Weed control	Butachlor/granules @ 5%	Hand weeding	Hand weeding
Harvesting date	5-10 Oct.	15-20 Feb.	15-20 May
Yield, q/ha	32-35	40 -42	60-65

*Adopted from AICRP=IFS (2008-2011).SR Centre, SKUAST-J

Constraints : There is a significant gap between the potential and realized yield because of the following:

- Unfavourable weather conditions.
- Abnormal soil condition.
- Varietal constraints:- Lack of high yielding varieties., Low harvest index, High susceptibility , Flower drop
- Indeterminate growth habit of most of the Rajmash varieties.
- Instability in performance.
- Seed Constraints:- Poor quality seed, Inadequate availability, High price, Untimely supply, Low SRR

- Agronomic Constraints:-Improper sowing time, Inadequate seed rate, Defective method of sowing, inadequate interculture, Mixed cropping or catch crop, Imbalance fertilization, Negligible use of Rhizobium culture, Weed infestation, Socio-economic factors.

Export –Import

With the rajmash acreage in Maharashtra declining steadily and Chinese rajmash chitra getting hold of the entire Asian market, trade in new rajmash chitra seems lucrative for stockists, opined marketmen. According to a recent market survey, rajmash trade is getting confusing. On the one hand, rajmash acreage in Maharashtra producing centres such as Pune, Satara, Wai, Khatav, Maini and Koregaon is witnessing a steady decrease during the past couple of years, while on the other hand, China, hugely benefiting from this, has captured almost 90 per cent of the world rajmash market, marketmen said. Though rajmash production in Maharashtra is seen normal, however, arrivals in country's northern region have plummeted as much of the produce in Maharashtra is being lapped up by retail packagers. Old stock of rajmash chitra is gradually coming down and this has firmed buying sentiment. Consequently, flat rajmash chitra has shot up to Rs 4150-4200 per quintal. In tandem, round rajmash chitra has also hardened at Rs 3900/3950. Seeing hardening of rajmash spot prices, buying flat and round grain Chinese rajmash chitra Oct-Nov shipment, pegged at Rs 3100 per quintal and Rs 2950 per quintal respectively, seems profitable.

Future Thrusts

The North Western Himalayan regions, comprising parts of Jammu and Kashmir and Himachal Pradesh , are rich sources of rajmash diversity for seed, colour, shape and plant type. The choice of parents in a breeding programme for hybridization is one of the most critical considerations, since the selection on the basis of performance does not provide clear information. Genetic parameters, such as combining ability, play a significant role in crop improvement, since they help in characterizing the nature and magnitude of genetic effects governing yield and component traits, besides pinpointing the promising parents to be used in the creation of genetic variability for eventual use in development of suitable varieties. Diallel analysis developed by Griffing (1956) offers an excellent means of obtaining information on differential parental combinations in terms of general and specific combination ability and nature and extent of gene action. The following measures may be undertaken for its futuristic research programme:

- Broadening of genetic base through breeding programme and biotechnological interventions
- Development of high yielding short duration, disease resistant (rust and blight resistant), synchronous maturity varieties of bean .
- Development of location specific agro-techniques for bean in the rice fallows of Andhra Pradesh, Tamil Nadu, Orissa, Karnataka andNorth eastern hill zones.

- Identification of more productive cropping systems involving pulses and replacement of cereal crops in the prevailing rice-wheat cropping systems with high yield varieties of pulses and refining the production technologies aimed at high input use efficiency
- Development of IPM module against major insect pests and diseases.
- Development of efficient plant type with high biomass, harvest index and high nitrogen fixing varieties, which will play a crucial role in sustainable agricultur
- Identification of efficient Rhizobium strains so as to economize on N use need to be pursued.
- Reducing storage losses and improving market information and infrastructure
- Linking MSP to market prices can bridge the gap between demand and supply.
- Coordination of research, extension and farmers to encourage farmer's participatory research.

Package of practices

Climate

Major *rajmash* producing areas are located in tropical and temperate regions with a temperature around 21°C. The optimum temperature for better growth is 16-24°C. Growth of plant ceases if temperature falls below 10°C. Temperatures above 35°C cause dropping of buds and flowers resulting in poor yield. It is highly susceptible to frost. The crop is generally raised in areas receiving 50-150 cm annual rainfall. Waterlogging at any stage adversely affects its yield. Rains cause flower drop and spread of leaf spot diseases.

Soil and land preparation

It can be grown successfully on any type of soil, from sandy loam to heavy loam to heavy clay. But well drained loams soils are the most suitable ones for its growth. Crop is sensitive to salinity. Soil pH around 5.2-5.8 is optimum. An electrical conductivity of 1 dS/m is threshold limit for frenchbean. Soil having high amount of organic matter promote more vegetative growth (Ahlawat, 1986).

Crop requires fine seedbed and adequate soil moisture for good germination. A deep ploughing followed by 2-3 harrowings and planking is adequate to obtain required tilth.

Varieties

K-198 (Triloki Rajmash): It has determinate bushy growth, white flowers, broad dark green leaves, bold and creamish yellow coloured seeds with good cooking quality and taste. The height is 45-55 cm and matures in 98- 100 days. It is resistant to bacterial blight, angular leaf spot and anthracnose in whole in Zone IV except in Sangla valley where it is moderately susceptible to anthracnose. It is tolerant to shattering. Average yield is 25-27 q/ha.

KRC-8 (Baspa) : It is a semi-dwarf variety and matures in 110-120 days. It is recommended for high hill areas of H.P. It is resistant to bean anthracnose. It has attractive magenta coloured bold grains with good cooking quality. It average yield is 18-20 q/ha.

HPR-12 (Jwala) : It is early maturing (75-85 days) and is erect and dwarf having compact plant type. Pod length is 8-10 cm and each pod contains 3-4 seeds. Its 100 seed weight is 38-40 g. It yields, on an average, 12-15 q/ha. It is resistance to shattering. The strain has attractive scarlet seed colour. Suitable for cultivation in Kullu, Barot,

Shimla and Chamba valleys between 1100-1800 m.

Him- : It is a dwarf and early maturing (80-90 days) variety. Pods are 10-13 cm long containing 4-5 seeds per pod. The seeds are light pinkish with dark red spots. Besides, this variety is good flavoured. Its average yield is 9 q/ha.

HPR-35 (Kanchan): It is semi-dwarf, determinate and early maturing variety (7-10 days earlier than Him 1 and Jawala) and recommended for mid and high hill areas. It is better suited for both under monoculture and intercropping with maize. It has mottled deep pink bold grains with good cooking quality. Its yield is 12-15 q/ha.

Contender : It is bushy type with light green, fleshy, thick and slightly curved pods without fibres and green colours. It is tolerant to mosaic and powdery mildew and its yield is 11-11.5 q/ha.

Pusa Parvati : Prolific bearer, pods mature in 40-45 days .It is tolerant to mosaic and powdery mildew yielding 10-11 q/ha.

Arka komal : Plant is erect and bushy. Pods are straight, flat, tender and green. It is very good for transport as keeping quality is good, and yields 12-13q/ha

Other varieties resistant to disease : The other varieties resistant to disease are given in table. 2.

Table 2 : Selected varieties resistant to diseases

Variety	Year of release	Zone	Yield (q/ha)	Characterstics/remarks
IPR 98-3-1	2007	Central	16.0	Tolerate to BCMV disease
IPR 98-5	2005	North eastern plain zone	17.0	Tolerate to BCMV disease
IPR 96-4	2002	-do-	15.0	Tolerate to BCMV disease and leaf crinkle.

Time of sowing

Rajma is grown in *kharif* and *rabi* seasons in different parts of the .country. In *kharif*, mid May – mid June is ideal. The spring crop can be sown from February – early March. In sub tropical regions of J&K, it can be grown during the last week of January to first fortnight of February, whereas in intermediate zones it is grown during March-April and in temperate zones, the 1st crop is taken during March-April while the 2nd crop is taken during May-June.

The optimum time of sowing in *rabi* season varies from state to state. It is first and second fortnight of November for central Uttar Pradesh and north Bihar. However, mid October is optimum for Vidarbha region (Maharashtra). For early varieties, October end is optimum, while late varieties can be sown up to mid November.

Seed treatment

To protect the crop from various diseases at early seedling stage, seed is treated with bavistin or Thiram @ 2g/kg of rajmash seed before sowing

Seed rate and Spacing

Seed rate varies with seed size. Bold seeded bush type varieties with a test weight of 350-450 g need 120-140 kg seed/ha, while in small seeded varieties, it varies from 80-100 kg/ha. For the pole type varieties, the seed rate is reduced to half that of bush type. The seed rate in intercropping may vary with row proportions.

Bush type rajma is generally sown in rows 60 cm apart with the plant to plant spacing of 10 cm whereas for pole type varieties the spacing is reduced to 20 cm. In northern plain, *rabi* rajma is sown in rows of 45 cm. The plant-to-plant spacing is 10-12 cm. For obtaining good yield, its plant population should be 2.5-3.0 lakh plants/ha. 8-10 cm is the optimum depth of sowing.

Manures and Fertilizers

Unlike other pulses, rajmash is inefficient in symbiotic nitrogen fixation (Ali and Lal, 1992) as it lacks nodulation due to the absence of NOD gene regulator (Kushwaha, 1994) even with native *Rhizobia* and commercially produced cultures. Hence, the nitrogen requirement of rajmash is different from other pulse crops and application of nitrogen through fertilizers is imperative for exploiting its yield potential. As nodulation is poor in rajmash, it requires more nitrogen and phosphorus for root development, nodulation and better plant growth and hence, responds even to application of fertilizer phosphorus (Ssali and Keya, 1986). Potassium application stimulates nitrogenase activity and causes accumulation of uricides in pod walls and greater partitioning of above ground nitrogen to seeds. Supplementary doses of potassium stimulate the transport of nitrogenous compounds to developing fruits and thereby increase seed yields (Thomas and Hungaria, 1988). The seed yield of rajmash (2079 kg/ha) and economic returns (Rs. 25,613 ha^{-1}) were significantly higher with application of 80:75: 30 kg N: P_2O_5:K_2O per ha during *kharif* at Dharwad (Veeresh, 2003).

Manuring

10-15 tones/ha of well rotten farm yard manure should be applied at the time of preparation of the land. This should be supplemented with by applying inorganic fertilizers. In temperate zones of J & K, N is required @20 kg/ha, Phosphorus @ 60kg /ha and Potash @30 kg /ha and Zinc sulphate should be applied once in three years @ 10 kg/ ha.

Half dose of N and full doses of P and K should be placed in the bands about 7-8 cm at the time of sowing, whereas, remaining half dose of N is top dressed at 30 days after sowing. Whereas, $ZnSO_4$ may be applied at the time of sowing or applied as a foliar application @ 0.1% for enhancing seed yield. When rajmash is grown as intercropped with maize crop then no additional fertilizer is provided to rajmash.

Irrigation

Rajma has shallow root system and hence moisture stress at any stage is detrimental to its performance. As a rainy season *(kharif)* crop, it does not require irrigation, when rainfall distribution is even throughout crop cycle. However, *rabi* crop requires irrigation. Irrigation at 25 days after sowing (DAS) is critical. In north-east plains zone, 3 irrigations at 25, 75 and 100 DAS and in central zone, 4 irrigations at 25, 50, 75 and 100 DAS are necessary for optimum crop performance.

Weed control

Rajma suffers severe competition from weeds in initial stages. The first 30-40 days after planting is the critical period for crop weed competition. One hand weeding at 30-35 days after sowing was beneficial. Pre-emergence application of pendimethalin @ 1.0 kg/ha or pre-plant incorporation of 1.0 kg/ha of fluchloralin have been found effective in controlling weeds.

Interculture

Two weedings and hoeings should be done to control weeds.

Pests and Diseases

Bihar hairy caterpillar, blister beetle, bean bug and aphids are important insect pests of the crop. Rot (collar, stem and pod) and bean golden mosaic virus are important diseases limiting rajmash productivity (Table 3).

Table 3 : Insect pest and diseases of French bean with their management

S .No.	Name of disease	Casual Organism/ Scientific name	Symptoms/damage type	Management	Remarks
Diseases					
1	Anthracnose of beans	*Colletotrichum lindemuthaiauum*	Blank sunken cankers with dark centres having red, yellow or orange margins appears and clearly visual on the pods	Seed treatment with bavistin (0.2%)	
2	Rot	*Rhizoctonia solani*	Rotting in root system with appearance of reddish brown cankers	Seed treatment with bavistin@ 2g/kg	
3	Bean common mosaic	Bean common Mosaic virus (BCMV)	Irregular shaped leaves with mosaic have, light - yellow and green areas. Leaves puckered, curled and malformed .	Eradication of infected plants. Use of Furadon or Disyston or Thimet 10-G granules @ 1.5 kg a.i/ha at the time of sowing	
4	Rust	*Uromyces phaseoli Uromyces appendiculatus*		Seed treatment with bavistin (0.2%) followed by spraying of amacozeb(2g/litre)	
5.	Web blight	*Rhizoctonia solani*	Appearance of water soaked, light tan colour spots on leaf and pods	Soil application of Trichoderma viride; Seed treatment with bavistin 2g/kg of seed	
Important peasts					
1	Blister beetle	*Cyaneolylta coerculea*	Grubs and adults feed voraciously on leaves and flowers.	Hand picking and destroying of adult beetles and spraying of malathion @ 2 ml/l of water.	
2	Hairy catterpiller	*Spilosoma oblique*	Larvae completely defoliate the plants.	Spraying of Malathion @ 2 ml/l of water	
3	Cutworm	*Agrotis ipsilon*)	At the an early stage of crop larvae cut the plants at ground level	Seed treatment with Chloropyriphos @ 2 gm/kg seed ; soil application with Phorate 10 G (1.0 kg a.i./ha.)	Main problem in temperate areas and hills,
4	Aphids	*Aphis fabad*	Vectors of the virus diseases	Spraying of Imidacloprid @ 0.30 ml Metasystox @1.5 ml per litre of water are effective.	

Harvesting and threshing

Harvesting of the crop should take place when leaves and pods turn yellowish brown and mast of the leaves fall. Avoid delay in harvesting as it may cause shattering of the pods. The harvested crop is kept for sun drying for 5-7 days, and thereafter threshed seeds are cleaned and properly stored in storage bins with moistur @ 9%.

Yield

Under optimum conditions, 2.0-2.5 t/ha of grain and 3.0-3.5 t/ha straw yield can be obtained.

References

Ahlawat, I.P.S., 1996, Response of French bean (Phaseolus vulgaris) varieties to plant density and phosphorus level. Indian J. Agron., 39: 246-248.

Ahlawat, I.P.S. and Sharma, R.P., 1989, Response of French bean genotypes to soil moisture regimes and phosphate fertilization. Indian J. Argon. 34: 70-74.

Anonymous, 2005, Food and Agriculture Organization of the United Nations. FAO STAT database. http:/ www.fao,org

Ali, M. and Lal, S., 1992, Technology for rajmash cultivation in the plains. Indian Fmg. 42: 25-26. Jasvinder Kaur, K.S. and Mehta, U., 1994, Nutritional evaluation of new varieties of kidney beans (Phaseolus vulgaris). Legume Res. 17: 17-21.

Kushwaha, B.L., 1994, Response of French bean (Phaseolus vulgaris L.) to nitrogen application in north Indian plains. Indian J. Agron., 39: 34-37.

Ssali, H. and Keya, S.O., 1986, The effect of phosphorus and nitrogen fertilizer level on nodulation, growth and dinitrogen fixation of three bean cultivars. Trop. Agric. 63: 105-109.

Thomas, R.J. and Hungaria, M., 1988, Effect of potassium on nitrogen fixation, nitrogen transport and nitrogen harvest index of bean. J. Pl. Nutr. 11: 175-188.

Veeresh, N.K., 2003, Response of French bean (Phaseolus vulgaris L.) to fertilizer levels in Northern Transitional Zone of Karnataka. M.Sc.(Agri.)Thesis, Univ. Agric. Sci., Dharwad. p. 37-79.

Barelli MAA, Vidigal MCG, Amaral ATJ, Filho PSV, Scapim CA (2000). Diallel analysis of the combining ability of common bean (Phaseolus vulgaris L.) cultivars. Braz. Arch. Biol. Technol., 43: 409-414.

Goncalves VMC, Silverio L, Elias HT, Filho PSV, Kvitschal MV, Retuci VS, Silva CR (2008). Combining ability and heterosis in common bean (Phaseolus vulgaris L.) cultivars. Pesq. Agrpec. Bras. Brasilia, 43(9): 1143-1150.

Griffing B (1956). Concept of general and specific combining ability in relation to diallel crossing systems. Aus. J. Biol. Sci., 9: 463-493.

Reddy,A.2009. Pulses Production Technology: Status and Way Forward. Economic & Political Weekly EPW December 26, 2009 vol xliv no 52:73-80.

❑❑❑

Chapter - 22

Arid Pulses : Mothbean, Cowpea and Horsegram

A. Henry & P.K. Roy
National Network Project on Arid Legumes, Central Arid Zone Research Institute, Jodhpur Rajasthan - 342 005

The Arid Legume includes clusterbean (*Cyamopsis tetragonoloba* (L.) Taub.); moth bean (*Vigna aconotifolia* (Jacq) Marechal; cowpea (*Vigna unguiculata* (L.) Walp) and horse gram (*Macrotylom uniflorum* (L.) Verdc). In India, these are grown extensively as rainfed crops in Rajasthan, Orissa, Maharashtra, Karnataka, Gujarat, Haryana, M.P. and to a limited extent in other states covering a total area of 51-80 lakh hectares (Table-1). Most of the production comes from harsh environments where both biotic and abiotic constraints prevail limiting the production.

Arid Legumes as a group of pulses are extremely important in the common vegetarian diet of the people in the area of its cultivation. They also offer a new dimension as they have become foreign exchange earners through export opportunities. For example, clusterbean (guar) contains 30-35 per cent guar gum and India is a world leader in the export of this commodity and earns more than Rs.1100 crores annually (Mathur and Henry, 2009). Similarly, moth bean is gaining great industrial importance because of *Bikaneri Bhujia* industry which is heavily based on this crop (Singh and Henry, 1985). Cowpea is very important in *Dahi bada* industry and making curry preparation while horsegram is a minor pulse and is utilized by the poorest of the poor in its growing areas as well used for kidney stone treatment (Henry *et.al.,* 2006; Vishwanatha *et. al.,* 2006). Clusterbean and moth bean together account for 3.5 million hectare areas in the great Indian Thar Desert and play a very important role in the economy of the desert farmers. Similarly, cowpea and horse gram are grown mostly in marginal and sub-marginal lands under rainfed situations with no inputs applied by the farmers. These two pulses are grown in various areas in the country and easily adapted to various conditions.

Table 1: Area, production and productivity of arid legumes in India

India/ State	Area(lakh ha)	Production(lakh t)	Productivity(kg ha^{-1})
Guar			
Rajasthan	19.85 (85.19%)	7.34(71.86%)	370
Haryana	1.20 (5.15%)	1.00(9.80%)	740
Gujarat	2.27 (9.27%)	1.15 (11.22%)	669
India	23.30	10.20	430
Moth bean			
Rajasthan	11.55 (86.0%)	2.41 (82.8%)	197.7
Maharashtra	1.25 (9.27%)	0.37 (12.7%)	296.2
Gujarat	0.60 (4.45%)	0.15 (5.40%)	264.1
India	13.52	2.91	215.2
Horse gram			
Karnataka	5.90 (34.6%)	2.15 (29.9%)	370
Maharashtra	3.01 (17.6%)	0.94 (13.0%)	314
Orrisa	2.73 (16.0%)	1.60 (22.2%)	587
Andhra Pradesh	2.72 (16.9%)	0.76 (10.5%)	280
Tamil Nadu	1.37 (8.1%)	0.63 (8.7%)	453
India	17.02	7.19	422.0

Moth bean or dew gram

In India, mothbean is largely grown in Rajasthan (>90%) and other states like Maharashtra, Gujarat, Punjab and Haryana as a rainfed crop. The distribution of this crop is confined to plains or very low elevation. It can tolerate high temperatures during growth without any adverse effect on the flowering and fruit development. It tolerates widely different soil textures but grows best in light soils. The bulk of the moth bean cultivation is confined to the dry lands of arid zone where seasonal rainfall ranges from 250 to 500 mm (Henry and Singh, 1985; Singh *et. al.* 1981).

Cowpea

Cowpea is predominantly a hot weather crop and does well in temperature range between 25 ^{0}C and 35 ^{0}C although it can survive up to 45 ^{0}C, if adequate soil moisture is maintained but under such conditions it sheds flowers. Temperatures lower than 25 ^{0}C adversely effect the growth of cow pea. In the Northern parts of India, where the temperatures during winter season can be quite low, it is cultivated during rainy season only. In the central and southern states where there is not much variation in the climate, it is cultivated both during winter and rainy season. Cowpea is also frequently sown mixed with other crops like maize, sorghum, and millet etc. Cowpeas is grown on a wide range of soil types from loamy sand to heavy expandable clays. Most of the cultivars do not tolerate water logging. Some of the cultivars/varieties are day length insensitive, while others require short days to mature within reasonable time. It is a dual purpose crop. It is capable of withstanding a much higher rainfall. Now early maturing types of cowpea occupy an important place even under rainfed conditions of Indian Arid Zone.

Horse gram

Horse gram grows well from terraced field of Himalayas to the adjoining districts of Thar desert, and the from plains of Maharashtra, Orissa, A.P. Karnataka, West Bengal, Tamil Nadu, Kerala to the coastal region of Konkan under various agro-ecological situations.

Horse gram is typically grown as a rainfed crop in very poor soils in which most of the other crops do not perform well. It is basically a short dry plant. However, photo-insensitive types are also available. In North-Western India, it is grown with the onset of monsoon on very poor marginal and sub-marginal lands. In Bihar, plateau region, horsegram is normally planted during third week of August to first week of September. In Orissa, it is grown as a *kharif* crop in shallow and red soils, gravelly soils and hard upland soils of Kalahandi, Sambalpur and Cuttack and as a *rabi* (post rainy season) crop in all the districts in sandy soils. It is cultivated after harvest of upland rice. In West-Bengal, it is grown on acid lateritic soils (pH 5.5). In Maharashtra, it is mostly grown during both *kharif* and *rabi* seasons. It is grown after the harvest of rice crop in *kharif* season. In Karnataka, it is generally grown in August-September. It is grown as a sole crop in poor red sandy soils as a late *kharif* season crop, relay crop in finger millet, in double cropping system after *kharif* crop of groundnut or sesamum, and also as main crop like finger millet, sorghum and pearl millet which can not be sown due to delayed monsoon. It is also extensively cultivated as an inter-crop, in the assured rainfall tracts of Karnataka comprising the eastern belt lying immediately to the east of the Western Ghats. Its location specific land races are very well adapted for growing from early June in H.P. to late *kharif* season in Bihar, Orissa, Maharashtra, Karnataka to October-November in Tamil Nadu. It has good feed and fodder value with valuable protein (23.30%). It has medicinal value and helpful in treatment of kidney and gall bladder stones, piles, bronchitis, cough and cold and urinary diseases. The areas where rainfall is low it can be an alternative crop to cowpea because it withstands drought better than cowpea.

Recommended varieties

Many institution have been working on developing high yielding varieties of various legumes for different agro-ecological situation (Table-2)

Table 2: High yielding varieties of arid legumes for different states

S.No / States / crop	Variety	Maturity(days)	Yield(kg/ha)	Special salient features
Cowpea				
1. Low rainfall zones.	RC-19	62-64	650-850	Sinchronized maturity.
2. T.N., Karnataka, A.P., Kerala	CoVu-702 Co (CP-7)	67-73	950-1250	MR to pod borer and LCV, Erect growth ovate trifoliate leaves, compact growth, light brown grain color.
3. Gujarat, Rajasthan	GC-5	75-80	1400-1600	Suited to *kharif* and summer season improved over GC-4. Adapted to Gujarat situations, bold seeds.
4. Karnataka, Tamilnadu	Vamban-1	90-100	950-1100	Bold and brown seed color, straight growth with less winy habit, long pods.
5. Karnataka, Tamilnadu	KBC-2	95-105	600-1200	Late maturing, unsinchronous maturity, seed light browns in color, rust tolerant, long pods.
6. Delhi, H.P., U.P.,Sampada) Maharashtra	V-585(Pusa	90-110	700-1000	Requires high rains late maturity, adapted to some what better conditions dual purpose, creamy color of seeds.
7. Delhi, H.P., U.P., Maharashtra, Karnataka, Tamilnadu, Jharkhand, Gujarat	GC-3	90-95	750-1150	Nationally adapted early and high yielding, short statured growth, pods are moderate in size, drought hardy and disease tolerant.
8. Haryana, Delhi, Rajasthan, Gujarat	V-130(Pusa Safed)	90-100	1100-1250	Resistant to CYMV, white seeded, semi-viny growth, unsinchronised maturity.
9. Rajasthan, M.P., Karnataka Delhi, T.N., Gujarat	V-240 (Rambha)	85-95	1200-1400	Suited to all cow pea growing areas of the country, tall growing, drought hardy. Dual purpose, red grains, dark green leaves color.
10. Rajasthan	RC-101	60-65	750-850	Extra early maturity, drought hardy, determinate

Contd.

				type, non-viny, white seeded, suited to low rain falls, high yielder than early maturing other varieties.
Mothbean				
1. Rajasthan, Haryana, Gujarat, Maharashtra	CAZRI Moth-3	62-64	900-1000	Erect, early drought hardy, escapes YMV diseases, heavy podding behaviour.
2. Rajasthan	RMO-257	63-66	650-700	Erect early improved in grain & fodder, medium early
3. Rajasthan	RMO- 423 (State release)	62-64	500-580	Erect, medium maturity drought hardy, may escape YMV, semi spreading behaviour, flowers in clusters, stems pigmented, a mutant from RMO-40.
4. Rajasthan, Gujarat	GMO-2	75-85	800-1100	Medium-early in maturity semi-erect, bold seeds with brown color, suited to inter cropping, escapes YMV infection.
5. Rajasthan, Haryana, Gujarat, Maharashtra	CAZRI Moth-2	65-67	1000-1200	Suited to moderate rainfalls, first variety developed by hybridization. (Jodiax RMO -40), brisk podding behaviour, high yield potential.
6. Rajasthan	RMO-435 (State release)	64-66	600-800	Erect, medium, early, good yield potential leaves green colored, mutant from RMO-40, Leaves green colored.
7. Rajasthan, Haryana, Gujarat, Maharashtra	CAZRI Moth-1	70-72	400-500	Semi-spreading, responding to inputs, suited to better environments, good regemerations potential, dual purpose, medium bold seeds, natural source of YMV, grain protein 25%.
8. Rajasthan	RMO-225	64-66	650-750	Erect, early, suited to low rainfalls, stem, more pigmented, field tolerance towards YMV.

Contd.

9. Rajasthan Gujarat	RMO-40	60-64	600-850	Erect, early, suited to low rainfalls, widely spreaded in low rainfall zones, short statured, less biomass.
Horsegram				
1. Uttranchal	VLG-10	115-135	1000-1200	Semi-erect, spreading type Seeds hold field tolerance towards anthracnose and stem rot.
2. Uttranchal & H.P.	VLG-8	95-112	1000-1300	Semi-erect, spreading type Seeds hold field tolerance towards anthracnose and stem rot.
3. A.P.,CRIDA-1- Maharashtra, T.N., Kerala, Karnataka	95-110 18 R	900-1100	High yielding,	20% more over PHG-9, heavy bearing, synchronized pod behaviour
4. Northern hills and Central zones.	VLG-15	100-110	400-600	30% higher yield over AK-21 check.
5. U.P., Uttranchal, Gujarat, Rajasthan.	VLG-19	95-112	490-700	29% higher yield over AK-21, Digestibility 83.4%, tannin content 1.00 mg/g.
6. Karnataka, A.P. and T.N. States	BJPL-1	90-95	600-800	27% higher grain yield over PHG-9, check.
7. Rajasthan, West Bengal, Gujarat, Jharkhand, H.P.	AK-42	90-95	600-850	High yields and medium late in maturity, suited to moderate rain falls, grain color brick red, straight plant.
8. Rajasthan, West Bengal, Gujarat, Jharkhand, H.P.	AK-21	85-90	650-850	Earliest maturing variety, high yield grin potential, compact, synchronized growth habit.
9. Karnataka	BGM-1 (KBH-1)	100-110	600-700	Bushy gives tendrils tolerant to YMV, seed color light brown.
10. Andhra Pradesh	Palem-2	105-112	650-750	Suited to A.P. situations, semi-spreading light yellow seeds bold pods.
11. Andhra Pradesh	Palem-1	100-110	600-700	Suited to A.P. situations, semi-spreading light yellow seeds.
12. Karnataka, Tamilnadu, Kerala	PHG-9	90-100	650-1000	Long duration suited to southern locations only, photo sensitive, some what winy.

Disease Management

The recent advances made in insect-pest management of arid legumes are highlighted here.

Moth bean

For management of yellow mosaic virus (YMV) disease, spray the crop with neem oil (0.2%) or neem seed kernel powder extract (5%) thrice starting from onset of disease at an interval of 15 days. Also treat the seed with imiddaclroprid 70 WS (5 g/kg seeds) than spray the crop with emidacloriprid 17.8 SL (0.05%) after 35 days of sowing.

Cowpea

For control of root rot use of compost (2.5 t/ha) + vermi compost (2.5 t/ha) reduced the incidence (75%) and increased the yield at At S.K. Nagar. The use of mustard compost (2.5 t/ha) + oil cake (0.5t/ha) was effective for control of root rot at Hisar. While at Durgapura, Amended soil with mustard compost (2.5 t/ha) + vermi compost (5 t/ha) and seed treatment with *T. hariinim* powder (4 gm/kg seeds) proved effective.

Horse gram

For management of yellow mosaic virus (YMV) the application of neem oil 3%, neem oil 5%, *azadiractin* 1.5 ml/litre, *dimethoate* 2 ml/litre and *triazophos* 2ml/litre were found effected in reducing the yellow mosaic disease incidence by managing the white fly population.

Management of insect-pests

Mothbean

Foliar spray of cow urine (100%) + neem leaf extract (5%) controlled white flies and was at par with insecticides.

Imidacloprid 17.8 SL (1 spray at 35 DAS) gave the better control of *Jassids and White flies* and also helped in management of yellow mosaic virus (YMV).

Coating with mustard oil @ 5 and 8 ml/kg seed while ground nut oil or Sesame oil and soybean oil @ 8ml/kg seed also proved effective for Bruchids Management in stored pests.

Cowpea

Application of Profenophos @ 0.05% and Indoxacarb 10.E.C. (50 gm a.i./ha) are recommended for control cowpea borer (*M. vitrata)* Bifenthrin 10E.C. (80 g a.i./ha) also proved effective.

Cartap hydrochloride (0.15%) and leaf extract (10%) proved effective for contral of Cowpea pod borer management.

Cropping systems involving arid legumes

The existing cropping systems are mainly subsistence oriented. Single crop systems using pearl millet and involving long fallow (Oct-June) are pre-dominant in arid region Pearl millet based mixed cropping systems with grain legumes and oilseed crop are quite their popular on small farms because of risk reducing effect. However, there is a need to develop cropping systems which aim at maximizing production in good and normal rainfall years and stabilizing it in sub normal rainfall years.

Single crop system: In arid areas mungbean-fallow system proved better alternative.

Double crop system sequence : In Rajasthan, double cropping in this tract is not common. However, years having more than 500 mm rainfall with an extended rainy season favour double cropping. The results on double cropping studies indicated that bajra-mustard double cropping was more remunerative than green gram-mustard (Singh *et. al.,* 1997).

Inter /mixed cropping system : Inter/mixed cropping is practiced to meet the domestic needs of the farmers for cereals, pulses, oilseeds etc. and to minimize the risk of failure of crops and lessen the incidence of pests and diseases. In arid regions particularly, it is an important practice to support subsistence oriented arable farming.

Intercropping with perennial forage grasses : Grass cultivation in arid zone is the optimum land use which brings in improvement in soil against wind erosion. Considering the farmers adherence to arable cropping, legumes can be grown advantageously as inter-crops with grasses in good rainfall years and this can be termed as opportunity cropping. This system of inter cropping or strip cropping is capable of controlling soil erosion by wind, improving soil physical condition by grass leys and soil fertility build up by nitrogen fixing legumes. A seven year study with four grain legumes (Mungbean, moth bean, clusterbean and cowpea) inter cropped in grasses planted in 80 cm regular and 40/80 cm paired row revealed that the mean green forage yield of *Cenchrus ciliaris* from pure and mixed stand were not significantly different during good rainfall years besides having a bonus grain yield from inter cropped legumes.

Intercropping of pearl millet and annual legumes for higher fodder production : Inter cropping of pearl millet and sorghum with annual legumes revealed that pearlmillet intercropping with legumes, in general, and with cowpea (HFC-42-1) in particular, gave higher forage production as compared to sorghum inter-cropped with legumes.

Intercropping of pearl millet with mungbean : From the point of view of the total productivity and gross returns, intercropping system of paired row of mungbean + one row of bajra in the inter-spaces appeared to be a paying proposition.

Intercropping of pearl millet in clusterbean and moth bean : The growing of one row of bajra in the inter-spaces of clusterbean and moth bean sown in paired row and fertilizing bajra with 15 kg N/ha resulted in the highest productivity as against the pure stand of clusterbean or mothbean.

Intercropping of grain legumes in sesame : Studies on mixed cropping of grain legumes with til revealed that clusterbean pure and mixture of til (1/3) + guar (2/3) gave higher gross returns than the other inter-cropping systems tried.

Intercropping of annual grain legumes with sunflower : Mothbean, clusterbean and cowpea grown as a inter-crop in sunflower planted in regular (60 cm) and paired row (40/80 cm) indicated that intercropping of mung, cowpea and moth bean did not affect the yield of sunflower adversely although clusterbean produced the highest grain yield among the inter crops, it offered severe competition to the base crops sunflower. Economic evaluation revealed that sunflower + cowpea gave the highest gross returns, followed by sunflower+ mungbean.

Intercropping of annual grain and fodder legumes in castor : Four grain legumes were tested as inter crops (mungbean, mothbean, clusterbean and cowpea) for fodder. Inter cropping of fodder legumes offered less competition to castor than grain legumes. Among grain legumes, cowpea offered less competition. Castor and mungbean combination gave the highest gross monetary returns MUE; followed by castor + cowpea for fodder and castor and cowpea for grain.

Intercropping of grain legumes in *cheena* (*Penicum milliaceum*) : Four grain legumes, *viz*; mungbean, mothbean, clusterbean and cowpea were inter cropped with *cheena* (PM-31)- a short duration minor millet as base crop, under late sown conditions. The base/inter crop ratio was 2:1. Sole cropping of clusterbean and cowpea resulted in the highest total grass monetary returns. Among the inter-cropping system, cheena + clusterbean gave the highest total productivity and monetary returns. (Saxena and Henry, 1988).

Suitable genotypes of grain legumes for inter/mixed cropping with pearlmillet

Intercropping proved to be a viable system for dry land crop production in the arid zone. Pearlmillet being the most important arid land crop of this region, it was felt essential to determine which of the leguminous crops and their varieties are the most compatible for inter/mixed cropping with pearl millet.

In moth bean, the combination of genotypes, viz; IPCMO-909, IPCMO-884 and Maru-Moth-1(JMM-259) were promising. However, genotype IPCMO-880 (Moth-880) resulted in higher yield levels of pearl millet. In green gram, the combination of genotype M-45 and pearl millet indicated a positive effect on both the components *viz*; base crop and companion crop (Daulay, 2000).

with better adaptability, resistance to drought, photo and thermo insensitive with efficient root system and nodulation. The crop-wise focus in the near future should be on the following.

Mothbean

- To identify/evolve a variety having a semi-erect plant habit, dark green leaves, early maturity, high yield potential and resistant to yellow mosaic virus.
- To identify suitable variety for inter/mixed cropping in pearl millet.

Cowpea

- To identify/evolve varieties which are more compact in their growth and which are relatively erect with a non-trailing habit. This will result in higher seed yield of the crop.
- To evolve varieties resistant to diseases and pests.

Horsegram

- Low productivity, presence of *trypsin* inhibitors in seed and poor palatability are some of the problems in this crop. There is need to develop new varieties to improve the yield and quality of horse gram.
- Being a leguminous crop, it fixes atmospheric nitrogen in the soil and improves fertility and health of the soil. The identification of suitable *rhizobial* strain for enhancing this ability is essential.

Agronomy

Though the information is available on agronomic aspects of these crops. However, with global warming and climatic change, it is necessary to intensify to research on various agronomic aspects under rainfed and irrigated conditions. Some suggestions are as follows:

- The production technology should be non-monetary input oriented and capable of giving high yield even in sub-normal rainfall years so that farmers are encouraged to cultivate these crops.
- The agronomic practices for maximizing production under irrigated and dry land areas needs to be worked out in view of changing climatic.
- Requirement of micro-nutrients in relation to nitrogen fixation by *Rhizobium* cultures is to be studied.
- A suitable system of water harvesting needs to be standardized for maximization of yield in good rainfall years and stabilization of production in low rainfall years.

Plant Pathology

The following efforts will result in the development of better disease resistant varieties :

- Inheritance studies for disease resistance.
- Identification of prevalence of different races of important diseases.
- Inititation of work on minor diseases in order to have information on varietal reactions and control measures before any one of these diseases become a major one.
- Bio-control of important diseases.

The cropwise specific diseases that need to the addressed on priority are :

Mothbean

- Yellow mosaic virus. It is regarded as the most serious and devastating viral disease.
- Bacterial leaf spot: It is common in Rajasthan and spreads rapidly during high humidity.
- Charcoal rot/seedling blight/root rot: It appears in hot and dry climate.
- Brown web blight: It appears during heavy rains and high temperature. The screening of above diseases in the breeding material will result in development of resistant varieties.

Cowpea

- Cowpea yellow mosaic, Anthracnose, root rot, cercospora leaf spot, bacterial blight and powdery mildew (Southern India) are important diseases of the crop.

Horsegram

- Yellow mosaic on horsegram is prevalent in Southern India resulting in enormous losses.
- Other diseases are anthracnose, powdery mildew, dry root rot, rust, aerial blight and leaf spot.

Entomology

Till now, these crops have been relatively free of major insect pests. Though several insect and mite pests have been reported to damage the crops, most of the pests are of are namely, minor importance. The insects pests are namely aphids, green leaf hopper, white fly (cowpea and guar), pod fly (horsegram), pod borers (cowpea), spotted pod border (cow pea) , leaf perforator(guar).To tackle them,it is essential to develop effective bio-control measures for different insects and pests, which may be of minor importance at present but may become a serious problem later. There is also a need to identify new insect pests and storage pests and evolve effective control measures.

References

Daulay, H.S. and Henry, A. 1986. Cropping system of pearl millet in arid zone of Rajasthan. Proc. International Pearl millet Workshop at ICRISAT (India). pp. 306-307.

Daulay, H.S. and Henry, A. 1987. Farming systems for managing marginal rainfed arid lands. Rangelands Resource and Management. Proc. National Rangeland Symposium, Nov. 9-12, 1987. (Ed. Singh, P. and Pathak, P.S.: 283-292.

Daulay, H.S.and Henry,A. 1987. Desert ecological environments and their management systems for optimizing crop/plant production. The Geographer 34 (2): 26-38.

Daulay, H.S. and Henry, A. 1988. Crop diversification and alternate land use system for rural land use planning in Indian Arid Zone. The Geographer 35 (2): 13-22.

Daulay, H.S. and Henry, A. 1988. Dry land cropping systems for arid lands. In : Challenges in Dry land Agriculture- A Global perspective. Proceedings of the International Conference on Dry land farming. A.marrillo/Bushland, Texas, U.S.A. pp. 781-783.

Daulay, H.S.; Henry, A. and Bhati, T.K. 1998. Compatibility of promising varieties of clusterbean for inter/ mixed cropping in pearl millet in dry land of Western Rajasthan. Forage Research, 24: 13-16.

Daulay, H.S.; Henry, A. and Bhati, T.K. 1999. Studies on suitable genotype of green gram for inter/mixed sowing in pearl millet on dry land of Western Rajasthan. Current Agriculture, 23: 93-96.

Daulay, H.S.; Henry, A. and Bhati, T.K. 2000. Screening of suitable genotypes of clusterbean, moth bean and green gram in legume based inter-cropping with pearl millet. Current Agriculture, 24: 83-88.

Daulay, H.S.; Henry, A.; Bhati, T.K. and Mathur, B.K. 2006. Compatibility of promising varieties of moth bean and cowpea for inter/mixed cropping in pearl millet in arid regions of Western Rajasthan. Journal of Arid Legumes 3(2):44-48.

Mathur, B.K. and Henry, A. 2006. Performance of grain legumes V/S Oilseeds and cereals in area, production and productivity in different agro-climatic zones of Western Rajasthan. Journal of Arid Legumes 3 (1):83-92.

Mathur, B.K. and Henry, A. 2008. District wise cropping trends of legumes crops vis-à-vis cereal and oilseed crops of arid region of Western Rajasthan. Journal of Arid Legumes 5(2):124-131.

Mathur, B.K. and Henry, A. 2009. District wise cost benefit analysis of major crops in arid regions of Western Rajasthan. In Golden Jubilee Symposium 3(1):83-92.

Meghwal, P.R. and Henry, A. 2009. Economic evaluation of different agro-horticulture systems under rainfed conditions of arid zone. In legumes in Dry Areas (Ed. Kumar, D. & et. al. 2009), Publ. by Indian Arid Legume Society and Scientific Publisher, Jodhpur (RAJ) : pp. 300-303.

Saxena, M.B.L. and Henry, A. 1988. Inter-cropping of grain legumes in Cheena (Panicum Miliaceum) under late sown conditions. Millet Newsletter.

Singh, H.P., Bhati, T.K. and Henry, A. 1997. Research Achievements (1971-1996). AICRP on Dry land Agriculture, CAZRI, Jodhpur, pp. 1-62.

❑❑❑